LYNCHBURG COLLEGE

SYMPOSIUM READINGS

CLASSICAL SELECTIONS ON GREAT ISSUES

VOLUME VII

THE NATURE OF THE UNIVERSE

Proclus
Euclid
Aristotle
Cleomedes
Simplicius
Archimedes
Plutarch
Copernicus
Galileo
Descartes
Newton
Bacon
Darwin
Agassiz
Clausius
Pasteur
Faraday
Einstein
Pierre and Marie Curie
Eve Curie
Heisenberg
Jeans
Cedering
Peirce
Poincaré
Kuhn
Keller
Hardy
Fisher

University Press of America,® Inc.
4720 Boston Way
Lanham, Maryland 20706

12 Hid's Copse Rd.
Cummor Hill, Oxford OX2 9JJ

Printed in the United States of America
British Library Cataloguing in Publication Information Available

Co-published by arrangement with Lynchburg College

ISBN 0-7618-0829-9 (pbk: alk. ppr.)

∞™ The paper used in this publication meets the minimum requirements of American National Standard for information Sciences—Permanence of Paper for Printed Library Materials, ANSI Z39.48—1984

CONTENTS

PREFACE

The Symposium Readings were first developed by a group of Lynchburg College faculty members to be used in a unique senior capstone course, the Senior Symposium, which has been required of all seniors at Lynchburg College since 1976. The symposium is, loosely speaking, a "great books—great issues" course whose aim is to integrate the fragmented knowledge acquired by students through the traditional "courses" offered by academic departments. A more complete description of the course and its philosophical underpinnings can be obtained by writing to the editor of the series.

More recently, another group of Lynchburg faculty proposed using the Symposium Readings to teach reading, writing, and speaking skills across the curriculum. The Lynchburg College Symposium Reading (LCSR) project evolved from their proposal. First funded by FIPSE, the project continues to evolve into an exciting part of the college curriculum. As various faculty experimented with the approach in their courses, it became apparent that a richer selection of readings would be required. Thus, the second edition. One might characterize the first edition as a largely male chorus, mostly white European males, punctuated by a few solo voices of women and minorities. The faculty who worked on the second edition has conscientiously tried to enrich the chorus, adding more voices of women, minorities, and third-world authors.

The editors do not claim that these ten volumes are in any way a "canon." What constitutes a "classic" in an evolving field such as science and technology is at best an educated guess. Academic interests change, and in fact, discussions of a third edition, with different selections, are already underway. But that lies well in the future, to be done by the next generation of faculty.

ACKNOWLEDGEMENTS

The editors acknowledge with appreciation the permissions granted by the following copyright holders:

From *The Structure of Scientific Revolutions* by Thomas Kuhn. Copyright © 1970, The University of Chicago Press.

From *Madame Curie* by Eve Curie. Copyright © 1937 by Doubleday, a division of Bantam, Doubleday, Dell Publishing Group, Inc. Used by permission of Doubleday, a division of Bantam, Doubleday, Dell Publishing Group, Inc.

From *Relflections on Gender and Science* by Evelyn Fox Keller. Copyright © 1985, Yale University Press.

From *The Design of Experiments* from *Statistical Methods, Experimental Design, and Scientific Inference* by R.A. Fisher edited by J.H. Bennett (1990). Oxford University Press.

"Letters from the Astronomers" from *Letters from the Floating World* by Siv Cedering, Copyright © 1984. Reprinted by permission of the University of Pittsburgh Press.

From *Discoveries and Opinions of Galileo* by Galileo Galilei. Copyright © 1957 by Stillman Drake. Used by permission of Doubleday, a division of Bantam Doubleday Dell Publishing Group, Inc.

From *Three Copernican Treatises*, "Commentariolus" by Nicholas Copernicus. Copyright © 1959, Dover Publications, Inc.

From *Mathematical Principles of Natural Philosophy and His System of the World* by Isaac Newton, edited by Florian Cajori, translated/edited by Andrew Motte (1947). Permission granted by the Regents of the University of California and the University of California Press.

From *Out of My Later Years* by Albert Einstein (1950). Philosophical Library, New York.

"The History of Quantum Theory" from *Physics and Philosophy* by Werner Heisenberg. Copyright © 1958 by Werner Heisenberg. Reprinted by permission of HarperCollins Publishers, Inc.

The editors further acknowledge with thanks the advice of faculty colleagues in the selection of these readings and the support of Dean Jacqueline Asbury. A special note of appreciation is due Mrs. Laurie Cassidy for her invaluable assistance in preparing these materials for publication.

INTRODUCTION

The mind of the most rational among us may be compared to a stormy ocean of passionate convictions based upon desire, upon which float perilously a few tiny boats carrying a cargo of scientifically tested beliefs. Nor is this to be altogether deplored: life has to be lived, and there is no time to test rationally all the beliefs by which our conduct is regulated. Without a certain wholesome rashness, no one could long survive. Scientific method, therefore, must in its very nature, be conned to the more solemn and official of our opinions.

Bertrand Russell

Human beings are an inquisitive lot. From the beginning of understanding, we have struggled to know our place in the cosmos—the universe in which we live and which in some mysterious way defines our being. The need to know, to understand, even to control our universe, provided the driving force for the particular way of knowing the world which we call science. Although the immediate appearance of nature as manifested in the position of a planet, the shape of a continent, the elusiveness of a breeze, or the violence of a black hole is temporal, scientists believe to the point of faith that the physical laws governing each of these are unchanging and are the same throughout the universe. Scientists believe that these laws are knowable and that they are called to ferret them out. What they learn provides an organizing framework for understanding who and what we are. Science has been incredibly successful in applying its understanding to the creation of new knowledge and understanding. For a wide range of questions dealing with aspects of the external world, no other way of knowing matches its success. But it is not the only way of knowing, and an important part of becoming an educated person is to be able to distinguish the questions which science can possibly answer from those which must be answered through other realms of knowledge.

The readings in this volume were selected because in some important way, each illustrates a particular aspect of the endeavor we call science. The selections may be used in a variety of ways. Many are appropriate for introductory courses in the history of Western civilization. All of the readings can be used to supplement textbooks in introductory science courses. Most of the readings are taken from physical science, because mathematics and astronomy, and their offspring physics, are *prior*. They *are* prior in the historical sense; they are also prior because they are arguably the most successful, in that they deal with the simplest systems, systems which readily lend themselves to mathematical analysis. The origins of what we recognize as science arise from the fusion of mathematics and astronomy which led to Newton's *Principia*. Several of the selections treat mathematics; one treats statistics, and three are taken from biology. A variety of sources from psychology can be found in Volume X of the series, *Human Nature,* and readings which discuss the impact of science and technology on society can be found in Volume VIII, *Science, Technology and Society.*

In some sense, any anthology is essentially a history of the subject. The readings in this volume are arranged roughly in chronological order and for that reason they may provide some crude idea of the history of thought in the sciences. However, the reader should not assume that this collection provides adequate insight into the history of physical thought. The selection of Greek writings is altogether too short to be of real use in a historical sense. The collection provides no sense of the important work done in Europe during the Medieval period, of the importance of translation of Greek texts into Latin, of the importance of the rise of the great European universities, or even of the importance of the Judeo-Christian idea that nature is ordered and knowable.

In the first section, **Greek Origins**, the reader will be able to glimpse the origins of modern science through the eyes of the Greeks. The Greek emphasis on rational thought led to the idea that order exists in the world and that humans can learn about that order by thinking carefully about it. These selections primarily treat Greek mathematics and astronomy, including speculations on the nature and causes of motion—speculations which were to dominate western views of motion for more than a millennium and which would later be rejected by Copernicus, Galileo, and Newton.

The next section, **The Scientific Revolution**, contains selections from the 16th and 17th centuries, readings which illustrate the origins of the Scientific Revolution. Copernicus argued for a sun-centered universe instead of an earth-centered universe; Galileo trained his new telescope on the heavens and became the first person to see the moon, the sun, planets and other heavenly objects with anything other than the unaided human eye. His discovery of the moons of Jupiter further eroded the previous belief that everything in the universe orbited the earth. Descartes believed that all of nature was subject to precise mathematical laws and his works were important to the development of mathematics and to the use of mathematics in attempts to describe nature. He tried mightily to develop a coherent method by which to study the world around us. Newton provided the great synthesis, connecting the falling of objects on earth to the motion of the moon about the earth, providing a simple explanation of such diverse phenomena as Kepler's laws of planetary motion, the orbit of Halley's comet and the precession of the perihelion of Mercury. Newton's work was so successful that it fostered the idea that dominated the 18th century—the enlightenment idea that science could solve all of humankind's problems. Francis Bacon refined the notion of "scientific method," as he articulated the primacy for science of inductive reasoning—reasoning from the particular of experiment to the general theory.

A variety of selections from the 19th and early 20th century provide glimpses of the growing maturity and complexity of scientific thought in the section entitled **The Growth and Maturation of Scientific Knowledge**. Darwin clearly had the greatest intellectual impact on the larger society and, like Galileo, his work fertilized a contentious and ongoing friction between science and religion. Louis Agassiz, a renowned teacher, also provided many of the seminal ideas of classical biology. Rudolph Clausius was among the first to articulate the trick which nature plays on those who would use heat as a source of energy. Researchers struggled for a century to understand why they could not build heat engines which would convert most of the heat energy to work. The answer rests in *The Second Law of Thermodynamics*, which describes the constraints nature places on the conversion of heat into other forms of energy. Louis Pasteur writes in an extraordinarily clear manner of the implications of his discovery of "germs" and the application of that discovery to medical practices. In a remarkable way, Michael Faraday illustrates one of the genuine fascinations which science holds for those who love it, the full and complex description of what appears to be a simple process. The reading, which describes the scientist's understanding of the burning of a candle, is actually one of Faraday's annual Christmas lectures for young people.

very high risk of reoffending', in addition to finite sentences for other violent offences.[271]

In the complaint before the Committee, the authors alleged that the discretionary sentence of preventive detention on the grounds of 'evidence of future dangerousness' breached Article 9(1) prohibition of arbitrary detention. The authors further submitted that the preventive detention regime under which their sentences would not come up for review until they had served a minimum of ten years, violated the right to periodic review of indefinite sentences guaranteed by Article 9, para 4 of the Covenant.[272] The authors also claimed that the ten-year non-parole period violated the prohibition of cruel and inhuman punishment (Article 7) and the right to persons in custody to be treated with humanity (Article 10, para 1). They also alleged that the sentences violated the 'essential aim of reformation and social rehabilitation required by Article 10, paragraph 3', since the treatment programmes aimed at reducing the risk of re-offending would not be made available until the close to the expiry of the ten-year period.[273]

The majority in *Rameka* found a violation of Article 9(4) in respect of Harris, who had been given a finite sentence of seven-and-a-half years but could not have his sentence reviewed before he had served out ten years, thus leaving an unreviewable period of over two years. However, the preventive detention scheme as applied to the authors was compatible with the Covenant as long as appropriate review was available to detainees.[274] In the instant case, compulsory annual reviews by the Parole Board upon the expiry of a ten-year non-parole period met that requirement. The majority declared the claims under Article 7 and Article 10 inadmissible as the authors had failed to specify in detail which courses they 'should be entitled to undertake at an earlier point of imprisonment'.[275] Disagreeing with the majority, four members of the Committee took the view that preventive detention based on a forecast made according to vague criteria was contrary to Article 9(1) of the ICCPR:

> In our view, the arbitrariness of such detention, even if the detention is lawful, lies in the assessment made of the possibility of the commission of a repeat offence. The science underlying the assessment in question is unsound. How can anyone seriously assert that there is a '20% likelihood' that a person will re-offend?[276]

Three other members of the Committee dissented on the holding of a violation of Article 9(4) in respect of Harris. In their view, the idea that the period of

271 Ibid para 2.6.
272 Ibid para 3.3.
273 Ibid para 3.6.
274 Ibid para 7.3.
275 Ibid para 6.4.
276 *Rameka and others v New Zeeland*. Individual Opinion of Committee members, Mr. Prafullachandra Natwarlal Bhagwati, Ms. Christine Chante, Mr. Gelled Hanahan and Mr. Hipolito Solara Irigoyen (dissenting in part), p 19.

Albert Einstein shook the scientific world to its very foundations in 1905, when he published several papers in *Annalen der Physick.* One of the papers offered by the then unknown German outlined the two simple assumptions underlying the special theory of relativity. Despite the aura of difficulty which surrounds this writing, it is a clear and fully comprehensible treatment of Einstein's understanding of space and time. Relativity provided one of the main streams of physical thought during the 20th century, especially in the area of cosmology. Concurrently, the Curies were doing their painstaking research into the mystery of radioactivity. A second paper by Einstein in 1905 provided much of the thrust which led to the modern understanding of the quantum nature of nature. Werner Heisenberg provides some insight into this second stream of thought which has dominated the 20th-century world of the very small. James Jeans and Siv Cedering provide additional perspective with their more recent commentaries on astronomy and astronomers.

The six readings in the next section, entitled **Science as a Way of Knowing**, along with Descartes, Newton, Bacon, and Einstein, can provide the reader with substantial insight into the heart of scientific methodology—how science differs from other ways of knowing the world. Most people who think about it believe that science as a way of knowing differs principally through its use of the so-called "scientific method." While philosophers study science and try to distill its "method," scientists go about their business, blithely unaware of the details of the "philosophy of science." Only the writings of Descartes, Peirce and Kuhn would likely be included in an anthology of the philosophy of science [and Peirce is rarely regarded as a philosopher of science]. Each of these readings illustrates how science is done and how scientists think about the bases of their theories. Peirce's pragmatism probably comes closest to the way most scientists actually proceed with their work. Poincaré touches on many aspects of scientific thought, and especially on the way in which mathematics joins with science. His "Mathematical Creation" describes in a wonderful way the creativity necessary to reach new insights in mathematics. Thomas Kuhn's *The Structure of Scientific Revolutions* is probably the most widely read and most influential book on the philosophical underpinnings of science. Evelyn Fox-Keller argues the importance of the language which actually describes science, particularly gender-based language. G.H. Hardy writes with great pride about his understanding of the joy of "pure" mathematics and Sir Ronald Fischer discusses the statistical nature of experimentation.

Finally, no collection of readings which purports to provide insight into human understanding of the universe would be in any way complete without some reference to the manner in which science has affected religion and vice versa. Prior to Copernicus and Galileo, most Western Europeans believed, as Aristotle had argued, that the Earth was the center of the universe. They also believed, as *Genesis* stated, that God created humans to inhabit that center. The Scientific Revolution removed humanity from that center and ultimately placed us orbiting an unremarkable [yellow dwarf] star two-thirds of the way along one arm of a galaxy which is itself ordinary and unremarkable. Darwin's theory allowed the possibility that the appearance of human life, or any life at all, on this planet may have been accidental. Whatever one believes about religion, one cannot avoid confronting the impact of evolutionary theory on those beliefs. Also, religious faith is based on ideas of the absolute, while science can change its beliefs at will, based on the experimental evidence. So it is not difficult to understand the tension between science and religion. One can argue, as many writers have, that Galileo, through his insistence on observation and objectivity drove the first wedge between science and religion, which prior to Galileo were not regarded as antithetical. Galileo's somewhat polemical "Letter to the Grand Duchess Christina" articulates his understanding of the separation of the two realms, as does Einstein's well-known essay, "Science and Religion." They, like Darwin, see no real conflict between science and religion.

Had I been present at the creation, I would have given some useful hints for the better ordering of the universe.

Alfonso X (1221-1284)

What science has to teach us . . .is not its techniques but its spirit: the irresistible need to explore. Perhaps the techniques of science may be practiced for a time without its spirit, in secret establishments, as the Egyptians practiced their priestcraft. But the inspiration of science for four hundred years has been opposite to this. It has created the values of our intellectual life and, with the arts, has taught them to our civilization. Science has nothing to be ashamed of even in the ruins of Nagasaki. The shame is theirs who appeal to other values than the human imaginative values which science has evolved. the shame is ours if we do not make science part of our world, intellectually as much as physically, so that we may at last hold these halves of the world together by the same values. For this is the lesson of science, that the concept is more profound than its laws, and the act of judging more critical than the judgement.

Jacob Bronowski, *Science and Human Values*, 1956

The only solid piece of scientific truth about which I feel totally confident is that we are profoundly ignorant about nature. . .It is this sudden confrontation with the depth and scope of ignorance that represents the most significant contribution of twentieth-century science to the human intellect.

Lewis Thomas

Science has a simple faith, which transcends utility. Nearly all men of science, all men of learning for that matter, and men of simple ways too, have it in some form and in some degree. It is the faith that it is the privilege of man to learn to understand, and that this is his mission. If we abandon that mission under stress we shall abandon it forever, for the stress will not cease. Knowledge for the sake of understanding, not merely to prevail, that is the essence of our being. None can define its limits, or set its ultimate boundaries.

Vannevar Bush, *Science is Not Enough*, 1967

Poetry is not the proper antithesis to prose, but to science. Poetry is opposed to science, and prose to metre. The proper and immediate object of science is the acquirement, or communication, of truth; the proper and immediate object of poetry is the communication of immediate pleasure.

Samuel Taylor Coleridge, *Definitions of Poetry*, 1811

But you tell me of an invisible planetary system where electrons gravitate around a nucleus. You explain this to me with an image. I realize then that you have been reduced to poetry: I shall never know. Have I the time to become indignant? You have changed theories. So that science that was to teach me everything ends up in a hypothesis, that lucidity founders in metaphor, that uncertainty is resolved in a work of art.

Albert Camus, *The Myth of Sisyphus*

Science is the search for truth—it is not a game in which one tries to beat his opponent, to do harm to others.

Linus Pauling, *No More War*, 1958

[Science is] the great antidote to the poison of enthusiasm and superstition.

Adam Smith, *The Wealth of Nations*, 1776

Science is a cemetery of dead ideas, even though life may issue from them.

Miguel de Unamuno, *The Tragic Sense of Life*, 1913

Science is, after all, only an expression of our ignorance of our own ignorance.

Samuel Butler, *Samuel Butler's Notebooks*, 1951

GREEK ORIGINS

Proclus

Euclid

Aristotle

Cleomedes

Simplicius

Archimedes

Plutarch

There are in fact two things, science and opinion. The former begets knowledge, the latter ignorance.

Hippocrates, *Law*

One must learn by doing the thing; though you think you know it, you have no certainty until you try.

Sophocles, *Trachiniae*

How comes it then that some things are moved continuously, though that which has caused them to move is no longer in contact with them, as, for instance, things thrown?

Aristotle, *Physics*

Euclid alone. . .
Has looked on Beauty bare. Fortunate they
Who, though once only and then but far away,
Have heard her massive sandal set on stone.

Edna St. Vincent Millay, 1923

Geometry enlightens the intellect and sets one's mind right. All its proofs are very clear and orderly. It is hardly possible for errors to enter into geometrical reasoning, because it is well arranged and orderly. Thus the mind that constantly applies itself to geometry is not likely to fall into error. In this convenient way, the person who knows geometry acquires intelligence. It has been assumed that the following statement was written upon Plato's door: "No one who is not a geometrician may enter our house.

Ibn Khaldun, *The Muqaddimadah*

Aristotle maintained that women have fewer teeth than men; although he was twice married, it never occurred to him to verify this statement by examining his wives' mouths.

Bertrand Russell, *The Impact of Science on Society*, 1952

GREEK ORIGINS

Proclus (410-485 A.D.) is a principal source on the history of mathematics and he is our chief source of information about Euclid. Surnamed Diadochus ("successor") because he succeeded Syrianus in the chair of philosophy at Athens, Proclus was born in Constantinople, but reared at Xanthus in Lycia. He studied grammar and philosophy at Alexandria before moving to Athens. There he devoted himself mainly to the writing of extensive commentaries on Plato and other philosophers. He wrote several commentaries on astronomy and mathematics, in addition to the one on the first book of Euclid's *Elements*, a part of which is reproduced in this section.

Euclid, the most noted of the ancient Greek mathematicians, lived in the third century B.C., though the actual dates of his birth and death are unknown. HE is said to have founded the mathematical school of Alexandria, and was reported by a contemporary and probably a student, Pappus, to have been unpretending, of mild and inoffensive temperament, and kind to all genuine student of mathematics. Although many of the principles of geometry had been discovered before Euclid, he brought them together in systematic form and presumably added theorems of his own until his *Elements* superseded all the predecessors. Indeed for the next twenty-two centuries the first six books of the *Elements* were used as an introduction to geometry almost everywhere in the Western world. On being asked by King Ptolemy I of Egypt if there were not an easier way to learn geometry than by studying the *Elements*, Euclid reportedly responded, "There is no royal road to geometry."

Aristotle (384-322 B.C.) has the reputation of being one of the greatest thinkers, philosophers, and scientific investigators of all time. He was born at Stagira in the Thracian peninsula of Chalcidice, and his father, Nicomachus, was court physician and friend of the Macedonian King Amyntas. At the age of eighteen, Aristotle went to Athens where he joined the Academy as a student of Plato. He remained there for twenty years, until Plato's death. In 343 he answered a call of Philip of Macedore to become the tutor of the King's 13 year old son who would become Alexander the Great. When Alexander set out on his military campaign in Asia, Artistotle returned to Athens where he founded a school at the Lyceum.

Aristotle became a prolific writer, and no field of knowledge seemed to escape his attention. In addition to a number of popular dialogues, modeled on those of Plato, he wrote extensively on logic, ethics, politics, metaphysics, and rhetoric and criticism as well as on natural science. In a way, he laid the foundation for all subsequent science. Yet his research and thinking were so profound and his prestige so great, they probably delayed for centuries acceptance of more modern views of the universe and the development of modern scientific research.

Almost nothing is known of the life of Cleomedes, but it is thanks to him that we know much of the work of Eratosthenes of Alexandria and Posidonius of Cyrene. Cleomedes probably wrote his great work *On the Orbits of the Heavenly Bodies* in the 2nd century A.D. Eratosthenes (c. 276-c.194 B.C.), like Aristotle, wrote on many subjects, but probably his most remarkable feat was in being the first man known to measure the circumference of the earth. According to one account, he set about this when he learned that at noon of the summer solstice the sun's rays completely lighted a well at Syene (Aswan), while at Alexandria the sun's rays were at a lower angle. He also worked out a "sieve" to discover all prime numbers. For several hears, he served as director of the great library of Alexandria. Posidonius (c. 135-c. 51 B.C.), of whom Cleomedes also wrote, was a leading Stoic philosopher and considered the most learned man of his time. After a number of years of travel and scientific research, he settled at Rhodes as a teacher. In fact he had been a teacher of Cicero.

The commentaries of Simplicius on certain of the works of Aristotle and of Epictetus have been invaluable for later scholars. A native of Cicilia, Simplicius was one of the last of the neoplatonists teaching in the school of philosophy at Athens. But in 529 A.D. the Roman emperor Justinian closed the school and forbade the teaching of philosophy. Simplicius, together with six colleagues, went over to Persia and received the protection of the Persian king. Two years later they were able to return to their homes under a treaty of peace that the Persian king concluded with Justinian that guaranteed these philosophers "enjoyment of liberty of conscience." It was then that Simplicius was able to do his serious writing.

"Give me a place to stand and I will move the earth." So is quoted Archimedes (c. 287-212 B.C.) in emphasizing the effectiveness of his lever. Born in the Greek city of Syracuse on the island of Sicily, he went to Alexandria for a few years of study and then spent the rest of his life in his native city. In addition to the lever, he is noted for developing a water screw for raising water from a stream for irrigating a field; the catapult, a weapon for hurling big stones at an enemy; and pulleys for lifting weights. He was an accomplished mathematician who determined a more exact value of pi and he anticipated Newton by nineteen-hundred years in some use of the methods of calculus. He also discovered basic laws of hydrostatics, notably in buoyancy and the displacement of water by an object according to volume or weight. This insight is said to have occurred to him while he was in the bath, and excitedly he ran out, without stopping to dress, into the street shouting, "Eureka! Eureka!" (I have found it). Archimedes died at the hands of a Roman soldier during the capture of Syracuse by Marcellus, though the general had declared that the great scientist and his house should be spared.

In his *Life of Marcellus*, Plutarch (writing over three hundred years after the event) took the occasion to explain how the ingenuity of Archimedes was instrumental in holding off the attacks of the Romans against the revolting city of Syracuse for over two years. This is taken from Plutarch's great work, *Parallel Live of Illustrious Greeks and Romans*.

1. Do you agree that scientific ideas appear and disappear in cycles? How do you account for that notion? What gave rise to the knowledge of arithmetic with the Phoenicians and of geometry with the Egyptians?

2. Why is Euclid called "the father of geometry" when Pythagoras preceded him by more than two hundred years?

3. What is Euclid's proof of the Pythagorean Theorem?

4. What is Proclus's criticism of Euclid's *Parallel Postulate*?

5. What is Aristotle's concept of motion? How does that compare with modern ideas? What does he mean by "natural and unnatural" movement and "natural and unnatural" states of rest?

6. For Aristotle, what is the significance of the number 3? (Might this have some bearing on the Christian conception of the Trinity?)

7. What was the difference in approach of Eratosthenes and Posidonius on the measurement of the circumference of the earth? Which do you think was more accurate?

8. What is Archimedes' law of the lever? What is its significance?

9. How did Archimedes contribute to the defense of Syracuse against the attacks of the Romans?

10. How do you account for the succession of great Greek mathematicians and scientists like Euclid, Aristotle, Eratosthenes, Archimedes, and the others, and yet for the Greeks to have failed to develop what we would call modern science?

COMMENTARY ON EUCLID'S ELEMENTS

PROCLUS DIADOCHUS

The History of Geometry to the Time of Euclid

We must next speak of the origin of geometry in the present world cycle. For, as the remarkable Aristotle tells us, the same ideas have repeatedly come to men at various periods of the universe. It is not, he goes on to say, in our time or in the time of those known to us that the sciences have first arisen, but they have appeared and again disappeared, and will continue to appear and to disappear, in various cycles, of which the number both past and future is countless. But since we must speak of the origin of the arts and sciences with reference to the present world cycle, it was, we say, among the Egyptians that geometry is generally held to have been discovered. It owed its discovery to the practice of land measurement. For the Egyptians had to perform such measurements because the overflow of the Nile would cause the boundary of each person's land to disappear. Furthermore, it should occasion no surprise that the discovery both of this science and of the other sciences proceeded from utility, since everything that is in the process of becoming advances from the imperfect to the perfect. The progress, then, from sense perception to reason and from reason to understanding is a natural one. And so, just as the accurate knowledge of numbers originated with the Phoenicians through the commerce and their business transactions, so geometry was discovered by the Egyptians for the reason we have indicated.

It was Thales, who, after a visit to Egypt, first brought this study to Greece. Not only did he make numerous discoveries himself, but laid the foundations for many other discoveries on the part of his successors, attacking some problems with greater generality and others more empirically. After him Mamercus, the brother of the poet Stesichorus, is said to have embraced the study of geometry, and in fact Hippias of Elis writes that he achieved fame in that study.

After these Pythagoras changed the study of geometry, giving it the form of a liberal discipline, seeking its first principles in ultimate ideas, and investigating its theorems abstractly and in a purely intellectual way. It was he who discovered the subject of proportions and the construction of the cosmic figures. After him Anaxagoras of Clazomenae devoted himself to many of the problems of geometry, as did Oenopides of Chios, who was a little younger than Anaxagoras. Plato in *The Rivals'* mentions them both as having achieved repute in mathematics. After them Hippocrates of Chios the discoverer of the quadrature of the lune, and Theodorus of Cyrene gained fame in geometry. For Hippocrates is the first man on record who also composed Elements.

Plato, who lived after Hippocrates and Theodorus, stimulated to a very high degree the study of mathematics and of geometry in particular because of his zealous interest in these subjects. For he filled his works with mathematical discussions, as is well known, and everywhere sought to awaken admiration for mathematics in students of philosophy.

At this time lived also Leodamas of Thasos, Archytas of Tarentum, and Theaetetus of Athens, all of whom increased the number of theorems and made progress towards a more scientific ordering thereof. Neoclides and his pupil,

Leon, who were younger than Leodamas, made many additions to the work of their predecessors. As a result the Elements composed by Leon are more carefully worked out both in the number and in the usefulness of the propositions proved. Leon also first discussed the *diorismi* (distinctions), that is the determination of the conditions under which the problem posed is capable of solution, and the conditions under which it is not.

A little younger than Leon was Eudoxus of Cnidus, an associate of Plato's school, who first increased the number of the general theorems, as they are called, and added three new proportions to the three then known. Using the method of analysis he greatly extended the theory of the section, a subject which had originated with Plato. Amyclas of Heraclea, a friend of Plato, Menaechmus, a pupil of Eudoxus who had also been associated with Plato, and Dinostratus, a brother of Menaechmus, brought the whole of geometry to an even higher degree of perfection. Theudius of Magnesia achieved a reputation for excellence both in mathematics and in the other parts of philosophy, for his Elements were excellently arranged and many of the (theretofore) limited propositions were put in more general form. Furthermore, Athenaeus of Cyzicus, a contemporary, distinguished himself in mathematics generally and in geometry in particular. These men spent their time together in the Academy and collaborated in their investigations. Hermotimus of Colophon extended the work that had been done by Eudoxus and Theaetetus, made many discoveries in the Elements, and put together some material on the subject of Loci. Philippus of Mende6 who, as a pupil of Plato, was inspired by him to the study of mathematics, conducted investigations along lines suggested by Plato, and in particular set before himself researches which he thought would contribute to the philosophy of Plato. Now those who have written histories trace the development of the science of geometry down to Philippus.

Euclid, who was not much younger than Hermotimus and Philippus, composed Elements, putting in order many of the theorems of Eudoxus, perfecting many that had been worked on by Theaetetus, and furnishing with rigorous proofs propositions that had been demonstrated less rigorously by his predecessors. Euclid lived in the time of the first Ptolemy, for Archimedes, whose life overlapped the reign of this Ptolemy too, mentions Euclid. Furthermore, there is a story that Ptolemy once asked Euclid whether there was any shorter way to a knowledge of geometry than by the study of the Elements. Whereupon Euclid answered that there was no royal road to geometry. He is, then, younger than Plato's pupils and older than Eratosthenes and Archimedes, who, as Eratosthenes somewhere remarks, were contemporaries.

By choice Euclid was a follower of Plato and connected with this school of philosophy. In fact he set up as the goal of the Elements as a whole the construction of the so-called Platonic figures.

There are, in addition, many other mathematical works by Euclid, written with remarkable accuracy and scientific insight, such as the *Optics*, the *Catoptrics*, works on the *Elements of Music*, and the book *On Divisions*.

But he is most to be admired for his *Elements of Geometry* because of the choice and arrangement of the theorems and problems made with regard to the elements. For he did not include all that he might have included, but only those theorems and problems which could fulfill the functions of elements. . . . If you seek to add or subtract anything, are you not unwittingly cast adrift from science and carried away toward falsehood and ignorance?

Now there are many things which seem to be grounded in truth and to follow from scientific principles but actually are at variance with these principles and deceive the more superficial. It was for this reason that Euclid set forth methods for intelligent discrimination in such matters, too. With these methods not only shall we be able to train beginners in this study to detect fallacies, but we shall be able to escape deception ourselves. The work in which he gives us this preparation he entitled *Pseudaria*.1 In it he recounts in order various types of fallacies, training us to understand each type by all sorts of theorems, setting the true alongside the false, and linking the refutation of the fallacy with empirical examples. This book, therefore, is for the purification and training of the understanding, while the *Elements* contain the complete and irrefutable guide to the scientific study of the subject of geometry.

* * * * *

The Divisions of Mathematics Pure and Applied

Such is the doctrine of the Pythagoreans and their fourfold division of mathematics. But others, among them Geminus, prefer to divide mathematics in another way. They consider mathematics on the one hand as concerned with things conceived by the mind, and on the other hand as concerned with and applied to things perceived by the senses. By things conceived by the mind they mean those which the psyche makes objects of contemplation by itself, when it completely divorces itself from forms connected with matter. In that division of mathematics which has to do with things conceived by the mind they place the two most basic and important branches, arithmetic and geometry. On the other hand, of that part of mathematics which devotes its attention to objects perceived by the senses they list six branches: mechanics, astronomy, optics, geodesy, canonics, and logistics.

They do not think, as some others do, that military science should be considered as one of the branches of mathematics, though, to be sure, it sometimes uses logistics, for example in the enumeration of companies,1 and at other times geodesy, for example in the division and measurement of areas. Similarly, there is even more reason why neither history nor medicine may be considered a part of mathematics, even though historians often use mathematical theorems, as when they give the location of climata or calculate the size and the diameter or circumference of cities. Physicians, too, by similar methods clarify many of the problems relevant to their field. For the importance of astronomy in medicine is made clear by Hippocrates and, in fact, by all who have discoursed on seasons and places. In the same way, then, the military tactician, without being a mathematician himself, uses the theorems of mathematics when, on occasion, he wishes to make the number of his troops appear as small as possible, and so arranges them in a circle, or, again, when he wishes to make the number appear as large as possible, and so arranges them in a square or pentagon or some other polygon.

While these are the branches of mathematics as a whole, geometry, in its turn, is divided into the theory of the plane and stereometry. For the study of points and lines cannot constitute a separate branch since the figures they form must be either plane or solid. Everywhere it is the task of geometry, in the case both of planes and solids, either to construct, or to compare or divide what has been constructed.

Similarly, arithmetic is divided into the study of linear, of plane, and of solid numbers. For it treats of the kinds of numbers in and of themselves, as they proceed from unity, the generation of plane numbers both similar and dissimilar, and the progression to the third dimension.

Geodesy and logistics are analogous to geometry and arithmetic, respectively, but are concerned with things perceived by the senses, not with numbers or figures as conceived by the mind. For it is not the business of geodesy to measure the cylinder or the cone, but to measure mounds as if they were cones, or wells as if they were cylinders. Geodesy deals, then, with straight lines, not as conceived by the mind, but as perceived by the senses, sometimes more sharply defined, as in the case of the rays of the sun, sometimes less sharply defined, as in the case of cords or a plumbline.

Similarly, one who employs logistics does not consider the properties of numbers by themselves, but always in connection with perceptible objects. Hence he gives to numbers a name after the objects that are being computed and thus speaks of *melites* and *phialites*. He does not admit an [absolute] minimum unit, as does the arithmetician, yet he does use a minimum unit in relation to some subject matter. Thus one man serves the logistician as the unit of a crowd of men.

Again, optics and canonics are derived from geometry and arithmetic, respectively. The science of optics makes use of lines as visual rays and makes use also of the angles formed by these lines. The divisions of optics are: (a) the study which is properly called optics and accounts for illusions in the perception of objects at a distance, for example, the apparent convergence of parallel lines or the appearance of square objects at a distance as circular; (b) catoptrics, a subject which deals, in its entirety, with every kind of reflection of light and embraces the theory of images; (c) scenography (scene-painting), as it is called, which shows how objects at various distances and of various heights may so be represented in drawings that they will not appear out of proportion and distorted in shape.

The science of canonics deals with the observed proportions of the notes of the musical scales and investigates the divisions of the monochord. It makes use, throughout, of sense perception and attributes greater importance to the ear than to the intellect, as Plato says.

In addition to these there is the science called mechanics which is a division of the study of material objects perceived by the senses. The science of mechanics embraces: *(a)* the manufacture of engines useful in war, for example, the engines of defense which Archimedes is said to have constructed against the besiegers of Syracuse; *(b)* the manufacture of wonderful devices, including those based on (1) air currents, e.g., devices such as Ctesibius and Hero describe, (2) weights (lack of equilibrium producing motion, and equilibrium producing rest, according to the definition in the *Timaeus)*, (3) ropes and cables, by means of which the motion of living beings may be imitated; *(c)* the study of equilibrium, in general, and of so-called centers of gravity; *(d)* sphere construction, for depicting the revolutions of the heavenly bodies, a field in which Archimedes worked; *(e)* in general, the whole subject of the kinetics of material bodies.

There remains astronomy, which is concerned with cosmic motions, the sizes and forms of the heavenly bodies, their illumination, their distances from the earth, and all other subjects of this sort. It makes wide use of sense perception but also has close connection with physical theory. Its branches are: (a) gnomonics, which is concerned with the measuring of the hours by the proper placing of gnomons; (b) meteoroscopy, which investigates the different elevations and distances of stars, and sets forth numerous other theorems of various sorts in the field of astronomy; (c) the science of dioptrics, which, with the use of the proper instruments, investigates the positions of the sun, moon, and the other stars.

The above is the account of the branches of mathematics that we have found in the ancient writers.

ELEMENTS

EUCLID

ELEMENTARY GEOMETRY

Postulates

Let the following be postulated:

1. To draw a straight line from any point to any point.
2. To produce a finite straight line continuously in a straight line.
3. To describe a circle with any centre and distance.
4. That all right angles are equal to one another.
5. That, if a straight line falling on two straight lines make the interior angles on the same side less than two right angles, the two straight lines if produced indefinitely, meet on that side on which are the angles less than the two right angles.

* * * * *

THE PYTHAGOREAN THEOREM

In right-angled triangles the square on the side subtending the right angle is equal to the squares on the sides containing the right angle.

Let *ABC* be a right-angled triangle having the angle *BAC* right;

I say that the square on *BC is* equal to the squares on *BA*, *,AC.*

For let there be described on *BC* the square *BDEC,* and on *BA*, *AC* the squares *GB*, *HC;*
through *A* let *AL* be drawn parallel to either *BD* or *CE,* and let *AD*, *FC* be joined.

Then, since each of the angles *BAC, BAG* is right, it follows that with a straight line *BA,* and at the point *A* on it, the two straight lines *AC, AG* not lying on the same side make the adjacent angles equal to two right angles; therefore *CA* is in a straight line with *AG.*

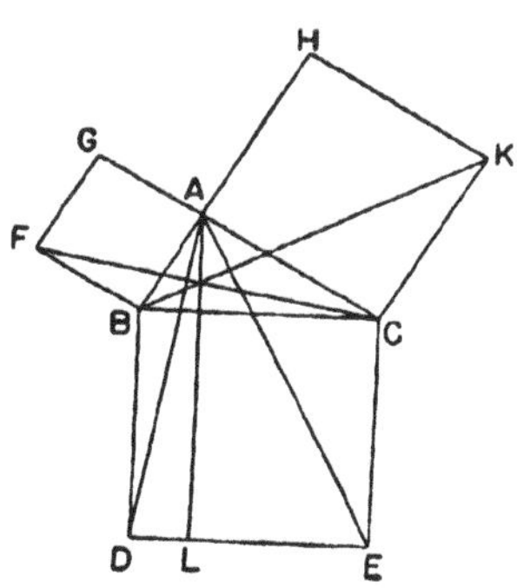

For the same reason
BA is also in a straight line with AH.

And, since the angle *DBC is* equal to the angle *FBA:*
for each is right:
let the angle *ABC* be added to each;

therefore the whole angle *DBA* is equal to the whole angle *FBC.*

And, since *DB* is equal to *BC,* and *FB* to *BA,* the two sides *AB, BD* are equal to the two sides *FB, BC* respectively;

and the angle *ABD* is equal to the angle *FBC;*

therefore the base *AD* is equal to the base *FC,*

and the triangle *ABD is* equal to the triangle *FBC.*

Now the parallelogram *BL* is double of the triangle *ABD*, for they have the same base *BD* and are in the same parallels *BD*, *AL*.

And the square *GB* is double of the triangle *FBC*,
for they again have the same base *FB* and are in the same parallels *FB*, *GC*.

But the doubles of equals are equal to one another.

Therefore the parallelogram *BL* is also equal to the square *GB*.

Similarly, if *AE*, *BK* be joined,
the parallelogram *CL* can also be proved equal to the square *HC*;

therefore the whole square *BDEC* is equal to the two squares *GB*,*HC*.

And the square *BDEC* is described on *BC*, and the squares *GB*, *HC* on *BA*, *AC*.

Therefore the square on the side *BC* is equal to the squares on the sides *BA*, *AC*.

Therefore etc. Q.E.D.

COMMENTARY ON EUCLID'S ELEMENTS

PROCLUS DIADOCHUS

The Parallel Postulate

This [fifth postulate] ought even to be struck out of the Postulates altogether; for it is a theorem involving many difficulties which Ptolemy, in a certain book, set himself to solve, and it requires for the demonstration of it a number of definitions as well as theorems. And the converse of it is actually proved by Euclid himself as a theorem. It may be that some would be deceived and would think it proper to place even the assumption in question among the postulates as affording, in the lessening of the two right angles, ground for such an instantaneous belief that the straight lines converge and meet. To such as these Geminus correctly replied that we have learned from the very pioneers of this science not to have any regard to mere plausible imaginings when it is a question of the reasonings to be included in our geometrical doctrine. For Aristotle says that it is as justifiable to ask scientific proofs of a rhetorician as to accept mere plausibilities from a geometer; and Simmias is made by Plato to say that he recognizes as quacks those who fashion for themselves proofs from probabilities. So in this case the fact that, when the right angles are lessened, the straight lines converge is true and necessary; but the statement that, since they converge more and more as they are produced, they will sometime meet is plausible but not necessary, in the absence of some argument showing that this is true in the case of straight lines. For the fact that some lines exist which approach indefinitely, but yet remain non-secant (ἀσύμπτωτοι), although it seems improbable and paradoxical, is nevertheless true and fully ascertained with regard to other species of lines. May not then the same thing be possible in the case of straight lines which happens in the case of the lines referred to? Indeed, until the statement in the Postulate is clinched by proof, the facts shown in the case of other lines may direct our imagination the opposite way. And, though the controversial arguments against the meeting of the straight lines should contain much that is surprising, is there not all the more reason why we should expel from our body of doctrine this merely plausible and unreasoned (hypothesis)?

It is then clear from this that we must seek a proof of the present theorem, and that it is alien to the special character of postulates. But how it should be proved, and by what sort of arguments the objections taken to it should be removed, we must explain at the point where the writer of the Elements is actually about to recall it and use it as obvious. It will be necessary at that stage to show that its obvious character does not appear independently of proof, but is turned by proof into matter of knowledge.

PHYSICS

ARISTOTLE

Book V
Chapter VI

BUT since we find a contrast not only between a given motion and its counter motion but also between experiencing that motion and being at rest from it, we must examine this matter also. For, while the opposite of a motion, in the full and proper sense, is the motion counter to it, absence of motion is also contrasted with motion, as being its non-accomplishment (for non-accomplishment is a kind of opposite to the 'might be'), and this holds in each particular category, e.g. local rest is contrasted with local motion.

This statement, however, is too general and needs qualification. Is abiding here opposed to moving hence or to moving hither? Evidently, since a movement must be between two terms A and B, movement from A to its opposite B will have for its opposite the abiding fixed in A, and movement from B to A will have for its opposite the abiding fixed in B.

At the same time these two fixities are also contrary to one another; for it would be absurd that there should be contrary movements and not also opposite states of rest, and there are such in fact, namely the states of rest in the opposite terms, *e.g.* abiding in health is opposed to abiding in sickness. The movement to which abiding in health is opposed is the movement from health to sickness; for it would be absurd to oppose it to the reverse movement from sickness to health (since movement towards the goal at which the subject is at rest is rather a process of 'coming to rest,' and 'coming to rest' is a process that coincides with the movement towards that goal), and it can only be one or the other of these two movements between health and sickness: no other opposite term can be in question, for that would involve the impossible consequence that rest in (say) 'whiteness' should be opposed to rest in health.

If a term has no opposite, then there can be no 'movement' from it or to it, but there may be a 'change' from its not being there to its being there and *vice versa.* In such a case you cannot say that it 'makes a stay' in being, or in not being, but only that there is absence of *change* from being or not-being. Now if there is some subject that changes from not being this or that into being it, then its not-changing from being that thing is the opposite of its not-changing from not being it. But if that which isn't the thing does not exist at all, one would be at a loss to express the opposite of the thing's being in existence and not changing from it, and one could hardly say whether such opposite were or were not a state of rest. If we are to speak of 'rest' in such a case, we must either say that 'rest' is not always opposed to 'movement,' or else that coming-to-be and ceasing-to-be are movements. It is clear, then, that since (as we have seen) these are not movements, we must not speak of rest, but call it something analogous to rest, namely unchangingness. Is there, then, an opposite to 'unchangingness in being'? Not if the non-existence opposed to existence is absolute; but otherwise the non-change of being a thing is opposed either to its non-change in non-existence or to its ceasing-to-be (which is a change from it, whereas genesis is a change to it).

One might. naturally ask why there are natural and unnatural *b* movings and abidings in locality, but not in other ways of passing from this to that. For instance, there is no natural or unnatural change of quality, for getting health is no more natural or unnatural than sickening, nor is growing white more or less natural than growing black; and in the same way neither are growth and shrinkage opposed to each other as one natural and the other unnatural, nor is one growth opposed to another in that way. And so again with coming-to-be and perishing, for neither is coming-to-be natural and perishing unnatural (for growing old is natural) nor do we distinguish between natural and unnatural coming-to-be. We may answer that if by 'unnatural' we mean 'enforced' then perishing may be opposed to perishing if the one is natural and the other enforced. Is there, then, such a thing as coming-into-being which is enforced and out of the naturally ordained way, and so contrary to natural coming-into-being, and are there forced growings and shrinkages, such as a youth's rapid growth to maturity due to luxury, or the rapid ripening of corn that is not packed solidly in the mould? And how about qualitive modifications ? It is the same with these, we may say: some may be considered forced and some natural; for instance recovery from a fever is a natural alteration if it occurs on a critical day, unnatural if the day is not critical. We shall, then, have perishings contrary in this respect rather to each other than each to an opposite coming-to-be. Why not ? For in a sense there are such, since one might be easy and another grievous, so that, if one way of perishing were not unqualifiedly opposed to another, yet they might be contrasted in so far as they have contrary qualities. It is, then, universally true that movements and states of rest are contrary in the manner just described. Take for instance the contrariety of upward and downward movement and of rest above and rest below. These are contrarieties in respect of place; and of the movements concerned the natural movement of fire is upward, of earth downward, and these movements of theirs are contrary; moreover, for fire upward movement is natural, downward movement unnatural, and its natural movement is certainly contrary to its unnatural movement. Also the same contrast of natural and unnatural applies to the 'staying where it is' of each element. 'Staying above' is contrary to motion from above downwards, and for earth staying above is an unnatural occurrence, while motion from above downwards is natural. And accordingly, one staying will be contrary to another—the unnatural staying contrary to the natural staying of the same thing (just as the unnatural movement was contrary to the natural movement of the same thing); for one of the two stayings—staying above or staying below as the ease may be—will be natural, the other unnatural.

The further question arises, whether every state of rest which does not exist at all times has a coming-into-being, and whether this coming-into-being can be identified with 'coming to a standstill.' Now when a thing is in an unnatural state of rest—when earth, for instance, stays aloft—that state of rest *b* must have a coming-into-being; and (if this is identified with 'coming to a standstill') that will mean that the earth in question was 'coming to a standstill' at a time when it was being forced upwards against its nature. But a thing that is coming to a standstill seems always to be moving with a quickening velocity, whereas what is forced against its nature is always losing velocity. So (if we identify the coming-to-be of an unnatural state of rest with 'coming to a stand-still,' and an unnatural movement cannot be said to come to a standstill), we shall have a thing which will *be* in a state of rest without ever having *come to be* in a state of rest—an absurd conclusion. Besides it is generally recognized that 'coming to a standstill' is either identical with a thing's moving to its proper place or a concomitant of that motion.

Again it may be questioned whether 'staying in a place' is really the contrary of 'moving out of it'; for when a thing starts moving out of its place or parting with some condition, it seems still to possess the condition it is losing, so that if the very being in it is the contrary of moving out of it towards the opposite, it combines two contraries. Or can it still be said to be at rest in some sort as long as any of it is left there ? At any rate, whenever anything is in process of shifting, some of it is in the state it occupied and is leaving, and some of it in the state it is changing into. So one counter movement is more perfectly contrary to the other than is the 'staying in place' which it abolishes.

Let this suffice for movement and staying still, and in what the unity of each consists, and which are contrary to which.

Yet another question occurs concerning the coming of things to a stand: Is there a state of rest opposed to such movements as are against nature and enforced? It would be absurd that there should not be; for the thing does stay where it is. On the other hand, this is the result of 'violence'; so we shall have something that is in a non-permanent state of rest without ever having *come to be* in that state. But clearly there must be such a state of rest; for just as things are moved against their nature, so a thing can be at rest against its nature.

Also since some things move both in accordance with their nature and against it, as fire goes up by nature and down by force, is its own down-going or that of the earth the contrary of its up-going (for it is earth that goes down by nature)? Obviously both its own down-going and that of earth are contrary to it, but not in the same sense. For one is a natural opposition of direction between things whose nature it is to go one this way and the other that. The other is an opposition between the natural up and the unnatural down of the same body. There is a similar opposition between abidings in position of the two bodies, though it may be said that, in a sense, a state of rest has for its opposite (not an opposite state of rest, but) a motion.

ON THE HEAVENS

ARISTOTLE

Book I
Chapter I

WE may say that the science of nature is for the most part plainly concerned with bodies and magnitudes and with their changing properties a and motions, as also with the principles which belong to that class of substance; for the sum of physically constituted entities consists of (a) bodies and magnitudes, (b) beings possessed of body and magnitude, (c) the principles or causes of these beings.

The continuous may be defined as that which is divisible into parts which are themselves divisible to infinity, body as that which is divisible in all ways. Magnitude divisible in one direction is a line, in two directions a surface, in three directions a body. There is no magnitude not included in these; for three are all, and "in three ways" is the same as "in all ways." It is just as the Pythagoreans say, the whole world and all things in it are summed up in the number three; for end, middle and beginning give the number of the whole, and their number is the triad. Hence it is that we have taken this number from nature, as it were one of her laws, and make use of it even for the worship of the gods. Our language too shows the same tendency, for of two things or people we say "both," not "all." This latter term we first employ when there are three in question; and in behaving thus, as I have said, we are accepting nature herself for our guide. Furthermore, "all," "the whole" and "the complete" do not differ from one another in form, but only, if at all, in their matter, and the subjects of which they are predicated. In this sense therefore body is the only complete magnitude, since it is the only one which is defined by extension in three directions, that is, which is an "all." Being divisible in three directions, it is divisible in all. Other magnitudes are divisible in one or two, for according to the number which can be appropriately applied to the different magnitudes, in so many directions is their divisibility and continuity. One is continuous in one direction, another in two, and the third in all.

We have seen that magnitudes which are divisible are continuous. Whether all continuous magnitudes are divisible has not emerged from the present inquiry. This much, however, is clear, that there is no further change to a fourth species of magnitude like the changes from length to surface and from surface to body. Otherwise body would not be the complete magnitude that we have shown it to be, for any advance on it could only be made where it fell short; but that which is complete cannot fall short; it extends in every possible direction.

Bodies which are parts of a whole are each complete in the sense here laid down, i.e., each has all the dimensions. But each one is limited by contact with the neighbouring part, so that in a sense each of such bodies is many. The whole of which these are parts, on the other hand, is of necessity complete, and, as its name suggests, wholly so, not complete in one respect and incomplete in another.

Chapter II

THE question of the nature of this Whole, whether it is of infinite magnitude or its total bulk is limited, must be left until later. We have now to speak of its formally distinct parts, and we may start from this, that all natural bodies and magnitudes are capable of moving of themselves in space; for nature we have defined as the principle of motion in them. Now all motion in space (locomotion) is either straight or circular or a compound of the two, for these are the only simple motions, the reason being that the straight and circular lines are the only simple magnitudes. By "circular motion" I mean motion around the centre, by "straight," motion up and down. "Up" means away from the centre, "down" towards the centre. It follows that all simple locomotion is either away from the centre or towards the centre or around the centre. This appears to follow consistently on what was said at the beginning: body was completed by the number three, and so now is its motion.

Of bodies some are simple, and some are compounds of the simple. By "simple" I mean all bodies which contain a principle of natural motion, like fire and earth and their kinds, and the other bodies of the same order. Hence motions also must be similarly divisible, some simple and others compound in one way or another; simple bodies will have simple motions and composite bodies composite motions, though the movement may be according to the prevailing element in the compound.

If we take these premises, (a) that there is such a thing as simple motion, (b) that circular motion is simple, (c) that simple motion is the motion of a simple body (for if a composite body moves with a simple motion, it is only by virtue of a simple body prevailing and imparting its direction to the whole), then it follows that there exists a simple body naturally so constituted as to move in a circle in virtue of its own nature. By force it can be brought to move with the motion of another, different body, but not naturally, if it is true that each of the simple bodies has one natural motion only. Moreover, granted that (a) unnatural motion is the contrary of natural, (b) a thing can have only one contrary, then circular motion, seeing it is one of the simple motions, must, if it is not the motion natural to the moved body, be contrary to its nature. Suppose now that the body which is moving in a circle be fire or some other of the four elements, then its natural motion must be contrary to the circular. But a thing can have only one contrary, and the contrary of upward is downward, and vice versa. Suppose on the other hand that this body which is moving in a circle contrary to its own nature is something other than the elements, there must be some other motion which is natural to it. But that is impossible: for if the motion were upward, the body would be fire or air, if downward, water or earth.

Furthermore, circular motion must be primary. That which is complete is prior in nature to the incomplete, and the circle is a complete figure, whereas no straight line can be so. An infinite straight line cannot, for to be complete it would have to have an end or completion, nor yet a finite, for all finite lines have something beyond them: any one of them is capable of being extended. Now if (a) a motion which is prior to another is the motion of a body prior in nature, (b) circular motion is prior to rectilinear, (c) rectilinear motion is the motion of the simple bodies (as e.g. fire moves in a straight line upwards and earthy bodies move downwards towards the centre), then circular motion also must of necessity be the motion of some simple body. (We have already made the reservation that the motion of composite bodies is determined by whatever simple body predominates in the mixture.) From all these premises therefore it clearly follows that there exists some physical substance besides the four in our sublunary world, and moreover that it is more divine than, and prior to, all these. The same can also be proved on the further assumption that all motion is either natural or unnatural, and that motion which is unnatural to one body is natural to another, as the motions up and down are natural or unnatural to fire and earth respectively; from these it follows that circular motion too, since it is unnatural to these elements, is natural to some other. Moreover, if circular motion is natural to anything, it will clearly be one of the simple and primary bodies of such a nature as to move naturally in a circle, as fire moves upward and earth downward. If on the other hand it be maintained that the revolutionary motion of the body which is carried round in a circle is unnatural, it is strange, in fact quite absurd, that being unnatural it should yet be the only continuous and eternal motion seeing that in the rest of nature what is unnatural is the quickest to fall into decay. And so, if fire be the body carried round, as some say, this motion will be no less unnatural to it than motion downwards for we see the natural motion of fire to be in a straight line away from the centre.

Thus the reasoning from all our premises goes to make us believe that there is some other body separate from those around us here, and of a higher nature in proportion as it is removed from the sublunary world.

ON THE ORBITS OF THE HEAVENLY BODIES

CLEOMEDES

(Translation by T. L. Heath)

I
Greek Astronomy

About the size of the earth the physicists, or natural philosophers, have held different views, but those of Posidonius and Eratosthenes are preferable to the rest. The latter shows the size of the earth by a geometrical method; the method of Posidonius is simpler. Both lay down certain hypotheses, and, by successive inferences from the hypotheses, arrive at their demonstrations.

Posidonius says that Rhodes and Alexandria lie under the same meridian. Now meridian circles are circles which are drawn through the poles of the universe and through the point which is above the head of any individual standing on the earth. The poles are the same for all these circles, but the vertical point is different for different persons. Hence we can draw an infinite number of meridian circles. Now Rhodes and Alexandria lie under the same meridian circle, and the distance between the cities is reputed to be 5,000 stades. Suppose this to be the case.

All the meridian circles are among the great circles in the universe, dividing it into two equal parts and being drawn through the poles. With these hypotheses, Posidonius proceeds to divide the zodiac circle, which is equal to the meridian circles, because it also divides the universe into two equal parts, into forty-eight parts, thereby cutting each of the twelfth parts of it (i.e., signs) into four. If, then, the meridian circle through Rhodes and Alexandria is divided into the same number of parts, forty-eight, as the zodiac circle, the segments of it are equal to the aforesaid segments of the zodiac. For, when equal magnitudes are divided into (the same number of) equal parts, the parts of the divided magnitudes must be respectively equal to the parts. This being so, Posidonius goes on to say that the very bright star called Canopus lies to the south, practically on the Rudder of Argo. The said star is not seen at all in Greece; hence Aratus does not even mention it in his *Phaenomena*. But, as you go from north to south, it begins to be visible at Rhodes, and, when seen on the horizon there, it sets again immediately as the universe revolves. But when we have sailed the 5,000 stades and are at Alexandria, this star, when it is exactly in the middle of the heaven, is found to be at a height above the horizon of one-fourth of a sign, that is, one forty-eighth part of the zodiac circle. It follows therefore,

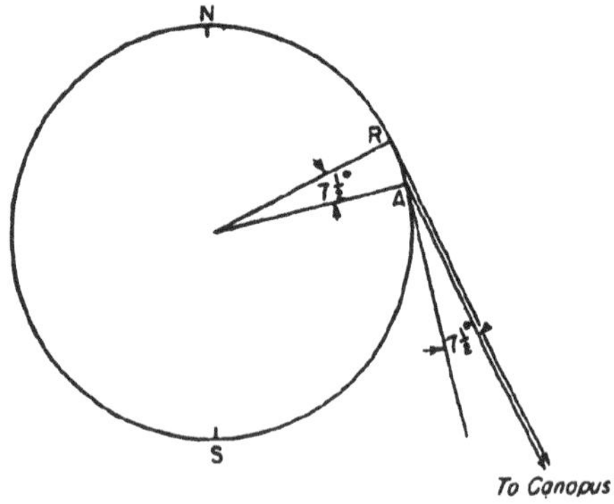

RA (Rhodes--Alexandria) = 5,000 stades. The star is assumed to be so distant that the lines to it from *R* and *A* may be considered parallel.

that the segment of the same meridian circle which lies above the distance between Rhodes and Alexandria is one forty-eighth part of the said circle, because the horizon of the Rhodians is distant from that of the Alexandrians by one forty-eighth of the zodiac circle. Since, then, the part of the earth under this segment is reputed to be 5,000 stades, the parts (of the earth) under the other (equal) segments (of the meridian circle) also measure 5,000 stades; and thus the great circle of the earth is found to measure 240,000 stades, assuming that from Rhodes to Alexandria is 5,000 stades; but, if not, it is in (the same) ratio to the distance. Such then is Posidonius' way of dealing with the size of the earth.

The method of Eratosthenes depends on a geometrical argument and gives the impression of being slightly more difficult to follow. But his statement will be made clear if we premise the following. Let us suppose, in this case too, first, that Syrene and Alexandria lie under the same meridian circle; secondly, that the distance between the two cities is 5,000 stades; and thirdly, that the rays sent down from different parts of the sun on different parts of the earth are parallel; for this is the hypothesis on which geometers proceed. Fourthly, let us assume that, as proved by the geometers, straight lines falling on parallel straight lines make the alternate angles equal, and fifthly, that the arcs standing on (i.e., subtended by) equal angles are similar, that is, have the same proportion and the same ratio to their proper circles—this, too, being a fact proved by the geometers. Whenever, therefore, arcs of circles stand on equal angles, if any one of these is (say) one tenth of its proper circle, all the other arcs will be tenth parts of their proper circles.

Any one who has grasped these facts will have no difficulty in understanding the method of Eratosthenes, which is this. Syrene and Alexandria lie, he says, under the same meridian circle. Since meridian circles are great circles in the universe, the circles of the earth which lie under them are necessarily also great circles. Thus, of whatever size this method shows the circle on the earth passing through Syrene and Alexandria to be, this will be the size of the great circle of the earth. Now Eratosthenes asserts, and it is the fact, that Syrene lies under the summer tropic. Whenever, therefore, the sun, being in the Crab at the summer solstice, is exactly in the middle of the heaven, the gnomons (pointers) of sundials necessarily throw no shadows, the position of the sun above them being exactly vertical; and it is said that this is true throughout a space three hundred stades in diameter. But in Alexandria, at the same hour, the pointers of sundials throw shadows, because Alexandria lies further to the north than Syrene. The two cities lying under the same meridian great circle, if we draw an arc from the extremity of the shadow to the base of the pointer of the sundial in Alexandria, the arc will be a segment of a great circle in the (hemispherical) bowl of the sundial, since the bowl of the sundial lies under the great circle (of the meridian). If now we conceive straight lines produced from each of the pointers through the earth, they will meet the centre of the earth. Since then the sundial at Syrene is vertically under the sun, if we conceive a straight line coming from the sun to the top of the pointer of the sun dial, the

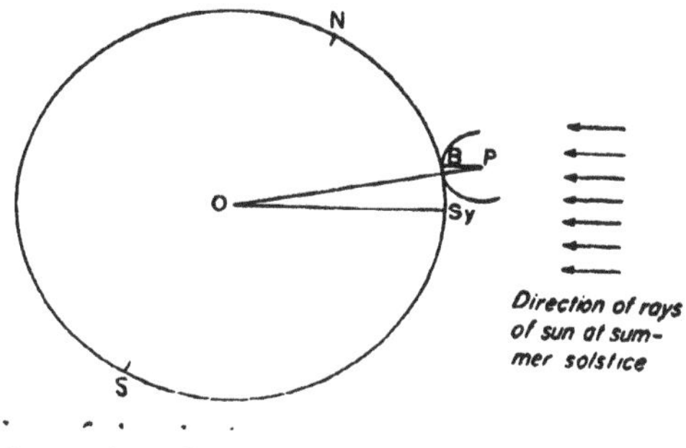

line reaching from the sun to the centre of the earth will be one straight line. If now we conceive another straight line drawn upwards from the extremity of the shadow of the pointer of the sundial in Alexandria, through the top of the pointer to the sun, this straight line and the aforesaid straight line will be parallel, since they are straight lines coming through from different parts of the sun to different parts of the earth. On these straight lines, there fore, which are parallel, there falls the straight line drawn from the centre of the earth to the pointer at Alexandria, so that the alternate angles which it makes are equal. One of these angles is that formed at the centre of the earth, at the intersection of the straight lines which were drawn from the sundials to the centre of the earth; the other is at the point of inter section of the top of the pointer at Alexandria and the straight line drawn from the extremity of its shadow to the sun through the point (the top) where it meets the pointer. Now on this latter angle stands the arc carried round from the extremity of the shadow of the pointer to its base while on the angle at the centre of the earth stands the arc reaching from Syrene to Alexandria. But the arcs are similar, since they stand on equal angles. Whatever ratio, therefore, the arc in the bowl of the sundial has to its proper circle, the arc reaching from Syrene to Alexandria has that ratio to its proper circle. But the arc in the bowl is found to be one-fiftieth of its proper circle. Therefore the distance from Syrene to Alexandria must necessarily be one-fiftieth part of the great circle of the earth. And the said distance is 5,000 stades; therefore the complete great circle measures 250,000 stades. Such is Eratosthenes' method.

THE SCOPE OF ASTRONOMY CONTRASTED WITH THAT OF PHYSICS

SIMPLICIUS

Translated by T. L. Heath

Alexander carefully quotes a certain explanation by Geminus taken from his summary of the *Meteorologica* of Posidonius. Geminus's comment, which is inspired by the views of Aristotle, is as follows:

> "It is the business of physical inquiry to consider the substance of the heaven and the stars, their force and quality, their coming into being and their destruction, nay, it is in a position even to prove the facts about their size, shape, and arrangement; astronomy, on the other hand, does not attempt to speak of anything of this kind, but proves the arrangement of the heavenly bodies by considerations based on the view that the heaven is a real (κόσμος) and further, it tells us of the shapes and sizes and distances of the earth, sun, and moon, and of eclipses and conjunctions of the stars, as well as of the quality and extent of their movements. Accordingly, as it is connected with the investigation of quantity, size, and quality of form or shape, it naturally stood in need, in this way, of arithmetic and geometry. The things, then, of which alone astronomy claims to give an account it is able to establish by means of arithmetic and geometry. Now in many cases the astronomer and the physicist will propose to prove the same point, e.g., that the sun is of great size or that the earth is spherical, but they will not proceed by the same road. The physicist will prove each fact by considerations of essence or substance, of force, of its being better that things should be as they are, or of coming into being and change; the astronomer will prove them by the properties of figures or magnitudes, or by the amount of movement and the time that is appropriate to it. Again, the physicist will in many cases reach the cause by looking to creative force; but the astronomer, when he proves facts from external conditions, is not qualified to judge of the cause, as when, for instance, he declares the earth or the stars to be spherical; sometimes he does not even desire to ascertain the cause, as when he discourses about an eclipse; at other times he invents by way of hypothesis, and states certain expedients by the assumption of which the phenomena will be saved. For example, why do the sun, the moon, and the planets appear to move irregularly? We may answer that, if we assume that their orbits are eccentric circles or that the stars describe an epicycle, their apparent irregularity will be saved; and it will be necessary to go further and examine in how many different ways it is possible for these phenomena to be brought about, so that we may bring our theory concerning the planets into agreement with that explanation of the causes which follows an admissible method. Hence we actually find a certain person, Heraclides of Pontus, coming forward and saying that, even on the assumption that the earth moves in a certain way, while the sun is in a certain way at rest, the apparent irregularity with reference to the sun can be saved. For it is no part of the business of an astronomer to know what is by nature suited to a position of rest, and

what sort of bodies are apt to move, but he introduces hypotheses under which some bodies remain fixed, while others move, and then considers to which hypotheses the phenomena actually observed in the heaven will correspond. But he must go to the physicist for his first principles, namely, that the movements of the stars are simple, uniform, and ordered, and by means of these principles he will then prove that the rhythmic motion of all alike is in circles, some being turned in parallel circles, others in oblique circles." Such is the account given by Geminus, or Posidonius in Geminus, of the distinction between physics and astronomy, wherein the commentator is inspired by the views of Aristotle.

ON THE EQUILIBRIUM OF PLANES

ARCHIMEDES

(Translation and paraphrase by T. L. Heath)

The Law of the Lever and Some Applications

"I postulate the following:

"1. Equal weights at equal distances are in equilibrium, and equal weights at unequal distances are not in equilibrium but incline towards the weight which is at the greater distance.

"2. If, when weights at certain distances are in equilibrium, something be added to one of the weights, they are not in equilibrium, but incline towards that weight to which the addition was made.

"3. Similarly, if anything be taken away from one of the weights, they are not in equilibrium but incline towards the weight from which nothing was taken.

"4. When equal and similar plane figures coincide if applied to one another, their centres of gravity similarly coincide.

"5. In figures which are unequal but similar the centres of gravity will be similarly situated. By points similarly situated in relation to similar figures I mean points such that, if straight lines be drawn from them to the equal angles, they make equal angles with the corresponding sides.

"6. If magnitudes at certain distances be in equilibrium, (other) magnitudes equal to them will also be in equilibrium at the same distances.

"7. In any figure whose perimeter is concave in (one and) the same direction the centre of gravity must be within the figure."

Proposition 1. Weights which balance at equal distances are equal. . . .

Proposition 2. Unequal weights at equal distances will not balance but will incline towards the greater weight. . . .

Proposition 3. Unequal weights will balance at unequal distances, the greater weight being at the lesser distance. . . .

Proposition 4. If two equal weights have not the same centre of gravity, the centre of gravity of both taken together is at the middle point of the line joining their centres of gravity. . . .

Proposition 5. If three equal magnitudes have their centres of gravity on a straight line at equal distances, the centre of gravity of the system will coincide with that of the middle magnitude. . . .

Corollary 1. The same is true of any odd number of magnitudes if those which are at equal distances from the middle one are equal, while the distances between their centres of gravity are equal.

Corollary 2. If there be an even number of magnitudes with their centres of gravity situated at equal distances on one straight line, and if the two middle ones be equal, while those which are equidistant from them (on each side) are equal respectively, the centre of gravity of the system is the middle point of the line joining the centres of gravity of the two middle ones.

LIFE OF MARCELLUS

PLUTARCH

Translation of Bernadotte Perrin (London, 1917)

ARCHIMEDES AND TECHNOLOGY

14. He [Marcellus] proceeded to attack the city [Syracuse] by land and sea, Appius leading up the land forces, and he himself having a fleet of sixty quinqueremes filled with all sorts of arms and missiles. Moreover, he had erected an engine of artillery on a huge platform supported by eight galleys fastened together, and with this he sailed up to the city wall, confidently relying on the extent and splendour of his equipment and his own great fame. But all this proved to be of no account in the eyes of Archimedes and in comparison with the engines of Archimedes. To these he had by no means devoted himself as work worthy of his serious effort, but most of them were mere accessories of a geometry practised for amusement, since in bygone days Hiero the king had eagerly desired and at last persuaded him to turn his art somewhat from abstract notions to material things, and by applying his philosophy somehow to the needs which make themselves felt, to render it more evident to the common mind.

For the art of mechanics, now so celebrated and admired, was first originated by Eudoxus and Archytas, who embellished geometry with its subtleties, and gave to problems incapable of proof by word and diagram, a support derived from mechanical illustrations that were patent to the senses. For instance, in solving the problem of finding two mean proportional lines, a necessary requisite for many geometrical figures, both mathematicians had recourse to mechanical arrangements, adapting to their purposes certain intermediate portions of curved lines and sections. But Plato was incensed at this, and inveighed against them as corrupters and destroyers of the pure excellence of geometry, which thus turned her back upon the incorporeal things of abstract thought and descended to the things of sense, making use, moreover, of objects which required such mean and manual labour. For this reason mechanics was made entirely distinct from geometry, and being for a long time ignored by philosophers, came to be regarded as one of the military arts.

And yet even Archimedes, who was a kinsman and a friend of King Hiero, wrote to him that with any given force it was possible to move any given weight; and emboldened, as we are told, by the strength of his demonstration, he declared that, if there were another world, and he could go to it, he could move this. Hiero was astonished, and begged him to put his proposition into execution, and show him some great weight moved by a slight force. Archimedes therefore fixed upon a three-masted merchantman of the royal fleet, which had been dragged ashore by the great labours of many men, and after putting on board many passengers and the customary freight, he seated himself at a distance from her, and without any great effort, but quietly setting in motion with his hand a system of compound pulleys, drew her towards him smoothly and evenly, as though she were gliding through the water. Amazed at this, then, and comprehending the power of his art, the king persuaded Archimedes to prepare for him offensive and defensive engines to be used in every kind of siege warfare. These he had never used himself, because he spent the greater part of his life in freedom from war and amid the festal rites of peace; but at the present time his apparatus stood the Syracusans in good stead, and, with the apparatus, its fabricator.

15. When, therefore, the Romans assaulted them by sea and land, the Syracusans were stricken dumb with terror; they thought that nothing could withstand so furious an onset by such forces. But Archimedes began to ply his engines, and shot against the land forces of the assailants all sorts of missiles and immense masses of stones, which came down with incredible din and speed; nothing whatever could ward off their weight, but they knocked down in heaps those who stood in their way, and threw their ranks into confusion. At the same time huge beams were suddenly projected over the ships from the walls, which sank some of them with great weights plunging down from on high; others were sieged at the prow by iron claws, or beaks like the beaks of cranes, drawn straight up into the air, and then plunged stern foremost into the depths, or were turned round and round by means of enginery within the city, and dashed upon the steep cliffs that jutted out beneath the wall of the city, with great destruction of the fighting men on board, who perished in the wrecks. Frequently, too, a ship would be lifted out of the water into mid-air, whirled hither and thither as it hung there, a dreadful spectacle, until its crew had been thrown out and hurled in all directions, when it would fall empty upon the walls, or slip away from the clutch that had held it. As for the engine which Marcellus was bringing up on the bridge of ships, and which was called "sambuca" from some resemblance it had to the musical instrument of that name, while it was still some distance off in its approach to the wall, a stone of ten talents' weight was discharged at it, then a second and a third; some of these, falling upon it with great din and surge of wave, crushed the foundation of the engine, shattered its frame-work, and dislodged it from the platform, so that Marcellus, in perplexity, ordered his ships to sail back as fast as they could, and his land forces to retire.

Then, in a council of war, it was decided to come up under the walls while it was still night, if they could; for the ropes which Archimedes used in his engines, since they imported great impetus to the missiles cast, would, they thought, send them flying over their heads, but would be ineffective at close quarters, where there was no space for the cast. Archimedes, however, as it seemed, had long before prepared for such an emergency engines with a range adapted to any interval and missiles of short flight, and, through many small and contiguous openings in the wall, short-range engines called "scorpions" could be brought to bear on objects close at hand without being seen by the enemy.

* * * * *

The simplest schoolboy is now familiar with truths for which Archimedes would have given his life.

Ernest Renan, *Souvenirs d'enfance et de juenesse*, 1883

Mathematics, rightly viewed, possesses not only truth, but supreme beauty—a beauty cold and austere, like that of sculpture, without appeal to any part of our weaker nature, without the gorgeous trappings of painting or music, yet sublimely pure, and capable of a stern perfection such as only the greatest art can show. . . .

Mathematics takes us still further from what is human, into the region of absolute necessity, to which not only the actual world, but every possible world must conform.

Bertrand Russell, *The Study of Mathematics*, 1902

Happy is he who gets to know the reason for things.

Virgil, *Georgics*

Practical sciences proceed by building up; theoretical sciences by resolving into components.

Thomas Aquinas, *Commentary on the Ethics*

THE SCIENTIFIC REVOLUTION

Copernicus

Galileo

Descartes

Newton

Bacon

To demonstrate that the appearances are saved by assuming the sun at the center and the earth in the heavens is not the same thing as to demonstrate that *in fact* the sun is in the center and the earth in the heavens. I believe that the first demonstration may exist, but I have very grave doubts about the second. . .

Cardinal Robert Bellarmine, 1615

What makes planets go around the sun? At the time of Kepler some people answered this problem by saying that there were angels behind them beating their wings and pushing the planets around an orbit. . .The answer is not far from the truth. The only difference is that the angels sit in a different direction and their wings push inwards.

Richard Feynman, *The Character of Physical Law*

The scientific revolution outshines everything since the rise of Christendom and reduces the Renaissance and Reformation to the rank of mere episodes . . . of Medieval Christendom.

Herbert Butterfield, *Origins of Modern Science*, 1949

COMMENTARIOLUS

NICHOLAS COPERNICUS

Nicholas Copernicus (1473-1543) was a priest, astronomer, and humanist who was founder of the heliocentric theory of the solar system.

Copernicus was born in the Polish town of Torun, in what was then West Prussia, of a Polish father and a German mother. He received his university education at Cracow in Poland and at Bologna and, in medicine, at Padua in Italy. He lectured on mathematics and astronomy at Rome and later studied canon law at Ferrara.

In 1497 he received a tenured appointment as canon of the cathedral at Fruenburg in East Prussia. In that same year, he made his first recorded astronomical observations. From his knowledge of ancient Greek and Roman writings, he was satisfied that Ptolemy's geocentric view of the universe was not the only one, and this encouraged him to formulate his own theory.

Aware of the opposition of anyone who held the current opinion, Copernicus was reluctant to publish his views. From 1510 to 1514, he summarized his theory in a manuscript, the *Commentariolus*, but he circulated it only among his friends. Not until 1540 did he consent to the publication of his complete work, *On the Revolutions of the Heavenly Bodies*, which he adroitly dedicated to Pope Paul III. It was reported that he received the first copy of the newly published book on the last day of his life, May 24, 1543. The book became a standard reference for problems in astronomical research. In effect, it led the way to the scientific revolution as a whole.

1. What motivations led Copernicus to place the sun rather than the earth at the center of the universe? In what senses did he not go far enough?

2. Why was the theory such a revolutionary one?

3. Why did Copernicus wait so long to publish his theory?

4. What were the effects of Copernicus' new theory?

5. Does Copernicus present his findings as a mere hypothesis? How does he distinguish between appearance and reality?

6. Why do you think Copernicus dedicated his great work to the Pope?

COMMENTARIOLUS

NICHOLAS COPERNICUS

SKETCH OF HIS HYPOTHESES FOR THE HEAVENLY MOTIONS

OUR ANCESTORS assumed, I observe, a large number of celestial spheres for this reason especially, to explain the apparent motion of the planets by the principle of regularity. For they thought it altogether absurd that a heavenly body, which is a perfect sphere, should not always move uniformly. They saw that by connecting and combining regular motions in various ways they could make any body appear to move to any position.

Callippus and Eudoxus, who endeavored to solve the problem by the use of concentric spheres, were unable to account for all the planetary movements; they had to explain not merely the apparent revolutions of the planets but also the fact that these bodies appear to us sometimes to mount higher in the heavens, sometimes to descend; and this fact is incompatible with the principle of concentricity. Therefore it seemed better to employ eccentrics and epicycles, a system which most scholars finally accepted.

Yet the planetary theories of Ptolemy and most other astronomers, although consistent with the numerical data, seemed likewise to present no small difficulty. For these theories were not adequate unless certain equants were also conceived; it then appeared that a planet moved with uniform velocity neither on its deferent nor about the center of its epicycle. Hence a system of this sort seemed neither sufficiently absolute nor sufficiently pleasing to the mind.

Having become aware of these defects, I often considered whether there could perhaps be found a more reasonable arrangement of circles, from which every apparent inequality would be derived and in which everything would move uniformly about its proper center, as the rule of absolute motion requires. After I had addressed myself to this very difficult and almost insoluble problem, the suggestion at length came to me how it could be solved with fewer and much simpler constructions than were formerly used, if some assumptions (which are called axioms) were granted me. They follow in this order.

Assumptions

1. There is no one center of all the celestial circles or spheres.
2. The center of the earth is not the center of the universe, but only of gravity and of the lunar sphere.
3. All the spheres revolve about the sun as their mid-point, and therefore the sun is the center of the universe.
4. The ratio of the earth's distance from the sun to the height of the firmament is so much smaller than the ratio of the earth's radius to its distance from the sun that the distance from the earth to the sun is imperceptible in comparison with the height of the firmament.
5. Whatever motion appears in the firmament arises not from any motion of the firmament, but from the earth's motion. The earth together with its circumjacent elements performs a complete rotation on its fixed poles in a daily motion, while the firmament and highest heaven abide unchanged.

6. What appear to us as motions of the sun arise not from its motion but from the motion of the earth and our sphere, with which we revolve about the sun like any other planet. The earth has, then, more than one motion.
7. The apparent retrograde and direct motion of the planets arises not from their motion but from the earth's. The motion of the earth alone, therefore, suffices to explain so many apparent inequalities in the heavens.

Having set forth these assumptions, I shall endeavor briefly to show how uniformity of the motions can be saved in a systematic way. However, I have thought it well, for the sake of brevity, to omit from this sketch mathematical demonstrations, reserving these for my larger work. But in the explanation of the circles I shall set down here the lengths of the radii; and from these the reader who is not unacquainted with mathematics will readily perceive how closely this arrangement of circles agrees with the numerical data and observations.

Accordingly, let no one suppose that I have gratuitously asserted, with the Pythagoreans, the motion of the earth; strong proof will be found in my exposition of the circles. For the principal arguments by which the natural philosophers attempt to establish the immobility of the earth rest for the most part on the appearances; it is particularly such arguments that collapse here, since I treat the earth's immobility as due to an appearance.

The Order of the Spheres

The celestial spheres are arranged in the following order. The highest is the immovable sphere of the fixed stars, which contains and gives position to all things. Beneath it is Saturn, which Jupiter follows, then Mars. Below Mars is the sphere on which we revolve; then Venus; last is Mercury. The lunar sphere revolves about the center of the earth and moves with the earth like an epicycle. In the same order also, one planet surpasses another in speed of revolution, according as they trace greater or smaller circles. Thus Saturn completes its revolution in thirty years, Jupiter in twelve, Mars in two and one-half, and the earth in one year; Venus in nine months, Mercury in three.

The Apparent Motions of the Sun

The earth has three motions. First, it revolves annually in a great circle about the sun in the order of the signs, always describing equal arcs in equal times; the distance from the center of the circle to the center of the sun is 1/25 of the radius of the circle. The radius is assumed to have a length imperceptible in comparison with the height of the firmament; consequently the sun appears to revolve witlr this motion, as if the earth lay in the center of the universe. However, this appearance is caused by the motion not of the sun but of the earth, so that, for example, when the earth is in the sign of Capricornus, the sun is seen diametrically opposite in Cancer, and so on. On account of the previously mentioned distance of the sun from the center of the circle, this apparent motion of the sun is not uniform, the maximum inequality being 2 1/6°. The line drawn from the sun through the center of the circle is invariably directed toward a point of the firmament about 10° west of the more brilliant of the two bright stars in the head of Gemini; therefore when the earth is opposite this point, and the center of the circle lies between them, the sun is seen at its greatest distance from the earth. In this circle, then, the earth revolves together with whatever else is included within the lunar sphere.

The second motion, which is peculiar to the earth, is the daily rotation on the poles in the order of the signs, that is, from west to east. On account of this rotation the entire universe appears to revolve with enormous speed. Thus does the earth rotate together with its circumjacent waters and encircling atmosphere.

The third is the motion in declination. For the axis of the daily rotation is not parallel to the axis of the great circle, but is inclined to it at an angle that intercepts a portion of a circumference, in our time about 23 ½°. Therefore, while the center of the earth always remains in the plane of the ecliptic, that is, in the circumference of the great circle, the poles of the earth rotate, both of them describing small circles about centers equidistant from the axis of the great circle. The period of this motion is not quite a year and is nearly equal to the annual revolution on the great circle. But the axis of the great circle is invariably directed toward the points of the firmament which are called the poles of the ecliptic. In like manner the motion in declination, combined with the annual motion in their joint effect upon the poles of the daily rotation, would keep these poles constantly fixed at the same points of the heavens, if the periods of both motions were exactly equal. Now with the long passage of time it has become clear that this inclination of the earth to the firmament changes. Hence it is the common opinion that the firmament has several motions in conformity with a law not yet sufficiently understood. But the motion of the earth can explain all these changes in a less surprising way. I am not concerned to state what the path of the poles is. I am aware that, in lesser matters, a magnetized iron needle always points in the same direction. It has nevertheless seemed a better view to ascribe the changes to a sphere, whose motion governs the movements of the poles. This sphere must doubtless be sublunar.

Equal Motion Should Be Measured Not by the Equinoxes but by the Fixed Stars

Since the equinoxes and the other cardinal points of the universe shift considerably, whoever attempts to derive from them the equal length of the annual revolution necessarily falls into error. Different determinations of this length were made in different ages on the basis of many observations. Hipparchus computed it as 365 1/4 days, and Albategnius the Chaldean as 365^d 5^h 46^m, that is, 13 3/5 ^{m} or 13 $1/3^m$ less than Ptolemy. Hispalensis increased Albategnius's estimate by the 20th part of an hour, since he determined the tropical year as 365^d 5^h 49^m.

Lest these differences should seem to have arisen from errors of observation, let me say that if anyone will study the details carefully, he will find that the discrepancy has always corresponded to the motion of the equinoxes. For when the cardinal points moved 1° in 100 years, as they were found to be moving in the age of Ptolemy, the length of the year was then what Ptolemy stated it to be. When however in the following centuries they moved with greater rapidity, being opposed to lesser motions, the year became shorter; and this decrease corresponded to the in«ease in precession. For the annual motion was completed in a shorter time on account of the more rapid recurrence of the equinoxes. Therefore the derivation of the equal length of the year from the fixed stars is more accurate. I used Spica Virginis and found that the year has always been 365 days, 6 hours, and about 10 minutes, which is also the estimate of the ancient Egyptians. The same method must be employed also with the other motions of the planets, as is shown by their apsides, by the fixed laws of their motion in the firmament, and by heaven itself with true testimony.

The Moon

The moon seems to me to have four motions in addition to the annual revolution which has been mentioned. For it revolves once a month on its deferent circle about the center of the earth in the order of the signs. The deferent carries the epicycle which is commonly called the epicycle of the first inequality or argument, but which I call the first or greater epicycle. In the upper portion of its circumference this greater epicycle revolves in the direction opposite to that of the deferent, and its period is a little more than a month. Attached to it is a second epicycle. The moon, finally, moving with this second epicycle, completes two revolutions a month in the direction opposite to that of the greater epicycle, so that whenever the center of the greater epicycle crosses the line drawn from the center of the great circle through the center of the earth (I call this line the diameter of the great circle), the moon is nearest to the center of the greater epicycle. This occurs at new and full moon; but contrariwise at the quadratures, midway between new and full moon, the moon is most remote from the center of the greater epicycle. The length of the radius of the greater epicycle is to the radius of the deferent as 1 1/18:10 and to the radius of the smaller epicycle as 4 3/4:1.

By reason of these arrangements the moon appears, at times rapidly, at times slowly, to descend and ascend; and to this first inequality the motion of the smaller epicycle adds two irregularities. For it withdraws the moon from uniform motion on the circumference of the greater epicycle, the maximum inequality being 12 1/4° of a circumference of corresponding size or diameter; and it brings the center of the greater epicycle at times nearer the moon, at times further from it, within the limits of the radius of the smaller epicycle. Therefore, since the moon describes unequal circles about the center of the greater epicycle, the first inequality varies considerably. In conjunctions and oppositions to the sun its greatest value does not exceed 4°56', but in the quadratures it increases to 6°36'. Those who employ an eccentric circle to account for this variation improperly treat the motion on the eccentric as unequal, and, in addition, fall into two manifest errors. For the consequence by mathematical analysis is that when the moon is in quadrature, and at the same time in the lowest part of the epicycle, it should appear nearly four times greater (if the entire disk were luminous) than when new and full, unless its magnitude increases and diminishes in no reasonable way. So too, because the size of the earth is sensible in comparison with its distance from the moon, the lunar parallax should increase very greatly at the quadratures. But if anyone investigates these matters carefully, he will find that in both respects the quadratures differ very little from new and full moon, and accordingly will readily admit that my explanation is the truer.

With these three motions in longitude, then, the moon passes through the points of its motion in latitude. The axes of the epicycles are parallel to the axis of the deferent, and therefore the moon does not move out of the plane of the deferent. But the axis of the deferent is inclined to the axis of the great circle or ecliptic; hence the moon moves out of the plane of the ecliptic. Its inclination is determined by the size of an angle which subtends 5° of the circumference of a circle. The poles of the deferent revolve at an equal distance from the axis of the ecliptic, in nearly the same manner as was explained regarding declination. But in the present case they move in the reverse order of the signs and much more slowly, the period of the revolution being nineteen years. It is the common opinion that the motion takes place in a higher sphere, to which the poles are attached as they revolve in the manner described. Such a fabric of motions, then, does the moon seem to have.

The Three Superior Planets
Saturn—Jupiter—Mars

Saturn, Jupiter, and Mars have a similar system of motions, since their deferents completely enclose the great circle and revolve in the order of the signs about its center as their common center. Saturn's deferent revolves in 30 years, Jupiter's in 12 years, and that of Mars in 29 months; it is as though the size of the circles delayed the revolutions. For if the radius of the great circle is divided into 25 units, the radius of Mars' deferent will be 38 units, Jupiter's 130 5/12, and Saturn's 230 1/6. By "radius of the deferent" I mean the distance from the center of the deferent to the center of the first epicycle. Each deferent has two epicycles, one of which carries the other, in much the same way as was explained in the case of the moon, but with a different arrangement. For the first epicycle revolves in the direction opposite to that of the deferent, the periods of both being equal. The second epicycle, carrying the planet, revolves in the direction opposite to that of the first with twice the velocity. The result is that whenever the second epicycle is at its greatest or least distance from the center of the deferent, the planet is nearest to the center of the first epicycle; and when the second epicycle is at the midpoints, a quadrant's distance from the two points just mentioned, the planet is most remote from the center of the first epicycle. Through the combination of these motions of the deferent and epicycles, and by reason of the equality of their revolutions, the aforesaid withdrawals and approaches occupy absolutely fixed places in the firmament, and everywhere exhibit unchanging patterns of motion. Consequently the apsides are invariable; for Saturn, near the star which is said to be on the elbow of Sagittarius; for Jupiter, 8° east of the star which is called the end of the tail of Leo; and for Mars, 6 1/2° west of the heart of Leo.

The radius of the great circle was divided above into 25 units. Measured by these units, the sizes of the epicycles are as follows. In Saturn the radius of the first epicycle consists of 19 units, 41 minutes; the radius of the second epicycle, 6 units, 34 minutes. In Jupiter the first epicycle has a radius of 10 units, 6 minutes; the second, 3 units, 22 minutes. In Mars the first epicycle, 5 units, 34 minutes; the second, 1 unit, 51 minutes. Thus the radius of the first epicycle in each case is three times as great as that of the second.

The inequality which the motion of the epicycles imposes upon the motion of the deferent is called the first inequality; it follows, as I have said, unchanging paths everywhere in the firmament. There is a second inequality, on account of which the planet seems from time to time to retrograde, and often to become stationary. This happens by reason of the motion, not of the planet, but of the earth changing its position in the great circle. For since the earth moves more rapidly than the planet, the line of sight directed toward the firmament regresses, and the earth more than neutralizes the motion of the planet. This regression is most notable when the earth is nearest to the planet, that is, when it comes between the sun and the planet at the evening rising of the planet. On the other hand, when the planet is setting in the evening or rising in the morning, the earth makes the observed motion greater than the actual. But when the line of sight is moving in the direction opposite to that of the planets and at an equal rate, the planets appear to be stationary, since the opposed motions neutralize each other; this commonly occurs when the angle at the earth between the sun and the planet is 120°. In all these cases, the lower the deferent on which the planet moves, the greater is the inequality. Hence it is smaller in Saturn than in Jupiter, and again greatest in Mars, in accordance with the ratio of the radius of the great circle to the radii of the deferents. The inequality attains its maximum for each planet when the line of sight to the planet is tangent to the circumference of the great circle. In this manner do these three planets move.

In latitude they have a twofold deviation. While the circumferences of the epicycles remain in a single plane with their deferent, they are inclined to the ecliptic. This inclination is governed by the inclination of their axes, which do not revolve, as in the case of the moon, but are directed always toward the same region of the heavens. Therefore the intersections of the deferent and ecliptic (these points of intersection are called the nodes) occupy eternal places in the firmament. Thus the node where the planet begins its ascent toward the north is, for Saturn, 8 1/2° east of the star which is described as being in the head of the eastern of the two Gemini; for Jupiter, 4° west of the same star; and for Mars, 6 1/2° west of Vergiliae. When the planet is at this point and its diametric opposite, it has no latitude. But the greatest latitude, which occurs at a quadrant's distance from the nodes, is subject to a large inequality. For the inclined axes and circles seem to rest upon the nodes, as though swinging from them. The inclination becomes greatest when the earth is nearest to the planet, that is, at the evening rising of the planet; at that time the inclination of the axis is, for Saturn 2 2/3°, Jupiter 1 2/5°, and Mars 1 5/6. On the other hand, near the time of the evening setting and morning rising, when the earth is at its greatest distance from the planet, the inclination is smaller, for Saturn and Jupiter by 5/12°, and for Mars by 1 2/3°. Thus this inequality is most notable in the greatest latitudes, and it becomes smaller as the planet approaches the node, so that it increases and decreases equally with the latitude.

The motion of the earth in the great circle also causes the observed latitudes to change, its nearness or distance increasing or diminishing the angle of the observed latitude, as mathematical analysis demands. This motion in libration occurs along a straight line, but a motion of this sort can be derived from two circles. These are concentric, and one of them, as it revolves, carries with it the inclined poles of the other. The lower circle revolves in the direction opposite to that of the upper, and with twice the velocity. As it revolves, it carries with it the poles of the circle which serves as deferent to the epicycles. The poles of the deferent are inclined to the poles of the circle halfway above at an angle equal to the inclination of these poles to the poles of the highest circle. So much for Saturn, Jupiter, and Mars and the spheres which enclose the earth.

Venus

There remain for consideration the motions which are included within the great circle, that is, the motions of Venus and Mercury. Venus has a system of circles like the system of the superior planets, but the arrangement of the motions is different. The deferent revolves in nine months, as was said above, and the greater epicycle also revolves in nine months. By their composite motion the smaller epicycle is everywhere brought back to the same path in the firmament, and the higher apse is at the point where I said the sun reverses its course. The period of the smaller epicycle is not equal to that of the deferent and greater epicycle, but has a constant relation to the motion of the great circle. For one revolution of the latter the smaller epicycle completes two. The result is that whenever the earth is in the diameter drawn through the apse, the planet is nearest to the center of the greater epicycle; and it is most remote, when the earth, being in the diameter perpendicular to the diameter through the apse, is at a quadrant's distance from the positions just mentioned. The smaller epicycle of the moon moves in very much the same way with relation to the sun. The ratio of the radius of the great circle to the radius of the deferent of Venus is 25:18; the greater epicycle has a value of 5/4 of a unit, and the smaller 1/4.

Venus seems at times to retrograde, particularly when it is nearest to the earth, like the superior planets, but for the opposite reason. For the regression of the superior planets happens because the motion of the earth is more rapid than theirs, but with Venus, because it is slower; and because the superior planets enclose the great circle, whereas Venus is enclosed within it. Hence Venus is never in opposition to the sun, since the earth cannot come between them, but it moves within fixed distances on either side of the sun. These distances are determined by tangents to the circumference drawn from the center of the earth, and never exceed 48° in our observations. Here ends the treatment of Venus' motion in longitude.

Its latitude also changes for a twofold reason. For the axis of the deferent is inclined at an angle of 2 ½°, the node whence the planet turns north being in the apse. However, the deviation which arises from this inclination, although in itself it is one and the same, appears twofold to us. For when the earth is on the line drawn through the nodes of Venus, the deviations on the one side are seen above, and on the opposite side below; these are called the reflexions. When the earth is at a quadrant's distance from the nodes, the same natural inclinations of the deferent appear, but they are called the declinations. In all the other positions of the earth, both latitudes mingle and are combined, each in turn exceeding the other; by their likeness and difference they are mutually increased and eliminated.

The inclination of the axis is affected by a motion in libration that swings, not on the nodes as in the case of the superior planets, but on certain other movable points. These points perform annual revolutions with reference to the planet. Whenever the earth is opposite the apse of Venus, at that time the amount of the libration attains its maximum for this planet, no matter where the planet may then be on the deferent. As a consequence, if the planet is then in the apse or diametrically opposite to it, it will not completely lack latitude, even though it is then in the nodes. From this point the amount of the libration decreases, until the earth has moved through a quadrant of a circle from the aforesaid position, and, by reason of the likeness of their motions, the point of maximum deviation has moved an equal distance from the planet. Here no trace of the deviation is found. Thereafter the descent of the deviation continues. The initial point drops from north to south, constantly increasing its distance from the planet in accordance with the distance of the earth from the apse. Thereby the planet is brought to the part of the circumference which previously was south. Now, however, by the law of opposition, it becomes north and remains so until the limit of the libration is again reached upon the completion of the circle. Here the deviation becomes equal to the initial deviation and once more attains its maximum. Thus the second semicircle is traversed in the same way as the first. Consequently this latitude, which is usually called the deviation, never becomes a south latitude. In the present instance, also, it seems reasonable that these phenomena should be produced by two concentric circles with oblique axes, as I explained in the case of the superior planets.

Mercury

Of all the orbits in the heavens the most remarkable is that of Mercury, which traverses almost untraceable paths, so that it cannot be easily studied. A further difficulty is the fact that the planet, following a course generally invisible in the rays of the sun, can be observed for a very few days only. Yet Mercury too will be understood, if the problem is attacked with more than ordinary ability.

Mercury, like Venus, has two epicycles which revolve on the deferent. The periods of the greater epicycle and deferent are equal, as in the case of Venus. The apse is located 14 1/2° east of Spica Virginis. The smaller epicycle revolves with twice the velocity of the earth. But by contrast with Venus, whenever the earth is above the apse or diametrically opposite to it, the planet is most remote from the center of the greater epicycle; and it is nearest, whenever the earth is at a quadrant's distance from the points just mentioned. I have said that the deferent of Mercury revolves in three months, that is, in 88 days. Of the 25 units into which I have divided the radius of the great circle, the radius of the deferent of Mercury contains 9 2/5. The first epicycle contains 1 unit, 41 minutes; the second epicycle is 1/3 as great, that is, about 34 minutes.

But in the present case this combination of circles is not sufficient, though it is for the other planets. For when the earth passes through the above-mentioned positions with respect to the apse the planet appears to move in a much smaller path than is required by the system of circles described above; and in a much greater path, when the earth is at a quadrant's distance from the positions just mentioned. Since no other inequality in longitude is observed to result from this, it may be reasonably explained by a certain approach of the planet to and withdrawal from the center of the deferent along a straight line. This motion must be produced by two small circles stationed about the center of the greater epicycle, their axes being parallel to the axis of the deferent. The center of the greater epicycle, or of the whole epicyclic structure, lies on the circumference of the small circle that is situated between this center and the outer small circle. The distance from this center to the center of the inner circle is exactly equal to the distance from the latter center to the center of the outer circle. This distance has been found to be 14 ½ minutes of one unit of the 25 by which I have measured the relative sizes of all the circles. The motion of the outer small circle performs two revolutions in a tropical year, while the inner one completes four in the same time with twice the velocity in the opposite direction. By this composite motion the centers of the greater epicycle are carried along a straight line, just as I explained with regard to the librations in latitude. Therefore, in the aforementioned positions of the earth with respect to the apse, the center of the greater epicycle is nearest to the center of the deferent; and it is most remote, when the earth is at a quadrant's distance from these positions. When the earth is at the midpoints, that is, 45° from the points just mentioned, the center of the greater epicycle joins the center of the outer small circle, and both centers coincide. The amount of this withdrawal and approach is 29 minutes of one of the abovementioned units. This, then, is the motion of Mercury in longitude.

Its motion in latitude is exactly like that of Venus, but always in the opposite hemisphere. For where Venus is in north latitude, Mercury is in south. Its deferent is inclined to the ecliptic at an angle of 7°. The deviation, which is always south, never exceeds 3/4°. For the rest, what was said about the latitudes of Venus may be understood here also, to avoid repetition.

Then Mercury runs on seven circles in all; Venus on five; the earth on three, and round it the moon on four; finally Mars, Jupiter, and Saturn on five each. Altogether, therefore, thirty-four circles suffice to explain the entire structure of the universe and the entire ballet of the planets.

THE STARRY MESSENGER

GALILEO GALILEI

Galileo Galilei (1564-1642), mathematician, astronomer, and physicist, was the first person to use a telescope for a study of the skies. He also was a founder of the experimental method in science, and as such, was one of the great pioneers of the scientific revolution.

He was born at Pisa, the son of a mathematician and musician. He received his early education at a monastery near Florence. For a time he studied music and art. Then he went on to the University of Pisa to study medicine. During his first year there, he supposedly observed the swinging of a lamp in the cathedral and this led later to his confirmation of the principle of the pendulum.

Withdrawing from Pisa on account of a lack of funds, he returned to Florence where he lectured at the Academy. During this time he invented the hydrostatic balance for determining the specific gravity of objects. A treatise on this subject won for him an appointment as lecturer in mathematics at Pisa. Three years later, mainly for financial reasons, he obtained appointment to a chair in mathematics at the University of Padua.

His researches into the theory of motion led to discrediting the Aristotelian notion that bodies of different weights fall at different speeds. There is no established basis for the story that he dropped weights from the leaning Tower of Pisa. He did roll metal balls down inclined planes in his experiments, inspired perhaps by Marlianus, who had rolled balls down inclined planes a century earlier.

With the use of a telescope with a power of 32 that he had built, he discovered the craters on the moon, sun spots, the satellites of Jupiter, that the Milky Way was made up of stars, the rings of Saturn, and the phases of Venus. These observations he published in *The Starry Messenger* in 1610, excerpts of which are given here.

Galileo long had supported, privately, the Copernican view of the solar system. In 1630 he made this explicit with the publication of *Dialogue Concerning the Two Chief World Systems* in 1632. Within a few months he was haled before the Inquisition in Rome. Found "vehemently suspect of heresy," he was forced to recant publicly and was sentenced to prison, though this was commuted to permanent house arrest.

While under arrest, he completed what probably was his greatest work, the *Discourse Concerning Two New Sciences*. Within two years after completing that writing in 1636 he was totally blind. He continued work until his health gave way completely in 1641. He died in January, 1642.

1. What discoveries did Galileo make with his telescope? Why were they so astounding?

2. What is the Milky Way? What are the nebulae?

3. What was distinctive about Galileo's scientific method?

4. What were "The Medicean Stars" and why did Galileo give them that name? How should theoretical scientific research be financed?

THE STARRY MESSENGER

GALILEO GALILEI

THE
STARRY MESSENGER
Revealing great, unusual, and re-
markable spectacles, opening these
to the consideration of every man,
and especially of philosophers and
astronomers;
AS OBSERVED BY GALILEO GALILEI
Gentleman of Florence
Professor of Mathematics in the
University of Padua,
WITH THE AID OF A
SPYGLASS
lately invented by him,
In the surface of the Moon, in innumerable
Fixed Stars, in Nebulae, and above all
in FOUR PLANETS
swiftly revolving about Jupiter at
differing distances and periods,
and known to no one before the
Author recently perceived them
and decided that they should
be named
THE MEDICEAN STARS

Venice
1610

TO THE MOST SERENE
COSIMO II DE' MEDICI
FOURTH GRAND DUKE OF TUSCANY

Surely a distinguished public service has been rendered by those who have protected from envy the noble achievements of men who have excelled in virtue, and have thus preserved from oblivion and neglect those names which deserve immortality. In this way images sculptured in marble or cast in bronze have been handed down to posterity; to this we owe our statues, both pedestrian and equestrian; thus have we those columns and pyramids whose expense (as the poet says) reaches to the stars; finally, thus cities have been built to bear the names of men deemed worthy by posterity of commendation to all the ages. For the nature of the human mind is such that unless it is stimulated by images of things acting upon it from without, all remembrance of them passes easily away.

Looking to things even more stable and enduring, others have entrusted the immortal fame of illustrious men not to marble and metal but to the custody of the Muses and to imperishable literary monuments. But why dwell upon these things as though human wit were satisfied with earthly regions and had not dared advance beyond? For, seeking further, and well understanding that all human monuments ultimately perish through the violence of the elements or by old age, ingenuity has in fact found still more incorruptible monuments over which voracious time and envious age have been unable to assert any rights. Thus turning to the sky, man's wit has inscribed on the familiar and everlasting orbs of most bright stars the names of those whose eminent and godlike deeds have caused them to be accounted worthy of eternity in the company of the stars. And so the fame of Jupiter, of Mars, of Mercury, Hercules, and other heroes by whose names the stars are called, will not fade before the extinction of the stars themselves.

Yet this invention of human ingenuity, noble and admirable as it is, has for many centuries been out of style. Primeval heroes are in possession of those bright abodes, and hold them in their own right. In vain did the piety of Augustus attempt to elect Julius Caesar into their number, for when he tried to give the name of "Julian" to a star which appeared in his time (one of those bodies which the Greeks call "comets" and which the Romans likewise named for their hairy appearance), it vanished in a brief time and mocked his too ambitious wish. But we are able, most serene Prince, to read Your Highness in the heavens far more accurately and auspiciously. For scarce have the immortal graces of your spirit begun to shine on earth when in the heavens bright stars appear as tongues to tell and celebrate your exceeding virtues to all time. Behold, then, four stars reserved to bear your famous name; bodies which belong not to the inconspicuous multitude of fixed stars, but to the bright ranks of the planets. Variously moving about most noble Jupiter as children of his own, they complete their orbits with marvelous velocity—at the same time executing with one harmonious accord mighty revolutions every dozen years about the center of the universe; that is, the sun.

Indeed, the Maker of the stars himself has seemed by clear indications to direct that I assign to these new planets Your Highness's famous name in preference to all others. For just as these stars, like children worthy of their sire, never leave the side of Jupiter by any appreciable distance, so (as indeed who does not know?) clemency, kindness of heart, gentleness of manner, splendor of royal blood, nobility in public affairs, and excellency of authority and rule have all fixed their abode and habitation in Your Highness. And who, I ask once more, does not know that all these virtues emanate from the benign star of Jupiter, next after God as the source of all things good? Jupiter; Jupiter, I say, at the instant of Your Highness's birth, having already emerged from the turbid mists of the horizon and occupied the midst of the heavens, illuminating the eastern sky from his own royal house, looked out from that exalted throne upon your auspicious birth and poured forth all his splendor and majesty in order that your tender body and your mind (already adorned by God with the most noble ornaments) might imbibe with their first breath that universal influence and power.

But why should I employ mere plausible arguments, when I may prove my conclusion absolutely? It pleased Almighty God that I should instruct Your Highness in mathematics, which I did four years ago at that time of year when it is customary to rest from the most exacting studies. And since clearly it was mine by divine will to serve Your Highness and thus to receive from near at hand the rays of your surpassing clemency and beneficence, what wonder is it that my heart is so inflamed as to think both day and night of little else than how I, who am indeed your subject not only by choice but by birth and lineage, may become known to you as most grateful and most anxious for your glory? And so, most serene Cosimo, having discovered under your patronage these stars unknown to every astronomer before me, I have with good right decided to designate them by the august name of your family. And if I am first to have investigated them, who can justly blame me if I likewise name them, calling them the Medicean Stars, in the hope that this name will bring as much honor to them as the names of other heroes have bestowed on other stars? For, to say nothing of Your Highness's most serene ancestors, whose everlasting glory is testified by the monuments of all history, your virtue alone, most worthy Sire, can confer upon these stars an immortal name. No one can doubt that you will fulfill those expectations, high though they are, which you have aroused by the auspicious beginning of your reign, and will not only meet but far surpass them. Thus when you have conquered your equals you may still vie with yourself, and you and your greatness will become greater every day.

Accept then, most clement Prince, this gentle glory reserved by the stars for you. May you long enjoy those blessings which are sent to you not so much from the stars as from God, their Maker and their Governor.

Your Highness's most devoted servant,
GALILEO GALILEI

PADUA, March 12, 1610

ASTRONOMICAL MESSAGE

Which contains and explains recent observations
made with the aid of a new spyglass
concerning the surface of the moon,
the Milky Way, nebulous stars, and
innumerable fixed stars,
as well as four planets never before seen, and
now named
THE MEDICEAN STARS

Great indeed are the things which in this brief treatise I propose for observation and consideration by all students of nature. I say great, because of the excellence of the subject itself, the entirely unexpected and novel character of these things, and finally because of the instrument by means of which they have been revealed to our senses.

Surely it is a great thing to increase the numerous host of fixed stars previously visible to the unaided vision, adding countless more which have never before been seen, exposing these plainly to the eye in numbers ten times exceeding the old and familiar stars.

It is a very beautiful thing, and most gratifying to the sight, to behold the body of the moon, distant from us almost sixty earthly radii, as if it were no farther away than two such measures—so that its diameter appears almost thirty times larger, its surface nearly nine hundred times, and its volume twenty-seven thousand times as large as when viewed with the naked eye. In this way one may learn with all the certainty of sense evidence that the moon is not robed in a smooth and polished surface but is in fact rough and uneven, covered everywhere, just like the earth's surface, with huge prominences, deep valleys, and chasms.

Again, it seems to me a matter of no small importance to have ended the dispute about the Milky Way by making its nature manifest to the very senses as well as to the intellect. Similarly it will be a pleasant and elegant thing to demonstrate that the nature of those stars which astronomers have previously called "nebulous" is far different from what has been believed hitherto. But what surpasses all wonders by far, and what particularly moves us to seek the attention of all astronomers and philosophers, is the discovery of four wandering stars not known or observed by any man before us. Like Venus and Mercury, which have their own periods about the sun, these have theirs about a certain star that is conspicuous among those already known, which they sometimes precede and sometimes follow, without ever departing from it beyond certain limits. All these facts were discovered and observed by me not many days ago with the aid of a spyglass which I devised, after first being illuminated by divine grace. Perhaps other things, still more remarkable, will in time be discovered by me or by other observers with the aid of such an instrument, the form and construction of which I shall first briefly explain, as well as the occasion of its having been devised. Afterwards I shall relate the story of the observations I have made.

About ten months ago a report reached my ears that a certain Fleming had constructed a spyglass by means of which visible objects, though very distant from the eye of the observer, were distinctly Seen as if nearby. Of this truly remarkable effect several experiences were related, to which some persons gave credence while others denied them. A few days later the report was confirmed to me in a letter from a noble Frenchman at Paris, Jacques Badovere, which caused me to apply myself wholeheartedly to inquire into the means by which I might arrive at the invention of a similar instrument. This I did shortly afterwards, my basis being the theory of refraction. First I prepared a tube of lead, at the ends of which I fitted two glass lenses, both plane on one side while on the other side one was spherically convex and the other concave. Then placing my eye near the concave lens I perceived objects satisfactorily large and near, for they appeared three times closer and nine times larger than when seen with the naked eye alone. Next I constructed another one, more accurate, which represented objects as enlarged more than sixty times. Finally, sparing neither labor nor expense, I succeeded in constructing for myself so excellent an instrument that objects seen by means of it appeared nearly one thousand times larger and over thirty times closer than when regarded with our natural vision.

It would be superfluous to enumerate the number and importance of the advantages of such an instrument at sea as well as on land. But forsaking terrestrial observations, I turned to celestial ones, and first I saw the moon from as near at hand as if it were scarcely two terrestrial radii away. After that I observed often with wondering delight both the planets and the fixed stars, and since I saw these latter to be very crowded, I began to seek (and eventually found) a method by which I might measure their distances apart. Here it is appropriate to convey certain cautions to all who intend to undertake observations of this sort, for in the first place it is necessary to prepare quite a perfect telescope, which will show all objects bright, distinct, and free from any haziness, while magnifying them at least four hundred times and thus showing them twenty times closer. Unless the instrument is of this kind it will be vain to attempt to observe all the things which I have seen in the heavens, and which will presently be set forth. Now in order to determine without much trouble the magnifying power of an instrument, trace on paper the contour of two circles or two squares of which one is four hundred times as large as the other, as it will be when the diameter of one is twenty times that of the other. Then, with both these figures attached to the same wall, observe them simultaneously from a distance, looking at the smaller one through the telescope and at the larger one with the other eye unaided. This may be done without inconvenience while holding both eyes open at the same time; the two figures will appear to be of the same size if the instrument magnifies objects in the desired proportion.

Such an instrument having been prepared, we seek a method of measuring distances apart. This we shall accomplish by the following contrivance.

Let ABCD be the tube and E be the eye of the observer. Then if there were no lenses in the tube, the rays would reach the object FG along the straight lines ECF and EDG. But when the lenses have been inserted, the rays go along the refracted lines ECH and EDI; thus they are brought closer together, and those which were previously directed freely to the object FG now include only the portion of it HI. The ratio of the distance EH to the line HI then being found, one may by means of a table of sines determine the size of the angle formed at the eye by the object HI, which we shall find to be but a few minutes of arc. Now, if to the lens CD we fit thin plates, some pierced with larger and some with smaller apertures, putting now one plate and now another over the lens as required, we may form at pleasure different angles subtending more or fewer minutes of arc, and by this means we may easily measure the intervals between stars which are but a few minutes apart, with no greater error than one or two minutes. And for the present let it suffice that we have touched lightly on these matters and scarcely more than mentioned them, as on some other occasion we shall explain the entire theory of this instrument.

Now let us review the observations made during the past two months, once more inviting the attention of all who are eager for true philosophy to the first steps of such important contemplations. Let us speak first of that surface of the moon which faces us. For greater clarity I distinguish two parts of this surface, a lighter and a darker; the lighter part seems to surround and to pervade the whole hemisphere, while the darker part discolors the moon's surface like a kind of cloud, and makes it appear covered with spots. Now those spots which are fairly dark and rather large are plain to everyone and have been seen throughout the ages; these I shall call the "large" or "ancient" spots, distinguishing them

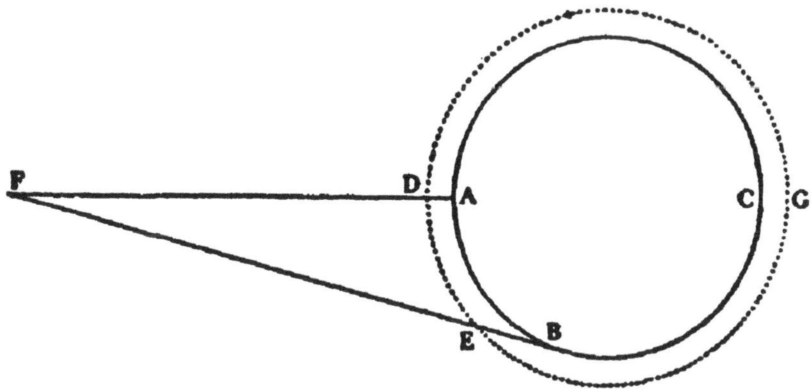

from others that are smaller in size but so numerous as to occur all over the lunar surface, and especially the lighter part. The latter spots have never been seen by anyone before me. From observations of these spots repeated many times I have been led to the opinion and conviction that the surface of the moon is not smooth, uniform, and precisely spherical as a great number of philosophers believe it (and the other heavenly bodies) to be, but is uneven, rough, and full of cavities and prominences, being not unlike the face of the earth, relieved by chains of mountains and deep valleys. The things I have seen by which I was enabled to draw this conclusion are as follows.

On the fourth or fifth day after the new moon, when the moon is seen with brilliant horns, the boundary which divides the dark part from the light does not extend uniformly in an oval line as would happen on a perfectly spherical solid, but traces out an uneven, rough, and very wavy line as shown in the figure below. Indeed, many luminous excrescences extend beyond the boundary into

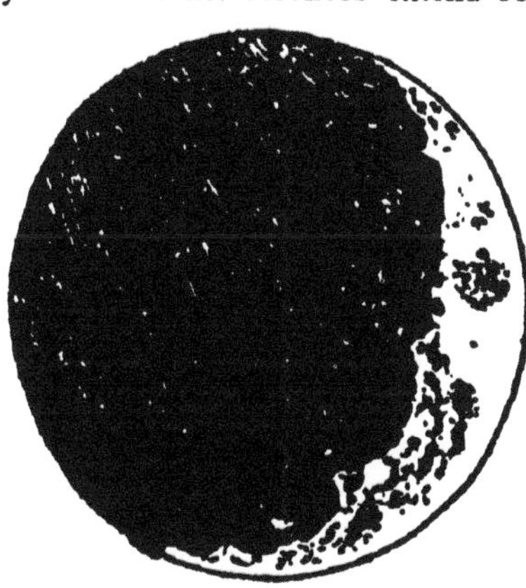

the darker portion, while on the other hand some dark patches invade the illuminated part. Moreover a great quantity of small blackish spots, entirely separated from the dark region, are scattered almost all over the area illuminated by the sun with the exception only of that part which is occupied by the large and ancient spots. Let us note, however, that the said small spots always agree in having their blackened parts directed toward the sun, while on the side opposite the sun they are crowned with bright contours, like shining summits. There is a similar sight on earth about sunrise, when we behold the valleys not yet flooded with light though the mountains surrounding them are already ablaze with glowing splendor on the side opposite the sun. And just as the shadows in the hollows on earth diminish in size as the sun rises higher, so these spots on the moon lose their blackness as the illuminated region grows larger and larger.

Again, not only are the boundaries of shadow and light in the moon seen to be uneven and wavy, but still more astonishingly many bright points appear within the darkened portion of the moon, completely divided and separated from the illuminated part and at a considerable distance from it. After a time these gradually increase in size and brightness, and an hour or two later they become joined with the rest of the lighted part which has now increased in size. Meanwhile more and more peaks shoot up as if sprouting now here, now there, lighting up within the shadowed portion; these become larger, and finally they too are united with that same luminous surface which extends ever further. An illustration of this is to be seen in the figure above. And on the earth, before the rising of the sun, are not the highest peaks of the mountains illuminated by the sun's rays while the plains remain in shadow? Does not the light go on spreading while the larger central parts of those mountains are becoming illuminated? And when the sun has finally risen, does not the illumination of plains and hills finally become one? But on the moon the variety of elevations and depressions appears to surpass in every way the roughness of the terrestrial surface, as we shall demonstrate further on.

At present I cannot pass over in silence something worthy of consideration which I observed when the moon was approaching first quarter, as shown in the previous figure. Into the luminous part there extended a great dark gulf in the neighborhood of the lower cusp. When I had observed it for a long time and had seen it completely dark, a bright peak began to emerge, a little below its center, after about two hours. Gradually growing, this presented itself in a triangular shape, remaining completely detached and separated from the lighted surface. Around it three other small points soon began to shine, and finally, when the moon was about to set, this triangular shape (which had meanwhile become more widely extended) joined with the rest of the illuminated region and suddenly burst into the gulf of shadow like a vast promontory of light, surrounded still by the three bright peaks already mentioned. Beyond the ends of the cusps, both above and below, certain bright points emerged which were quite detached from the remaining lighted part, as may be seen depicted in the same figure. There were also a great number of dark spots in both the horns, especially in the lower one; those nearest the boundary of light and shadow appeared larger and darker, while those more distant from the boundary were not so dark and distinct. But in all cases, as we have mentioned earlier, the blackish portion of each spot is turned toward the source of the sun's radiance, while a bright rim surrounds the spot on the side away from the sun in the direction of the shadowy region of the moon. This part of the moon's surface, where it is spotted as the tail of a peacock is sprinkled with azure eyes, resembles those glass vases which have been plunged while still hot into cold water and have thus acquired a crackled and wavy surface, from which they receive their common name of "ice-cups."

As to the large lunar spots, these are not seen to be broken in the above manner and full of cavities and prominences; rather, they are even and uniform, and brighter patches crop up only here and there. Hence if anyone wished to revive the old Pythagorean opinion that the moon is like another earth, its brighter part might very fitly represent the surface of the land and its darker region that of the water. I have never doubted that if our globe were seen from afar when flooded with sunlight, the land regions would appear brighter and the watery regions darker.

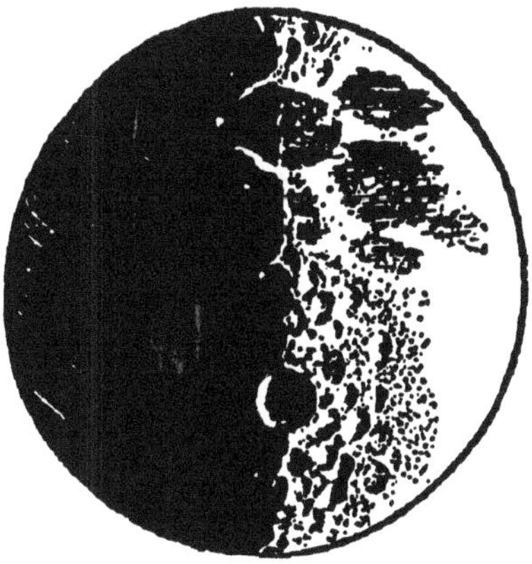

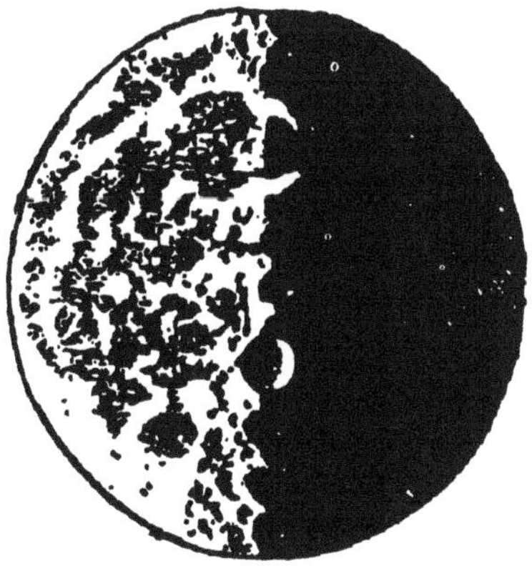

The large spots in the moon are also seen to be less elevated than the brighter tracts, for whether the moon is waxing or waning, there are always seen, here and there along its boundary of light and shadow, certain ridges of brighter hue around the large spots (and we have attended to this in preparing the diagrams); the edges of these spots are not only lower, but also more uniform, being uninterrupted by peaks or ruggedness.

Near the large spots the brighter part stands out particularly in such a way that before first quarter and toward last quarter, in the vicinity of a certain spot in the upper (or northern) region of the moon, some vast prominences arise both above and below as shown in the figures reproduced below. Before last quarter this same spot is seen to be walled about with certain blacker contours which, like the loftiest mountaintops, appear darker on the side away from the sun and brighter on that which faces the sun. (This is the opposite of what happens in the cavities, for there the part away from the sun appears brilliant, while that which is turned toward the sun is dark and in shadow.) After a time, when the lighted portion of the moon's surface has diminished in size and when all (or nearly all) the said spot is covered with shadow, the brighter ridges of the mountains gradually emerge from the shade. This double aspect of the spot is illustrated in the ensuing figures.

There is another thing which I must not omit, for I beheld it not without a certain wonder; this is that almost in the center of the moon there is a cavity larger than all the rest, and perfectly round in shape. I have observed it near both first and last quarters, and have tried to represent it as correctly as possible in the second of the above figures. As to light and shade, it offers the same appearance as would a region like Bohemia if that were enclosed on all sides by very lofty mountains arranged exactly in a circle. Indeed, this area on the moon is surrounded by such enormous peaks that the bounding edge adjacent to the dark portion of the moon is seen to be bathed in sunlight before the boundary of light and shadow reaches halfway across the same space. As in other spots, its shaded portion faces the sun while its lighted part is toward the dark side of the moon; and for a third time I draw attention to this as a very cogent proof of

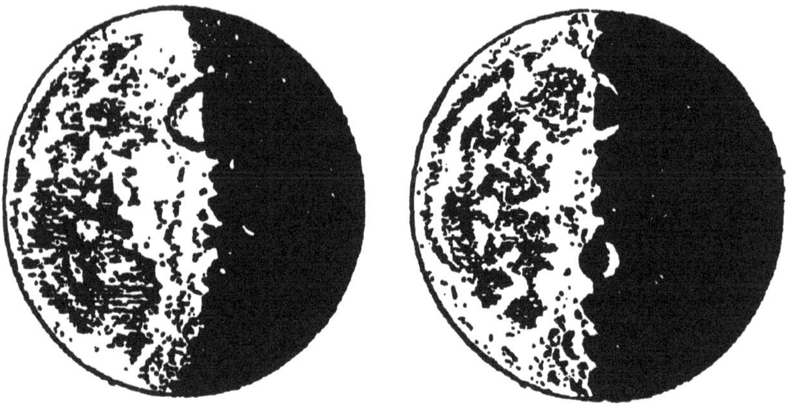

the ruggedness and unevenness that pervades all the bright region of the moon. Of these spots, moreover, those are always darkest which touch the boundary line between light and shadow, while those farther off appear both smaller and less dark, so that when the moon ultimately becomes full (at opposition to the Sun), the shade of the cavities is distinguished from the light of the places in relief by a subdued and very tenuous separation.

The things we have reviewed are to be seen in the brighter region of the moon. In the large spots, no such contrast of depressions and prominences is perceived as that which we are compelled to recognize in the brighter parts by the changes of aspect that occur under varying illumination by the sun's rays throughout the multiplicity of positions from which the latter reach the moon. In the large spots there exist some holes rather darker than the rest, as we have shown in the illustrations. Yet these present always the same appearance, and their darkness is neither intensified nor diminished, although with some minute difference they appear sometimes a little more shaded and sometimes a little lighter according as the rays of the sun fall on them more or less obliquely. Moreover, they join with the neighboring regions of the spots in a gentle linkage, the boundaries mixing and mingling. It is quite different with the spots which occupy the brighter surface of the moon; these, like precipitous crags having rough and jagged peaks, stand out starkly in sharp contrasts of light and shade. And inside the large spots there are observed certain other zones that are brighter, some of them very bright indeed. Still, both these and the darker parts present always the same appearance; there is no change either of shape or of light and shadow; hence one may affirm beyond any doubt that they owe their appearance to some real dissimilarity of parts. They cannot be attributed merely to irregularity of shape, wherein shadows move in consequence of varied illuminations from the sun, as indeed is the case with the other, smaller, spots which occupy the brighter part of the moon and which change, grow, shrink, or disappear from one day to the next, as owing their origin only to shadows of prominences.

But here I foresee that many persons will be assailed by uncertainty and drawn into a grave difficulty, feeling constrained to doubt a conclusion already explained and confirmed by many phenomena. If that part of the lunar surface which reflects sunlight more brightly is full of chasms (that is, of countless prominences and hollows), why is it that the western edge of the waxing moon, the eastern edge of the waning moon, and the entire periphery of the full moon are not seen to be uneven, rough, and wavy? On the contrary they look as precisely round as if they were drawn with a compass; and yet the whole periphery consists of that brighter lunar substance which we have declared to be filled with heights and chasms. In fact not a single one of the great spots extends to the extreme periphery of the moon, but all are grouped together at a distance from the edge.

Now let me explain the twofold reason for this troublesome fact, and in turn give a double solution to the difficulty. In the first place, if the protuberances and cavities in the lunar body existed only along the extreme edge of the circular periphery bounding the visible hemisphere, the moon might (indeed, would necessarily) look to us almost like a toothed wheel, terminated by a warty or wavy edge. Imagine, however, that there is not a single series of prominences arranged only along the very circumference, but a great many ranges of mountains together with their valleys and canyons disposed in ranks near the edge of the moon, and not only in the hemisphere visible to us but everywhere near the boundary line of the two hemispheres. Then an eye viewing them from afar will not be able to detect the separation of prominences by cavities, because the intervals between the mountains located in a given circle or a given chain will be hidden by the interposition of other heights situated in yet other ranges. This will be especially true if the eye of the observer is placed in the same straight line with the summits of these elevations. Thus on earth the summits of several mountains close together appear to be situated in one plane if the spectator is a long way off and is placed at an equal elevation. Similarly in a rough sea the tops of the waves seem to lie in one plane, though between one high crest and another there are many gulfs and chasms of such depth as not only to hide the hulls but even the bulwarks, masts, and rigging of stately ships. Now since there are many chains of mountains and chasms on the moon in addition to those around its periphery, and since the eye, regarding these from a great distance, lies nearly in the plane of their summits, no one need wonder that they appear as arranged in a regular and unbroken line.

To the above explanation another may be added; namely, that there exists around the body of the moon, just as around the earth, a globe of some substance denser than the rest of the aether. This may serve to receive and reflect the sun's radiations without being sufficiently opaque to prevent our seeing through it, especially when it is not illuminated Such a globe, lighted by the sun's rays, makes the body of the moon appear larger than it really is, and if it were thicker it would be able to prevent our seeing the actual body of the moon. And it actually is thicker near the circumference of the moon; I do not mean in an absolute sense, but relatively to the rays of our vision, which cut it obliquely there. Thus it may obstruct our vision, especially when it is lighted, and cloak the lunar periphery that is exposed to the sun. This may be more clearly understood from the figure below, in which the body of the moon, ABC, is surrounded by the vaporous globe DEG.

The eyesight from F reaches the moon in the central region, at A for example, through a lesser thickness of the vapors DA, while toward the extreme edges a deeper stratum of vapors, EB, limits and shuts out our sight. One indication of this is that the illuminated portion of the moon appears to be larger in circumference than the rest of the orb, which lies in shadow. And perhaps this

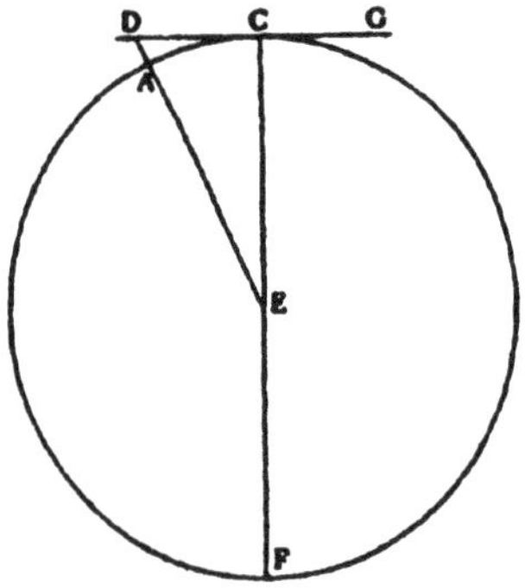

same cause will appeal to some as reasonably explaining why the larger spots on the moon are nowhere seen to reach the very edge probable though it is that some should occur there. Possibly they are invisible by being hidden under a thicker and more luminous mass of vapors.

That the lighter surface of the moon is everywhere dotted with protuberances and gaps has, I think, been made sufficiently clear from the appearances already explained. It remains for me to speak of their dimensions, and to show that the earth's irregularities are far less than those of the moon. I mean that they are absolutely less, and not merely in relation to the sizes of the respective globes. This is plainly demonstrated as follows.

I had often observed, in various situations of the moon with respect to the sun, that some summits within the shadowy portion appeared lighted, though lying some distance from the boundary of the light. By comparing this separation to the whole diameter of the moon, I found that it sometimes exceeded one-twentieth of the diameter. Accordingly, let CAF be a great circle of the lunar body, E its center, and CF a diameter, which is to the diameter of the earth as two is to seven.

Since according to very precise observations the diameter of the earth is seven thousand miles, CF will be two thousand, CE one thousand, and one-twentieth of CF will be one hundred miles. Now let CF be the diameter of the great circle which divides the light part of the moon from the dark part (for because of the very great distance of the un from the moon, this does not differ appreciably from a great circle), and let A be distant from C by one-twentieth of this. Draw the radius EA, which, when produced, cuts the tangent line GCD (representing the illuminating ray) in the point D. Then the arc CA, or rather the straight line CD, will consist of one hundred units whereof CE contains one thousand, and the sum of the squares of DC and CE will be 1,010,000. This is equal to the square of DE; hence ED will exceed 1,004, and AD will be more than four of those units of which CE contains one thousand. Therefore the altitude AD on the moon, which represents a summit reaching up to the solar ray GCD and standing at the distance CD from C, exceeds four miles. But on the earth we have no mountains which reach to a perpendicular height of even one mile. Hence it is quite clear that the prominences on the moon are loftier than those on the earth.

Here I wish to assign the cause of another lunar phenomenon well worthy of notice. I observed this not just recently, but many years ago, and pointed it out to some of my friends and pupils, explaining it to them and giving its true cause. Yet since it is rendered more evident a nd easier to observe with the aid of the telescope, I think it not unsuitable for introduction in this place, especially as it shows more clearly the connection between the moon and the earth.

When the moon is not far from the sun, just before or after new moon, its globe offers itself to view not only on the side where it is adorned with shining horns, but a certain faint light is also seen to mark out the periphery of the dark part which faces away from the sun, separating this from the darker background of the aether. Now if we examine the matter more closely, we shall see that not only does the extreme limb of the shaded side glow with this uncertain light, but the entire face of the moon (including the side which does not receive the glare of the sun) is whitened by a not inconsiderable gleam. At first glance only a thin luminous circumference appears, contrasting with the darker sky coterminous with it; the rest of the surface appears darker from its contact with the shining horns which distract our vision. But if we place ourselves so as to interpose a roof or chimney or some other object at a considerable distance from the eye, the shining horns may be hidden while the rest of the lunar globe remains exposed to view. It is then found that this region of the moon, though deprived of sunlight, also shines not a little. The effect is heightened if the gloom of night has already deepened through departure of the sun, for in a darker field a given light appears brighter.

Moreover, it is found that this secondary light of the moon (so to speak) is greater according as the moon is closer to the sun. It diminishes more and more as the moon recedes from that body until, after the first quarter and before the last, it is seen very weakly and uncertainly even when observed in the darkest sky. But when the moon is within sixty degrees of the sun it shines remarkably, even in twilight; so brightly indeed that with the aid of a good telescope one may distinguish the large spots. This remarkable gleam has afforded no small perplexity to philosophers, and in order to assign a cause for it some have offered one idea and some another. Some would say it is an inherent and natural light of the moon's own; others, that it is imparted by Venus; others yet, by all the stars together; and still others derive it from the sun, whose rays they would have permeate the thick solidity of the moon. But statements of this sort are refuted and their falsity evinced with little difficulty. For if this kind of light were the moon's own, or were contributed by the stars, the moon would retain it and would display it particularly during eclipses, when it is left in an unusually dark sky. This is contradicted by experience, for the brightness which is seen on the moon during eclipses is much fainter and is ruddy, almost copper-colored, while this is brighter and whitish. Moreover the other light is variable and movable, for it covers the face of the moon in such a way that the place near the edge of the earth's shadow is always seen to be brighter than the rest of the moon; this undoubtedly results from contact of the tangent solar rays with some denser zone which girds the moon about. By this contact a sort of twilight is diffused over the neighboring regions of the moon, just as on earth a sort of crepuscular light is spread both morning and evening; but with this I shall deal more fully in my book on the system of the world.

To assert that the moon's secondary light is imparted by Venus is so childish as to deserve no reply. Who is so ignorant as not to understand that from new moon to a separation of sixty degrees between moon and sun, no part of the moon which is averted from the sun can possibly be seen from Venus? And it is likewise unthinkable that this light should depend upon the sun's rays penetrating the thick solid mass of the moon, for then this light would never dwindle, inasmuch as one hemisphere of the moon is always illuminated except during lunar eclipses. And the light does diminish as the moon approaches first quarter, becoming completely obscured after that is passed.

Now since the secondary light does not inherently belong to the moon, and is not received from any star or from the sun, and since in the whole universe there is no other body left but the earth, what must we conclude? What is to be proposed? Surely we must assert that the lunar body (or any other dark and sunless orb) is illuminated by the earth. Yet what is there so remarkable about this? The earth, in fair and grateful exchange, pays back to the moon an illumination similar to that which it receives from her throughout nearly all the darkest gloom of night.

Let us explain this matter more fully. At conjunction the moon occupies a position between the sun and the earth; it is then illuminated by the sun's rays on the side which is turned away from the earth. The other hemisphere, which faces the earth, is covered with darkness; hence the moon does not illuminate the surface of the earth at all. Next, departing gradually from the sun, the moon comes to be lighted partly upon the side it turns toward us, and its whitish horns, still very thin, illuminate the earth with a faint light. The sun's illumination of the moon increasing now as the moon approaches first quarter, a refection of that light to the earth also increases. Soon the splendor on the moon extends into a semicircle, and our nights grow brighter; at length the entire visible face of the moon is irradiated by the sun's resplendent rays, and at full moon the whole surface of the earth shines in a flood of moonlight. Now the moon, waning, sends us her beams more weakly, and the earth is less strongly lighted; at length the moon returns to conjunction with the sun, and black night covers the earth.

In this monthly period, then, the moonlight gives us alternations of brighter and fainter illumination; and the benefit is repaid by the earth in equal measure. For while the moon is between us and the sun (at new moon), there lies before it the entire surface of that hemisphere of the earth which is exposed to the sun and illuminated by vivid rays. The moon receives the light which this reflects, and thus the nearer hemisphere of the moon—that is, the one deprived of sunlight—appears by virtue of this illumination to be not a little luminous. When the moon is ninety degrees away from the sun it sees but half the earth illuminated (the western half), for the other (the eastern half) is enveloped in night. Hence the moon itself is illuminated less brightly from the earth, and as a result its secondary light appears fainter to us. When the moon is in opposition to the sun, it faces a hemisphere of the earth that is steeped in the gloom of night, and if this position occurs in the plane of the ecliptic the moon will receive no light at all, being deprived of both the solar and the terrestrial rays. In its various other positions with respect to the earth and sun, the moon receives more or less light according as it faces a greater or smaller portion of the illuminated hemisphere of the earth. And between these two globes a relation is maintained such that whenever the earth is most brightly lighted by the moon, the moon is least lighted by the earth, and vice versa.

Let these few remarks suffice us here concerning this matter, which will be more fully treated in our *System of the world.* In that book, by a multitude of arguments and experiences, the solar reflection from the earth will be shown to be quite real—against those who argue that the earth must be excluded from the dancing whirl of stars for the specific reason that it is devoid of motion and of light. We shall prove the earth to be a wandering body surpassing the moon in splendor, and not the sink of all dull refuse of the universe; this we shall support by an infinitude of arguments drawn from nature.

Thus far we have spoken of our observations concerning the body of the moon. Let us now set forth briefly what has thus far been observed regarding the fixed stars. And first of all, the following fact deserves consideration: The stars, whether fixed or wandering, appear not to be enlarged by the telescope in the same proportion as that in which it magnifies other objects, and even the moon itself. In the stars this enlargement seems to be so much less that a telescope which is sufficiently powerful to magnify other objects a hundredfold is scarcely able to enlarge the stars four or five times. The reason for this is as follows.

When stars are viewed by means of unaided natural vision, they present themselves to us not as of their simple (and, so to speak, their physical) size, but as irradiated by a certain fulgor and as fringed with sparkling rays, especially when the night is far advanced. From this they appear larger than they would if stripped of those adventitious hairs of light, for the angle at the eye is determined not by the primary body of the star but by the brightness which extends so widely about it. This appears quite clearly from the fact that when stars first emerge from twilight at sunset they look very small, even if they are of the first magnitude; Venus itself, when visible in broad daylight, is so small as scarcely to appear equal to a star of the sixth magnitude. Things fall out differently with other objects, and even with the moon itself; these, whether seen in daylight or the deepest night, appear always of the same bulk. Therefore the stars are seen crowned among shadows, while daylight is able to remove their headgear; and not daylight alone, but any thin cloud that interposes itself between a star and the eye of the observer. The same effect is produced by black veils or colored glasses, through the interposition of which obstacles the stars are abandoned by their surrounding brilliance. A telescope similarly accomplishes the same result. It removes from the stars their adventitious and accidental rays, and then it enlarges their simple globes (if indeed the stars are naturally globular) so that they seem to be magnified in a lesser ratio than other objects. In fact a star of the fifth or sixth magnitude when seen through a telescope presents itself as one of the first magnitude.

Deserving of notice also is the difference between the appearances of the planets and of the fixed stars. The planets show their globes perfectly round and definitely bounded, looking like little moons, spherical and flooded all over with light; the fixed stars are never seen to be bounded by a circular periphery, but have rather the aspect of blazes whose rays vibrate about them and scintillate a great deal. Viewed with a telescope they appear of a shape similar to that which they present to the naked eye, but sufficiently enlarged so that a star of the fifth or sixth magnitude seems to equal the Dog Star, largest of all the fixed stars. Now, in addition to stars of the sixth magnitude, a host of other stars are perceived through the telescope which escape the naked eye; these are so numerous as almost to surpass belief. One may, in fact, see more of them than all the stars included among the first six magnitudes. The largest of these, which we may call stars of the seventh magnitude, or the first magnitude of invisible stars, appear through the telescope as larger and brighter than stars of the second magnitude when the latter are viewed with the naked eye. In order to give one or two proofs of their almost inconceivable number, I have adjoined pictures of two constellations. With these as samples, you may judge of all the others.

In the first I had intended to depict the entire constellation of Orion, but I was overwhelmed by the vast quantity of stars and by limitations of time, so I have deferred this to another occasion. There are more than five hundred new stars distributed among the old ones within limits of one or two degrees of arc. Hence to the three stars in the Belt of Orion and the six in the Sword which were previously known, I have added eighty adjacent stars discovered recently, preserving the intervals between them as exactly as I could. To distinguish the known or ancient stars, I have depicted them larger and have outlined them doubly; the other (invisible) stars I have drawn smaller and without the extra line. I have also preserved differences of magnitude as well as possible.

In the second example I have depicted the six stars of Taurus known as the Pleiades (I say six, inasmuch as the seventh is hardly ever visible) which lie within very narrow limits in the sky. Near them are more than forty others, invisible, no one of which is much more than half a degree away from the original six. I have shown thirty-six of these in the diagram; as in the case of Orion I have preserved their intervals and magnitudes, as well as the distinction between old stars and new.

Third, I have observed the nature and the material of the Milky Way. With the aid of the telescope this has been scrutinized so directly and with such ocular certainty that all the disputes which have vexed philosophers through so many ages have been resolved, and we are at last freed from wordy debates about it. The galaxy is, in fact, nothing but a congeries of innumerable stars grouped together in clusters. Upon whatever part of it the telescope is directed, a vast crowd of stars is immediately presented to view. Many of them are rather large and quite bright, while the number of smaller ones is quite beyond calculation.

But it is not only in the Milky Way that whitish clouds are seen; several patches of similar aspect shine with faint light here and there throughout the aether, and if the telescope is turned upon any of these it confronts us with a tight mass of stars. And what is even more remarkable, the stars which have been called "nebulous" by every astronomer up to this time turn out to be groups of very small stars arranged in a wonderful manner Although each star separately escapes our sight on account of its smallness or the immense distance from us, the mingling of their rays gives rise to that gleam which was formerly believed to be some denser part of the aether that was capable of reflecting rays from stars or from the sun. I have observed some of these constellations and have decided to depict two of them.

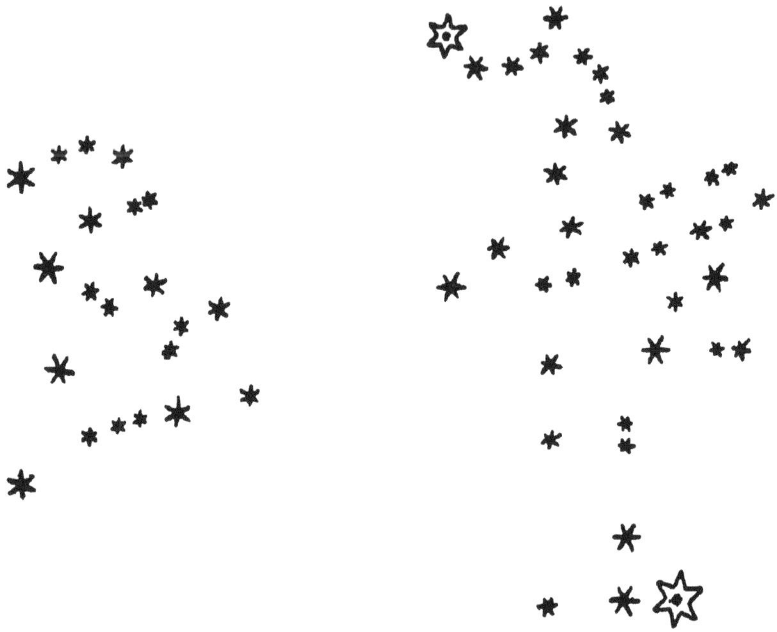

In the first you have the nebula called the Head of Orion, in which I have counted twenty-one stars. The second contains the nebula called Praesepe, which is not a single star but a mass of more than forty starlets. Of these I have shown thirty-six, in addition to the Aselli, arranged in the order shown.

We have now briefly recounted the observations made thus far with regard to the moon, the fixed stars, and the Milky Way. There remains the matter which in my opinion deserves to be considered the most important of all—the disclosure of four PLANETS never seen from the creation of the world up to our own time, together with the occasion of my having discovered and studied them, their arrangements, and the observations made of their movements and alterations during the past two months. I invite all astronomers to apply themselves to examine them and determine their periodic times, something which has so far been quite impossible to complete, owing to the shortness of the time. Once more, however, warning is given that it will be necessary to have a very accurate telescope such as we have described at the beginning of this discourse.

On the seventh day of January in this present year 1610, at the first hour of night, when I was viewing the heavenly bodies with a telescope, Jupiter presented itself to me; and because I had prepared a very excellent instrument for myself, I perceived (as I had not before, on account of the weakness of my previous instrument) that beside the planet there were three starlets, small indeed, but very bright. Though I believed them to be among the host of fixed stars, they aroused my curiosity somewhat by appearing to lie in an exact straight line parallel to the ecliptic, and by their being more splendid than others of their size. Their arrangement with respect to Jupiter and each other was the following:

East ✱ ✱ ○ ✱ **West**

That is, there were two stars on the eastern side and one to the west. The most easterly star and the western one appeared larger than the other. I paid no attention to the distances between them and Jupiter, for at the outset I thought them to be fixed stars, as I have said. But returning to the same investigation on January eighth—led by what, I do not know—I found a very different arrangement. The three starlets were now all to the west of Jupiter, closer together, and at equal intervals from one another as shown in the following sketch:

East ○ ✱ ✱ ✱ **West**

At this time, though I did not yet turn my attention to the way the stars had come together, I began to concern myself with the question how Jupiter could be east of all these stars when on the previous day it had been west of two of them. I commenced to wonder whether Jupiter was not moving eastward at that time, contrary to the computations of the astronomers, and had got in front of them by that motion. Hence it was with great interest that I awaited the next night. But I was disappointed in my hopes, for the sky was then covered with clouds everywhere.

On the tenth of January, however, the stars appeared in this position with respect to Jupiter:

East ✱ ✱ ○ **West**

that is, there were but two of them, both easterly, the third (as I supposed) being hidden behind Jupiter. As at first, they were in the same straight line with Jupiter and were arranged precisely in the line of the zodiac. Noticing this, and knowing that there was no way in which such alterations could be attributed to Jupiter's motion, yet being certain that these were still the same stars I had observed (in fact no other was to be found along the line of the zodiac for a long way on either side of Jupiter), my perplexity was now transformed into amazement. I was sure that the apparent changes belonged not to Jupiter but to the observed stars, and I resolved to pursue this investigation with greater care and attention.

And thus, on the eleventh of January, I saw the following disposition:

East ✱ ✱ ○ **West**

There were two stars, both to the east, the central one being three times as far from Jupiter as from the one farther east. The latter star was nearly double the size of the former, whereas on the night before they had appeared approximately equal.I had now decided beyond all question that there existed in the heavens three stars wandering about Jupiter as do Venus and Mercury about the sun, and this became plainer than daylight from observations on similar occasions which followed. Nor were there just three such stars; four wanderers complete their revolutions about Jupiter, and of their alterations as observed more precisely later on we shall give a description here. Also I measured the distances between them by means of the telescope, using the method explained before. Moreover I recorded the times of the observations, especially when more than one was made during the same night—for the revolutions of these planets are so speedily completed that it is usually possible to take even their hourly variations.

Thus on the twelfth of January at the first hour of night I saw the stars arranged in this way:

The most easterly star was larger than the western one, though both were easily visible and quite bright. Each was about two minutes of arc distant from Jupiter. The third star was invisible at first, but commenced to appear after two hours; it almost touched Jupiter on the east, and was quite small. All were on the same straight line directed along the ecliptic.

On the thirteenth of January four stars were seen by me for the first time, in this situation relative to Jupiter:

Three were westerly and one was to the east; they formed a straight line except that the middle western star departed slightly toward the north. The eastern star was two minutes of arc away from Jupiter, and the intervals of the rest from one another and from Jupiter were about one minute. All the stars appeared to be of the same magnitude, and though small were very bright, much brighter than fixed stars of the same size.

* * * * *

On the twenty-sixth of February, midway in the first hour of night, there were only two stars: One was to the east, ten minutes from Jupiter; the other to the west, six minutes away. The eastern one was somewhat smaller than the western. But at the fifth hour three stars were seen:

In addition to the two already noticed, a third was discovered to the west near Jupiter; it had at first been hidden behind Jupiter and was now one minute away. The eastern one appeared farther away than before, being eleven minutes from Jupiter.

This night for the first time I wanted to observe the progress of Jupiter and its accompanying planets along the line of the zodiac in relation to some fixed star, and such a star was seen to the east, eleven minutes distant from the easterly starlet and a little removed toward the south, in the following manner:

On the twenty-seventh of February, four minutes after the first hour, the stars appeared in this configuration:

The most easterly was ten minutes from Jupiter; the next, thirty seconds; the next to the west was two minutes thirty seconds from Jupiter, and the most westerly was one minute from that. Those nearest Jupiter appeared very small, while the end ones were plainly visible, especially the westernmost. They marked out an exactly straight line along the course of the ecliptic. The progress of these planets toward the east is seen quite clearly by reference to the fixed star mentioned, since Jupiter and its accompanying planets were closer to it, as may be seen in the figure above. At the fifth hour, the eastern star closer to Jupiter was one minute away.

At the first hour on February twenty-eighth, two stars only were seen; one easterly, distant nine minutes from Jupiter, and one to the west, two minutes away. They were easily visible and on the same straight line. The fixed star, perpendicular to this line, now fell under the eastern planet as in this figure:

At the fifth hour a third star, two minutes east of Jupiter, was seen in this position:

On the first of March, forty minutes after sunset, four stars all to the east were seen, of which the nearest to Jupiter was two minutes away, the next was one minute from this, the third two seconds from that and brighter than any of the others; from this in turn the most easterly was four minutes distant, and it was smaller than the rest. They marked out almost a straight line, but the third one counting from Jupiter was a little to the north. The fixed star formed an equilateral triangle with Jupiter and the most easterly star, as in this figure:

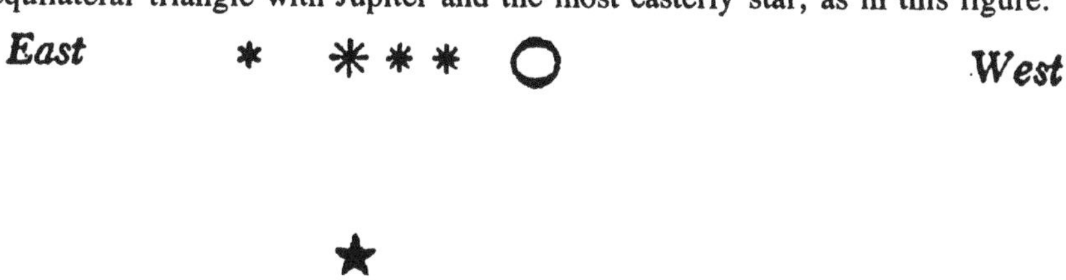

On March second, half an hour after sunset, there were three planets, two to the east and one to the west, in this configuration:

The most easterly was seven minutes from Jupiter and thirty seconds from its neighbor; the western one was two minutes away from Jupiter. The end stars were very bright and were larger than that in the middle, which appeared very small. The most easterly star appeared a little elevated toward the north from the straight line through the other planets and Jupiter. The fixed star previously mentioned was eight minutes from the western planet along the line drawn from it perpendicularly to the straight line through all the planets, as shown above.

I have reported these relations of Jupiter and its companions with the fixed star so that anyone may comprehend that the progress of those planets, both in longitude and latitude, agrees exactly with the movements derived from planetary tables.

Such are the observations concerning the four Medicean planets recently first discovered by me, and although from these data their periods have not yet been reconstructed in numerical form, it is legitimate at least to put in evidence some facts worthy of note. Above all, since they sometimes follow and sometimes precede Jupiter by the same intervals, and they remain within very limited distances either to east or west of Jupiter, accompanying that planet in both its retrograde and direct movements in a constant manner, no one can doubt that they complete their revolutions about Jupiter and at the same time effect all together a twelve-year period about the center of the universe. That they also revolve in unequal circles is manifestly deduced from the fact that at the greatest elongation from Jupiter it is never possible to see two of these planets in conjunction, whereas in the vicinity of Jupiter they are found united two, three, and sometimes all four together. It is also observed that the revolutions are swifter in those planets which describe smaller circles about Jupiter, since the stars closest to Jupiter are usually seen to the east when on the previous day they appeared to the west, and vice versa, while the planet which traces the largest orbit appears upon accurate observation of its returns to have a semimonthly period.

Here we have a fine and elegant argument for quieting the doubts of those who, while accepting with tranquil mind the revolutions of the planets about the sun in the Copernican system, are mightily disturbed to have the moon alone revolve about the earth and accompany it in an annual rotation about the sun. Some have believed that this structure of the universe should be rejected as impossible. But now we have not just one planet rotating about another while both run through a great orbit around the sun; our own eyes show us four stars which wander around Jupiter as does the moon around the earth, while all together trace out a grand revolution about the sun in the space of twelve years.

And finally we should not omit the reason for which the Medicean stars appear sometimes to be twice as large as at other times, though their orbits about Jupiter are very restricted. We certainly cannot seek the cause in terrestrial vapors, as Jupiter and its neighboring fixed stars are not seen to change size in the least while this increase and diminution are taking place. It is quite unthinkable that the cause of variation should be their change of distance from the earth at perigee and apogee, since a small circular rotation could by no means produce this effect, and an oval motion (which in this case would have to be nearly straight) seems unthinkable and quite inconsistent with the appearances. But I shall gladly explain what occurs to me on this matter, offering it freely to the judgment and criticism of thoughtful men. It is known that the interposition of terrestrial vapors makes the sun and moon appear large, while the fixed stars and planets are made to appear smaller. Thus the two great luminaries are seen larger when close to the horizon, while the stars appear smaller and for the most part hardly visible. Hence the stars appear very feeble by day and in twilight, though the moon does not, as we have said. Now from what has been said above, and even more from what we shall say at greater length in our *system,* it follows that not only the earth but also the moon is surrounded by an envelope of vapors, and we may apply precisely the same judgment to the rest of the planets. Hence it does not appear entirely impossible to assume that around Jupiter also there exists an envelope denser than the rest of the aether, about which the Medicean planets revolve as does the moon about the elemental sphere. Through the interposition of this envelope they appear larger when they are in perigee by the removal, or at least the attenuation, of this envelope.

Time prevents my proceeding further, but the gentle reader may expect more soon.

FINIS

Up stood the learned Galileo
Glanced briefly at the sun
And said: "Almighty God was wrong
In Genesis, Chapter One!"
  Now that was rash, my friends, it is no matter small
  For heresy will spread today like foul diseases.
  Change Holy Writ, forsooth? What will be left at all?
  Why: each of us would say and do just what he pleases!

Bertolt Brecht, *Galileo*, 1940

When you throw it with your arm, what is it that stays with the ball when it has left your hand, except the motion received from your arm which is conserved in it?

Galileo Galilei, *Dialogue Concerning the Two Chief World Systems*, 1632

DISCOURSE ON METHOD

DESCARTES

Rene Descartes (1596-1650) philosopher, mathematician, physicist, has been called by Laurence Lafleur the one "more than any other person responsible for the ideas and methods which rather sharply differentiate modern times from the medieval." He has been called by Hegel and others, "the father of modern philosophy."

With the famous Cartesian imperative, "I think, therefore I am," as his starting point, Descartes tried to develop a coherent system of thought based on principles which he determined to be necessarily true. He saw all nature as being subject to mathematical laws. He invented analytical geometry. He also made major contributions in mechanics, optics, and meteorology.

Descartes was born at LaHaye, Touraine, in France. His elementary education was a ten-year stint at the hands of the Jesuits at the Royal College at La Flèche. Then he took a degree in law, though he never practiced. After serving in the army of the Prince of Orange in Holland, then in the army of the Prince of Bavaria, and further travels around Europe, he returned to Paris and began writing. But he was troubled by the controversies surrounding his ideas.

In 1628 he settled in Holland. There he remained, with only occasional visits back to France, and some visits to England and elsewhere, until 1649. There he did his major work, though the condemnation of Galileo in 1633 caused Descartes to shelve a volume on which he was working that supported Galileo's doctrines.

Descartes died in Stockholm where he had gone to serve as tutor for Queen Christina.

T.V. Smith, noted philosopher and teacher at the University of Chicago and Syracuse University, and onetime Congressman from Illinois, has related how the reading of Descartes' *Discourse on Method* changed his whole life. One long, winter night when, as a young lad, he was riding in a freight train caboose across Texas, he came across the book which the brakeman had left open on his table while he was out working the train. Here Smith found many of the questions, and many of the challenges to prevailing opinion which he often had felt, but never had been able to articulate. This led him into philosophy as a career.

1. What does Descartes mean by method?

2. How does the method of Descartes differ from that of Bacon? What are the advantages and disadvantages of each?

3. What does Descartes see as the utility of mathematics in scientific reasoning?

4. What was Descartes' appraisal of his own contributions to thought?

DISCOURSE ON METHOD

DESCARTES

DISCOURSE ON THE METHOD OF RIGHTLY CONDUCTING THE REASON AND SEEKING THE TRUTH IN THE SCIENCES

PART I

GOOD SENSE is, of all things among men, the most equally distributed; for every one thinks himself so abundantly provided with it, that those even who are the most difficult to satisfy in everything else, do not usually desire a larger measure of this quality than they already possess. And in this it is not likely that all are mistaken: the conviction is rather to be held as testifying that the power of judging aright and of distinguishing Truth from Error, which is properly what is called Good Sense or Reason, is by nature equal in all men; and that the diversity of our opinions, consequently, does not arise from some being endowed with a larger share of Reason than others, but solely from this, that we conduct our thoughts along different ways, and do not fix our attention on the same objects. For to be possessed of a vigorous mind is not enough; the prime requisite is rightly to apply it. The greatest minds, as they are capable of the highest excellencies, are open likewise to the greatest aberrations; and those who travel very slowly may yet make far greater progress, provided they keep always to the straight road, than those who, while they run, forsake it.

For myself, I have never fancied my mind to be in any respect more perfect than those of the generality; on the contrary, I have often wished that I were equal to some others in promptitude of thought, or in clearness and distinctness of imagination, or in fulness and readiness of memory. And besides these, I know of no other qualities that contribute to the perfection of the mind; for as to the Reason or Sense, inasmuch as it is that alone which constitutes us men, and distinguishes us from the brutes, I am disposed to believe that it is to be found complete in each individual; and on this point to adopt the common opinion of philosophers, who say that the difference of greater and less holds only among the *accidents*, and not among the *forms* or *natures* of *individuals* of the same *species*.

I will not hesitate, however, to avow my belief that it has been my singular good fortune to have very early in life fallen in with certain tracks which have conducted me to considerations and maxims, of which I have formed a Method that gives me the means, as I think, of gradually augmenting my knowledge, and of raising it by little and little to the highest point which the mediocrity of my talents and the brief duration of my life will permit me to reach. For I have already reaped from it such fruits that, although I have been accustomed to think lowly enough of myself, and although when I look with the eye of a philosopher at the varied courses and pursuits of mankind at large, I find scarcely one which does not appear vain and useless, I nevertheless derive the highest satisfaction from the progress I conceive myself to have already made in the search after truth, and cannot help entertaining such expectations of the future as to believe that if, among the occupations of men as men, there is any one really excellent and important, it is that which I have chosen.

After all, it is possible I may be mistaken; and it is but a little copper and glass, perhaps, that I take for gold and diamonds. I know how very liable we are to delusion in what relates to ourselves, and also how much the judgments of our friends are to be suspected when given in our favour. But I shall endeavour in this Discourse to describe the paths I have followed, and to delineate my life as in a picture, in order that each one may be able to judge of them for himself, and that in the general opinion entertained of them, as gathered from current report. I myself may have a new help towards instruction to be added to those I have been in the habit of employing.

My present design, then, is not to teach the Method which each ought to follow for the right conduct of his Reason, but solely to describe the way in which I have endeavoured to conduct my own. They who set themselves to give precepts must of course regard themselves as possessed of greater skill than those to whom they prescribe; and if they err in the slightest particular, they subject themselves to censure. But as this Tract is put forth merely as a history, or, if you will, as a tale, in which, amid some examples worthy of imitation, there will be found, perhaps, as many more which it were advisable not to follow, I hope it will prove useful to some without being hurtful to any, and that my openness will find some favour with all.

From my childhood, I have been familiar with letters; and as I was given to believe that by their help a clear and certain knowledge of all that is useful in life& might be acquired, I was ardently desirous of instruction. But as soon as I had finished the entire course of study, at the close of which it is customary to be admitted into the order of the learned, I completely changed my opinion. For I found myself involved in so many doubts and errors, that I was convinced I had advanced no farther in all my attempts at learning, than the discovery at every turn of my own ignorance. And yet I was studying in one of the most celebrated Schools in Europe, in which I thought there must be learned men, if such were anywhere to be found. I had been taught all that others learned there; and not contented with the sciences actually taught us, I had, in addition, read all the books that had fallen into my hands, treating of such branches as are esteemed the most curious and rare. I knew the judgment which others had formed of me; and I did not find that I was considered inferior to my fellows, although there were among them some who were already marked out to fill the places of our instructors. And, in fine, our age appears to me as flourishing, and as fertile in powerful minds as and preceding one. I was thus led to take the liberty of judging of ail other men by myself, and of concluding that there was no science in existence that was of such a nature as I had previously been given to believe.

I still continued, however, to hold in esteem the studies of the Schools. I was aware that the Languages taught in them are necessary to the understanding of the writings of the ancients; that the grace of Fable stirs the mind; that the memorable deeds of History elevate it; and, if read with discretion, aid in forming the judgment; that the perusal of all excellent books is, as it were, to interview with the noblest men of past ages, who have written them, and even a studied interview, in which are discovered to us only their choicest thoughts; that Eloquence has incomparable force and beauty; that Poesy has its ravishing graces and delights; that in the Mathematics there are many refined discoveries eminently suited to gratify the inquisitive, as well as further all the arts and lessen the labour of man; that numerous highly useful precepts and exhortations to virtue are contained in treatises on Morals; that Theology points out the path to heaven; that Philosophy affords the means of discoursing with an appearance of truth on all matters, and commands the admiration of the more simple; that jurisprudence, Medicine, and the other Sciences, secure for their cultivators honours and riches; and in fine, that it is useful to bestow some attention upon all, even upon those abounding the most in superstition and error, that we may be in a position to determine their real value, and guard against being deceived.

But I believed that I had already given sufficient time to Languages, and likewise to the reading of the writings of the ancients, to their Histories and Fables. For to hold converse with those of other ages and to travel, are almost the same thing. It is useful to know something of the manners of different nations, that we may be enabled to form a more correct judgment regarding our own, and be prevented from thinking that everything contrary to our customs is ridiculous and irrational,—a conclusion usually come to by those whose experience has been limited to their own country. On the other hand, when too much time is occupied in travelling, we become strangers to our native country; and the over curious in the customs of the past are generally ignorant of those of the present. Besides, fictitious narratives lead us to imagine the possibility of many events that are impossible; and even the most faithful histories, if they do not wholly misrepresent matters, or exaggerate their importance to render the account of them more worthy of perusal, omit, at least, almost always the meanest and least striking of the attendant circumstances; hence it happens that the remainder does not represent the truth, and that such as regulate their conduct by examples drawn from this source, are apt to fall into the extravagances of the knight errants of Romance, and to entertain projects that exceed their powers.

I esteemed Eloquence highly, and was in raptures with Poesy; but I thought that both were gifts of nature rather than fruits of study. Those in whom the faculty of Reason is predominant, and who most skillfully dispose their thoughts with a view to render them clear and intelligible, are always the best able to persuade others of the truth of what they lay down, though they should speak only in the language of Lower Brittany, and be wholly ignorant of the rules of Rhetoric; and those whose minds are stored with the most agreeable fancies, and who can give expression to them with the greatest embellishment and harmony, are still the best poets, though unacquainted with the Art of Poetry.

I was especially delighted with the Mathematics, on account of the certitude and evidence of their reasonings: but I had not as yet a precise knowledge of their true use; and thinking that they but contributed to the advancement of the mechanical arts, I was astonished that foundations, so strong and solid, should have had no loftier superstructure reared on them. On the other hand, I compared the disquisitions of the ancient Moralists to very towering and magnificent palaces with no better foundation than sand and mud: they laud the virtues very highly, and exhibit them as estimable far above anything on earth; but they give us no adequate criterion of virtue, and frequently that which they designate with so fine a name is but apathy, or pride, or despair, or parricide.

I revered our Theology, and aspired as much as any one to reach heaven: but being given assuredly to understand that the way is not less open to the most ignorant than to the most learned, and that the revealed truths which lead to heaven are above our comprehension, I did not presume to subject them to the impotency of my Reason; and I thought that in order competently to undertake their examination, there was need of some special help from heaven, and of being more than man.

Of philosophy I will say nothing, except that when I saw that it had been cultivated for many ages by the most distinguished men, and that yet there is not a single matter within its sphere which is not still in dispute, and nothing, therefore, which is above doubt, I did not presume to anticipate that my success would be greater in it than that of others; and further. when I considered the number o f conflicting opinions touching a single matter that may be upheld by learned men, while there can be but one true, I reckoned as well-nigh false all that was only probable.

As to the other Sciences, inasmuch as these borrow their principles from Philosophy, I judged that no solid superstructures could be reared on foundations so infirm; and neither the honour nor the gain held out by them was sufficient to determine me to their cultivation: for I was not, thank Heaven, in a condition which compelled me to make merchandise of Science for the bettering of my fortune; and though I might not profess to scorn glory as a Cynic, I yet made very slight account of that honour which I hoped to acquire only through fictitious titles. And, in fine, of false Sciences I thought I knew the worth sufficiently to escape being deceived by the professions of an alchemist, the predictions of an astrologer, the impostures of a magician, or by the artifices and boasting of any of those who profess to know things of which they are ignorant.

For these reasons, as soon as my age permitted me to pass from under the control of my instructors, 1 entirely abandoned the study of letters, and resolved no longer to seek any other science than the knowledge of myself, or of the great book of the world. I spent the remainder of my youth in travelling, in visiting courts and armies, in holding intercourse with men of different dispositions and ranks in collecting varied experience, in proving myself in the different situations into which fortune threw me, and, above all, in making such reflection on the matter of my experience as to secure my improvement. For it occurred to me that I should find much more truth in the reasonings of each individual with reference to the affairs in which he is personally interested, and the issue of which must presently punish him if he has judged amiss, than in those conducted by a man of letters in his study, regarding speculative matters that are of no practical moment, and followed by no consequences to himself, farther, perhaps, than that they foster his vanity the better the more remote they are from common sense; requiring, as they must in this case, the exercise of greater ingenuity and art to render them probable. In addition, I had always a most earnest desire to know how to distinguish the true from the false, in order that I might be able clearly to discriminate the right path in life, and proceed in it with confidence.

It is true that, while busied only in considering the manners of other men, I found here, too, scarce any ground for settled conviction, and remarked hardly less contradiction among them than in the opinions of the philosophers. So that the greatest advantage I derived from the study consisted in this, that, observing many things which, however extravagant and ridiculous to our apprehension, are yet by common consent received and approved by other great nations, I learned to entertain too decided a belief in regard to nothing of the truth of which I had been persuaded merely by example and custom: and thus I gradually extricated myself from many errors powerful enough to darken our Natural Intelligence, and incapacitate us in great measure from listening to Reason. But after I had been occupied several years in thus studying the book of the world, and in essaying to gather some experience, I at length resolved to make myself an object of study, and to employ all the powers of my mind in choosing the paths I ought to follow; an undertaking which was accompanied with greater success than it would have been had I never quitted my country or my books.

PART II

I WAS then in Germany, attracted thither by the wars in that country, which have not yet been brought to a termination; and as I was returning to the army from the coronation of the Emperor, the setting in of winter arrested me in a locality where, as I found no society to interest me, and was besides fortunately undisturbed by any cares or passions, I remained the whole day in seclusion, with full opportunity to occupy my attention with my own thoughts. Of these one of the very first that occurred to me was, that there is seldom so much perfection in works composed of many separate parts, upon which different hands have been employed, as in those completed by a single master. Thus it is observable that the buildings which a single architect has planned and executed, are generally more elegant and commodious than those which several have attempted to improve, by making old walls serve for purposes for which they were not originally built. Thus also, those ancient cities which, from being at first only villages, have become, in course of time, large towns, are usually but ill-laid out compared with the regularly constructed towns which a professional architect has freely planned on an open plain; so that although the several buildings of the former may often equal or surpass in beauty those of the latter, yet when one observes their indiscriminate juxtaposition, there a large one and here a small, and the consequent crookedness and irregularity of the streets, one is disposed to allege that chance rather than any human will guided by reason, must have led to such a arrangement. And if we consider that nevertheless there have been at all times certain officers whose duty it was to see that private buildings contributed to public ornament, the difficulty of reaching high perfection with but the materials of others to operate on, will be readily acknowledged. In the same way I fancied that those nations which, starting from a semi-barbarous state and advancing to civilisation by slow degrees, have had their laws successively determined, and, as it were, forced upon them simply by experience of the hurtfulness of particular crimes and disputes, would by this process come to be possessed of less perfect institutions than those which, from the commencement of their association as communities, have followed the appointments of some wise legislator. It is thus quite certain that the constitution of the true religion, the ordinances of which are derived from God, must be incomparably superior to that of every other. And, to speak of human affairs, I believe that the past preeminence of Sparta was due not to the goodness of each of its laws in particular, for many of these were very strange, and even opposed to good morals, but to the circumstance that, originated by a single individual, they all tended to a single end. In the same way I thought that the sciences contained in books, (such of them at least as are made up of probable reasonings, without demonstrations,) composed as they are of the opinions of many different individuals massed together, are farther removed from truth than the simple inferences which a man of good sense using his natural and unprejudiced judgment draws respecting the matters of his experience. And because we have all to pass through a state of infancy to manhood, and have been of necessity, for a length of time, governed by our desires and preceptors, (whose dictates were frequently conflicting, while neither perhaps always counselled us for the best,) I farther concluded that it is almost impossible that our judgments can be so correct or solid as they would have been, had our Reason been mature from the moment of our birth, and had we always been guided by it alone.

It is true, however, that it is not customary to pull down all the houses of a town with the single design of rebuilding them differently, and thereby rendering the streets more handsome; but it often happens that a private individual takes down his own with the view of erecting it anew, and that people are even sometimes constrained to this when their houses are in danger of falling from age, or when the foundations are insecure. With this before me by way of example, I was persuaded that it would indeed be preposterous for a private individual to think of reforming a state by fundamentally changing it throughout, and overturning it in order to set it up amended; and the same I thought was true of any similar project for reforming the body of the Sciences, or the order of teaching them established in the Schools: but as for the opinions which up to that time I had embraced, I thought that I could not do better than resolve at once to sweep them wholly away, that I might afterwards be in a position to admit either others more correct, or even perhaps the same when they had undergone the scrutiny of Reason. I firmly believed that in this way I should much better succeed in the conduct of my life, than if I built only upon old foundations, and leant upon principles which, in my youth, I had taken upon trust. For although I recognised various difficulties in this undertaking, these were not, however, without remedy, nor once to be compared with such as attend the slightest reformation in public affairs. Large bodies, if once overthrown, are with great difficulty set up again, or even kept erect when once seriously shaken, and the fall of such is always disastrous. Then if there are any imperfections in the constitutions of states, (and that many such exist the diversity of constitutions is alone sufficient to assure us,) custom has without doubt materially smoothed their inconveniences, and has even managed to steer altogether clear of, or insensibly corrected a number which sagacity could not have provided against with equal effect; and, in fine, the defects are almost always more tolerable than the change necessary for their removal; in the same manner that highways which wind among mountains, by being much frequented, become gradually so smooth and commodious, that it is much better to follow them than to seek a straighter path by climbing over the tops of rocks and descending to the bottoms of precipices.

Hence it is that I cannot in any degree approve of those restless and busy meddlers who, called neither by birth nor fortune to take part in the management of public affairs, are yet always projecting reforms; and if I thought that this Tract contained aught which might justify the suspicion that I was a victim of such folly, I would by no means permit its publication. I have never contemplated anything higher than the reformation of my own opinions, and basing them on a foundation wholly my own. And although my own satisfaction with my work has led me to present here a draft of it, I do not by any means therefore recommend to every one else to make a similar attempt. Those whom God has endowed with a larger measure of genius will entertain, perhaps, designs still more exalted; but for the many I am much afraid lest even the present undertaking be more than they can safely venture to imitate. The single design to strip one's self of all past beliefs is one that ought not to be taken by every one. The majority of men is composed of two classes, for neither of which would this be at all a befitting resolution: in the first place, of those who with more than a due confidence in their own powers, are precipitate in their judgments and want the patience requisite for orderly and circumspect thinking; whence it happens, that if men of this class once take the liberty to doubt of their accustomed opinions, and quit the beaten highway, they will never be able to thread the byeway that would lead them by a shorter course, and will lose themselves and continue to wander for life; in the second place, of those who, possessed of sufficient sense or modesty to determine that there are others who excel them in the power of discriminating between truth and error, and by whom they may be instructed, ought rather to content themselves with the opinions of such than trust for more correct to their own Reason.

For my own part, I should doubtless have belonged to the latter class, had I received instruction from but one master, or had I never known the diversities of opinion that from time immemorial have prevailed among men of the greatest learning. But I had become aware, even so early as during my college life, that no opinion, however absurd and incredible, can be imagined, which has not been maintained by some one of the philosophers; and afterwards in the course of my travels I remarked that all those whose opinions are decidedly repugnant to ours are not on that account barbarians and savages, but on the contrary that many of these nations make an equally good, if not a better, use of their Reason than we do. I took into account also the very different character which a person brought up from infancy in France or Germany exhibits, from that which, with the same mind originally, this individual would have possessed had he lived always among the Chinese or with savages, and the circumstance that in dress itself the fashion which pleased us ten years ago, and which may again, perhaps, be received into favour before ten years have gone, appears to us at this moment extravagant and ridiculous. I was thus led to infer that the ground of our opinions is far more custom and example than any certain knowledge. And, finally, although such be the ground of our opinions, I remarked that a plurality of suffrages is no guarantee of truth where it is at all of difficult discovery, as in such cases it is much more likely that it will be found by one than by many. I could, however, select from the crowd no one whose opinions seemed worthy of preference, and thus I found myself constrained, as it were, to use my own Reason in the conduct of my life.

But like one walking alone and in the dark, I resolved to proceed so slowly and with such circumspection, that if I did not advance far, I would at least guard against falling. I did not even choose to dismiss summarily any of the opinions that had crept into my belief without having been introduced by Reason, but first *of* all took sufficient time carefully to satisfy myself of the general nature of the task I was setting myself, and ascertain the true Method by which to arrive at the knowledge of whatever lay within the compass of my powers.

Among the branches of Philosophy, I had, at an earlier period, given some attention to Logic, and among those of the Mathematics to Geometrical Analysis and Algebra,—three Arts or Sciences which ought, as I conceived, to contribute something to my design. But, on examination, I found that, as for Logic, its syllogisms and the majority of its other precepts are of avail rather in the communication of what we already know, or even as the Art of Lully, in speaking without judgment of things of which we are ignorant, than in the investigation of the unknown; and although this Science contains indeed an number of correct and very excellent precepts, there are, nevertheless, so many others, and these either injurious or superfluous, mingled with the former, that it is almost quite as difficult to effect a severance of the true from the false as it is to extract a Diana or a Minerva from a rough block of marble. Then as to the Analysis of the ancients and the Algebra of the moderns, besides that they embrace only matters highly abstract, and, to appearance, of no use, the former is so exclusively restricted to the consideration of figures, that it can exercise the Understanding only on condition of greatly fatiguing the Imagination; and, in the latter, there is so complete a subjection to certain rules and formulas, that there results an art full of confusion and obscurity calculated to embarrass, instead of a science fitted to cultivate the mind. By these considerations I was induced to seek some other Method which would comprise the advantages of the three and be exempt from their defects. And as a multitude of laws often only hampers justice, so that a state is best governed when, with few laws, these are rigidly administered; in like manner, instead of the great number of precepts of which Logic is composed, I believed that the four following would prove perfectly sufficient for me, provided I took the firm and unwavering resolution never in a single instance to fail in observing them.

The *first* was never to accept anything for true which I did not clearly know to be such; that is to say, carefully to avoid precipitancy and prejudice, and to comprise nothing more in my judgment than what was presented to my mind so dearly and distinctly as to exclude all ground of doubt.

The *second*, to divide each of the difficulties under examination into as many parts as possible, and as might be necessary for its adequate solution.

The *third*, to conduct my thoughts in such order that, by commencing with objects the simplest and easiest to know, I might ascend by little and little, and, as it were, step by step, to the knowledge of the more complex; assigning in thought a certain order even to those objects which in their own nature do not stand in a relation of antecedence and sequence.

And the *last*, in every case to make enumerations so complete, and reviews so general, that I might be assured that nothing was omitted.

The long chains of simple and easy reasonings by means of which geometers are accustomed to reach the conclusions of their most difficult demonstrations, had led me to imagine that all things, to the knowledge of which man is competent, are mutually connected in the same way, and that there is nothing so far removed from us as to be beyond our reach, or so hidden that we cannot discover it, provided only we abstain from accepting the false for the true, and always preserve in our thoughts the order necessary for the deduction of one truth from another. And I had little difficulty in determining the objects with which it was necessary to commence, for I was already persuaded that it must be with the simplest and easiest to know, and, considering that of all those who have hitherto sought truth in the Sciences, the mathematicians alone have been able to find any demonstrations, that is, any certain and evident reasons, I did not doubt but that such must have been the rule of their investigations. I resolved to commence, therefore, with the examination of the simplest objects, not anticipating however, from this any other advantage than that to be found in accustoming my mind to the love and nourishment of truth, and to a distaste for all such reasonings as were unsound. But I had no intention on that account of attempting to master all the particular Sciences commonly denominated Mathematics: but observing that however different their objects, they all agree in considering only the various relations or proportions subsisting among those objects, I thought it best for my purpose to consider these proportions in the most general form possible, without referring them to any objects in particular, except such as would most facilitate the knowledge of them, and without by any means restricting them to these, that afterwards I might thus be the better able to apply them to every other class of objects to which they are legitimately applicable. Perceiving further, that in order to understand these relations I should sometimes have to consider them one by one, and sometimes only to bear them in mind, or embrace them in the aggregate, I thought that, in order the better to consider them individually, I should view them as subsisting between straight lines, than which I could find no objects more simple or capable of being more distinctly represented to my imagination and senses; and on the other hand, that in order to retain them in the memory, or embrace an aggregate of many, I should express them by certain characters the briefest possible. In this way I believed that I could borrow all that was best both in Geometrical Analysis and in Algebra, and correct all the defects of the one by help of the other.

And, in point of fact, the accurate observance of these few precepts gave me, I take the liberty of saying, such ease in unravelling all the questions embraced in these two sciences, that in the two or three months I devoted to their examination, not only did I reach solutions of questions I had formerly deemed exceedingly difficult, but even as regards questions of the solution of which I continued ignorant, I was enabled, as it appeared to me, to determine the means whereby, and the extent to which, a solution was possible; results attributable to the circumstance that I commenced with the simplest and most general truths, and that thus each truth discovered was a rule available in the discovery of subsequent ones. Nor in this perhaps shall I appear too vain if it be considered that, as the truth on any particular pint is one, whoever apprehends the truth, knows all that on that point can be known. The child, for example, who has been instructed in the elements of Arithmetic, and has made a particular addition, according to rule, may be assured that he has found, with respect to the sum of the numbers before him, all that in this instance is within the reach of human genius. Now, in conclusion, the Method which teaches adherence to the true order, and an exact enumeration of all the conditions of the thing sought includes all that gives certitude to the rules of Arithmetic.

But the chief ground of my satisfaction with this Method, was the assurance I had of thereby exercising my reason in all matters, if not with absolute perfection, at least with the greatest attainable by me: besides, 1 was conscious that by its use my mind was becoming gradually habituated to clearer and more distinct conceptions of its objects; and I hoped also, from not having restricted this Method to any particular matter, to apply it to the difficulties of the other Sciences, with not less success than to those of Algebra. I should not, however, on this account have ventured at once on the examination of all the difficulties of the Sciences which presented themselves to me, for this would have been contrary to the order prescribed in the Method, but observing that the knowledge of such is dependent on principles borrowed from Philosophy, in which I found nothing certain, I thought it necessary first of all to endeavour to establish its principles. And because I observed, besides, that an inquiry of this kind was of all others of the greatest moment, and one in which precipitancy and anticipation in judgment were most to be dreaded, I thought that I ought not to approach it till I had reached a more mature age,(being at that time but twenty-three,) and had first of all employed much of my time in preparation for the work, as well by eradicating from my mind all the erroneous opinions I had up to that moment accepted, as by amassing variety of experience to afford materials for my reasonings, and by continually exercising myself in my chosen Method with a view to increased skill in its application.

PART III

AND, finally, as it is not enough, before commencing to rebuild the house in which we live, that it be pulled down, and materials and builders provided, or that we engage in the work ourselves, according to a plan which we have beforehand carefully drawn out, but as it is likewise necessary that we be furnished with some other house in which we may live commodiously during the operations, so that I might not remain irresolute in my actions, while my Reason compelled me to suspend my judgment, and that I might not be prevented from living thenceforward in the greatest possible felicity, I formed a provisory code of Morals, composed of three or four maxims, with which I am desirous to make you acquainted.

The *first* was to obey the laws and customs of my country, adhering firmly to the Faith in which, by the grace of God, I had been educated from my childhood, and regulating my conduct in every other matter according to the most moderate opinions, and the farthest removed from extremes, which should happen to be adopted in practice with general consent of the most judicious of those among whom I might be living. For, as I had from that time begun to hold my own opinions for nought because I wished to subject them all to examination, I was convinced that I could not do better than follow in the meantime the opinions of the most judicious; and although there are some perhaps among the Persians and Chinese as judicious as among ourselves, expediency seemed to dictate that I should regulate my practice conformably to the opinions of those with whom I should have to live; and it appeared to me that, in order to ascertain the real opinions of such, I ought rather to take cognizance of what they practised than of what they said, not only because, in the corruption of our manners, there are few disposed to speak exactly as they believe, but also because very many are not aware of what it is that they really believe; for, as the act of mind by which a thing is believed is different from that by which we know that we believe it, the one act is often found without the other. Also, amid many opinions held in equal repute, I chose always the most moderate, as much for the reason that these are always the most convenient for practice, and probably the best, (for all excess is generally vicious,) as that, in the event of my falling into error, I might be at less distance from the truth than if, having chosen one of the extremes, it should turn out to be the other which I ought to have adopted. And I placed in the class of extremes especially all promises by which somewhat of our freedom is abridged; not that I disapproved of the laws which, to provide against the instability of men of feeble resolution, when what is sought to be accomplished is some good, permit engagements by vows and contracts binding the parties to persevere in it, or even, for the security of commerce, sanction similar engagements where the purpose sought to be realized is indifferent: but because I did not find anything on earth which was wholly superior to change, and because, for myself in particular, I hoped gradually to perfect my judgments, and not to suffer them to deteriorate, I would have deemed it a grave sin against good sense, if, for the reason that I approved of something at a particular time, I therefore bound myself to hold it for good at a subsequent time, when perhaps it had ceased to be so, or I had ceased to esteem it such.

My second maxim was to be as firm and resolute in my actions as I was able, and not to adhere less steadfastly to the most doubtful opinions, when once adopted, than if they had been highly certain; imitating in this the example of travellers who, when they have lost their way in a forest, ought not to wander from side to side, far less remain in one place, but proceed constantly towards the same side in as straight a line as possible, without changing their direction for slight reasons, although perhaps it might be chance alone which at first determined the selection; for in this way, if they do not exactly reach the point they desire, they will come at least in the end to some place that will probably be preferable to the middle of a forest. In the same way, since in action it frequently happens that no delay is permissible, it is very certain that, when it is not in our power to determine what is true, we ought to act according to what is most probable; and even although we should not remark a greater probability in one opinion than in another, we ought notwithstanding to choose one or the other, and afterwards consider it, in so far as it relates to practice, as no longer dubious, but manifestly true and certain, since the reason by which our choice has been determined is itself possessed of these qualities. This principle was sufficient thenceforward to rid me of all those repentings and pangs of remorse that usually disturb the consciences of such feeble and uncertain minds as, destitute of any clear and determinate principle of choice, allow themselves one day to adopt a course of action as the best, which they abandon the next, as the opposite.

My *third* maxim was to endeavour always to conquer myself rather than fortune, and change my desires rather than the order of the world, and in general, accustom myself to the persuasion that, except our own thoughts, there is nothing absolutely in our power; so that when we have done our best in respect of things external to us, all wherein we fail of success is to be held, as regards us, absolutely impossible: and this single principle seemed to me sufficient to prevent me from desiring for the future anything which I could not obtain, and thus render me contented; for since our will naturally seeks those objects alone which the understanding represents as in some way possible of attainment, it is plain, that if we consider all external goods as equally beyond our power, we shall no more regret the absence of such goods as seem due to our birth, when deprived of them without any fault of ours, than our not possessing the kingdoms of China or Mexico; and thus making, so to speak, a virtue of necessity, we shall no more desire health in disease, or freedom in imprisonment, than we now do bodies incorruptible as diamonds, or the wings of birds to fly with. But I confess there is need of prolonged discipline and frequently repeated meditation to accustom the mind to view all objects in this light; and I believe that in this chiefly consisted the secret of the power of such philosophers as in former times were enabled to rise superior to the influence of fortune, and amid suffering and poverty, enjoy a happiness which their gods might have envied. For, occupied incessantly with the consideration of the limits prescribed to their power by nature, they became so entirely convinced that nothing was at their disposal except their own thoughts, that this conviction was of itself sufficient to prevent their entertaining any desire of other objects; and over their thoughts they acquired a sway so absolute, that they had some ground on this account for esteeming themselves more rich and more powerful, more free and more happy, than other men who, whatever be the favours heaped on them by nature and fortune, if destitute of this philosophy, can never command the realization of all their desires.

In fine, to conclude this code of Morals, I thought of reviewing the different occupations of men in this life, with the view of making choice of the best. And, without wishing to offer any remarks on the employments of others, I may state that it was my conviction that I could not do better than continue in that in which I was engaged, viz., in devoting my whole life to the culture of my Reason, and in making the greatest progress I was able in the knowledge of truth, on the principles of the Method which I had prescribed to myself. This Method, from the time I had begun to apply it, had been to me the source of satisfaction so intense as to lead me to believe that more perfect or more innocent could not be enjoyed in this life; and as by its means I daily discovered truths that appeared to me of some importance, and of which other men were generally ignorant, the gratification thence arising so occupied my mind that I was wholly indifferent to every other object. Besides, the three preceding maxims were founded singly on the design of continuing the work of self instruction. For since God has endowed each of us with some Light of Reason by which to distinguish truth from error, I could not have believed that I ought for a single moment to rest satisfied with the opinions of another, unless I had resolved to exercise my own judgment in examining these whenever I should be duly qualified for the task. Nor could I have proceeded on such opinions without scruple, had I supposed that I should thereby forfeit any advantage for attaining still more accurate, should such exist. And, in fine, I could not have restrained my desires, nor remained satisfied, had I not followed a path in which I thought myself certain of attaining all the knowledge to the acquisition of which I was competent, as well as the largest amount of what is truly good which I could ever hope to secure. Inasmuch as we neither seek nor shun any object except in so far as our understanding represents it as good or bad, all that is necessary to right action is right judgment, and to the best action the most correct judgment—that is, to the acquisition of all the virtues with all else that is truly valuable and within our reach; and the assurance of such an acquisition cannot fail to render us contented.

Having thus provided myself with these maxims, and having placed them in reserve along with the truths of Faith, which have ever occupied the first place in my belief, I came to the conclusion that I might with freedom set about ridding myself of what remained of my opinions. And, inasmuch as I hoped to be better able success fully to accomplish this work by holding intercourse with mankind, than by remaining longer shut up in the retirement where these thoughts had occurred to me, I betook me again to travelling before the winter was well ended. And, during the nine subsequent years, I did nothing but roam from one place to another, desirous of being a spectator rather than an actor in the plays exhibited on the theatre of the world; and, as I made it my business in each matter to reflect particularly upon what might fairly be doubted and prove a source of error, I gradually rooted out from my mind all the errors which had hitherto crept into it. Not that in this I imitated the Sceptics who doubt only that they may doubt, and seek nothing beyond uncertainty itself; for, on the contrary, my design was singly to find ground of assurance, and cast aside the loose earth and sand, that I might reach the rock or the clay. In this, as appears to me, I was successful enough; for, since I endeavoured to discover the falsehood or incertitude of the propositions I examined, not by feeble conjectures, but by clear and certain reasonings, I met with nothing so doubtful as not to yield some conclusion of adequate certainty, although this were merely the inference, that the matter in question contained nothing certain. And, just as in pulling down an old house, we usually reserve the ruins to contribute towards the erection, so, in destroying such of my opinions as I judged to be ill founded, I made a variety of observations and acquired an amount of experience of which I availed myself in the establishment of more certain. And further, I continued to exercise myself in the Method I had prescribed; for, besides taking care in general to conduct all my thoughts according to its rules, I reserved some hours from time to time which I expressly devoted to the employment of the Method in the solution of Mathematical difficulties, or even in the solution likewise of some questions belonging to other Sciences, but which, by my having detached them from such principles of these Sciences as were of inadequate certainty, were rendered almost Mathematical: the truth of this will be manifest from the numerous examples contained in this volume. And thus, without in appearance living otherwise than those who, with no other occupation than that of spending their lives agreeably and innocently, study to sever pleasure from vice, and who, that they may enjoy their leisure without ennui, have recourse to such pursuits as are honourable, I was nevertheless prosecuting my design, and making greater progress in the knowledge of truth, than I might, perhaps, have made had I been engaged in the perusal of books merely, or in holding converse with men of letters.

These nine years passed away, however, before I had come to any determinate judgment respecting the difficulties which form matter of dispute among the learned, or had commenced to seek the principles of any Philosophy more certain than the vulgar. And the examples of many men of the highest genius, who had, in former times, engaged in this inquiry, but, as appeared to me, without success, led me to imagine it to be a work of so much difficulty, that I would not perhaps have ventured on it so soon had I not heard it currently rumoured that I had already completed the inquiry. I know not what were the grounds of this opinion; and, if my conversation contributed in any measure to its rise, this must have happened rather from my having confessed my ignorance with greater freedom than those are accustomed to do who have studied a little, and expounded, perhaps, the reasons that led me to doubt of many of those things that by others are esteemed certain, than from my having boasted of any system of Philosophy. But, as I am of a disposition that makes me unwilling to be esteemed different from what I really am, I thought it necessary to endeavour by all means to render myself worthy of the reputation accorded to me; and it is now exactly eight years since this desire constrained me to remove from all those places where interruption from any of my acquaintances was possible, and betake myself to this country, in which the long duration of the war has led to the establishment of such discipline, that the armies maintained seem to be of use only in enabling the inhabitants to enjoy more securely the blessings of peace; and where, in the midst of a great crowd actively engaged in business, and more careful of their own affairs than curious about those of others, I have been enabled to live without being deprived of any of the conveniences to be had in the most populous cities, and yet as solitary and as retired as in the midst of the most remote deserts.

PART IV

I AM in doubt as to the propriety of making my first meditations in the place above mentioned matter of discourse; for these are so metaphysical, and so uncommon, as not, perhaps, to be acceptable to every one. And yet, that it may be determined whether the foundations that I have laid are sufficiently secure, I find myself in a measure constrained to advert to them. I had long before remarked that, in (relation to) practice, it is sometimes necessary to adopt, as if above doubt, opinions which we discern to be highly uncertain, as has been already said; but as I then desired to give my attention solely to the search after truth, I thought that a procedure exactly the opposite was called for, and that I ought to reject as absolutely false all opinions in regard to which I could suppose the least ground for doubt, in order to ascertain whether after that there remained aught in my belief that was wholly indubitable. Accordingly, seeing that our senses sometimes deceive us, I was willing to suppose that there existed nothing really such as they presented to us; and because some men err in reasoning, and fall into paralogisms, even on the simplest matters of Geometry, I, convinced that I was as open to error as any other, rejected as false all the reasonings I had hitherto taken for demonstrations; and finally, when I considered that the very same thoughts (presentations) which we experience when awake may also be experienced when we are asleep, while there is at that time not one of them true, I supposed that all the objects (presentations) that had ever entered into my mind when awake, had in them no more truth than the illusions of my dreams. But immediately upon this I observed that, whilst I thus wished to think that all was false, it was absolutely necessary that I, who thus thought, should be somewhat; and as I observed that this truth, *I think, hence I am,* was so certain and of such evidence, that no ground of doubt, however extravagant, could be alleged by the Sceptics capable of shaking it, I concluded that I might, without scruple, accept it as the first principle of the Philosophy of which I was in search.

In the next place, I attentively examined what I was, and as I observed that I could suppose that I had no body, and that here was no world nor any place in which I might be; but that I could not therefore suppose that I was not; and that, on the contrary, from the very circumstance that I thought to doubt of the truth of other things, it most clearly and certainly followed that I was; while, on the other hand, if I had only ceased to think, although all the other objects which I had ever imagined had been in reality existent, I would have had no reason to believe that I existed; I thence concluded that I was a substance whose whole essence or nature consists only in thinking, and which, that it may exist, has need of no place, nor is dependent on any material thing; so that "I," that is to say, the mind by which I am what I am, is wholly distinct from the body, and is even more easily known than the latter, and is such, that although the latter were not, it would still continue to be all that it is.

After this I inquired in general into what is essential to the truth and certainty of a proposition; for since 1 had discovered one which I knew to be true I thought that I must likewise be able to discover the ground of this certitude. And as I observed that in the words *I think, hence I am,* there is nothing at all which gives me assurance of their truth beyond this, that I see very clearly that in order to think it is necessary to exist, I concluded that I might take, as a general rule the principle, that all the things which we very clearly and distinctly conceive are true, only observing, however, that there is some difficulty in rightly determining the objects which we distinctly conceive.

In the next place, from reflecting on the circumstance that I doubted, and that consequently my being was not wholly perfect, (for I clearly saw that it was a greater perfection to know than to doubt,) I was led to inquire whence I had learned to think of something more perfect than myself; and I clearly recognised that I must hold this notion from some Nature which in reality was more perfect. As for the thoughts of many other objects external to me, as of the sky, the earth, light, heat, and a thousand more, I was less at a loss to know whence these came; for since I remarked in them nothing which seemed to render them superior to myself, I could believe that, if these were true, they were dependencies on my own nature, in so far as it possessed a certain perfection, and, if they were false, that I held them from nothing, that is to say, that they were in me because of a certain imperfection of my nature. But this could not be the case with the idea of a Nature more perfect than myself; for to receive it from nothing was a thing manifestly impossible; and, because it is not less repugnant that the more perfect should be an effect of, and dependence on the less perfect, than that something should proceed from nothing, it was equally impossible that I could hold it from myself: accordingly, it but remained that it had been placed in me by a Nature which was in reality more perfect than mine and which even possessed within itself all the perfections of which I could form any idea; that is to say, in a single word, which was God. And to this I added that, since I knew some perfections which I did not possess, I was not the only being in existence, (I will here, with your permission, freely use the terms of the schools); but, on the contrary, that there · as of necessity some other more perfect Being upon whom I was dependent, and from whom I had received all that I possessed; for if I had existed alone, and independently of every other being, so as to have had from myself all the perfection, however little which I actually possessed, I should have been able, for the same reason, to have had from myself the whole remainder of perfection, of the want of which I was conscious, and thus could of myself have become infinite, eternal, immutable, omniscient, all-powerful, and, in fine, have possessed all the perfection which I could recognise in God. For in order to know the nature of God. (whose existence has been established by the preceeding reasonings,) as far as my own nature permitted, I had only to consider in reference to all the properties of which I found in my mind some idea, whether their possession was a mark of perfection; and I was assured that no one which indicated any imperfection was in him, and that none of the rest was awanting. Thus I perceived that doubt, inconstancy, sadness, and such like, could not be found in God, since I myself would have been happy to be free from them. Besides, I had ideas of many sensible and corporeal things; for although I might suppose that I was dreaming, and that all which I saw or imagined was false, I could not, nevertheless, deny that the ideas were in reality in my thoughts. But, because I had already very clearly recognised in myself that the intelligent nature is distinct from the corporeal, and as I observed that all composition is an evidence of dependency, and that a state of dependency is manifestly a state of imperfection, I therefore determined that it could not be a perfection in God to be compounded of these two natures, and that consequently he was not so compounded; but that if there were any bodies in the world, or even any intelligences, or other natures that were not wholly perfect, their existence depended on his power in such a way that they could not subsist without him for a single moment.

I was disposed straightway to search for other truths; and when I had represented to myself the object of the geometers, which I conceived to be a continuous body, or a space indefinitely extended in length, breadth, and height or depth, divisible into divers parts which admit of different figures and sizes, and of being moved or transposed in all manner of ways, (for all this the geometers suppose to be in the object they contemplate,) I went over some of their simplest demonstrations. And, in the first place, I observed, that the great certitude which by common consent is accorded to these demonstrations, is founded solely upon this, that they are clearly conceived in accordance with the rules I have already laid down. In the next place, I perceived that there was nothing at all in these demonstrations which could assure me of the existence of their object: thus, for example, supposing a triangle to be given, I distinctly perceived that its three angles were necessarily equal to two right angles, but I did not on that account perceive anything which could assure me that any triangle existed: while, on the contrary, recurring to the examination of the idea of a Perfect Being, I found that the existence of the Being was comprised in the idea in the same way that the equality of its three angles to two right angles is comprised in the idea of a triangle, or as in the idea of a sphere, the equidistance of all points on its surface from the centre, or even still more clearly; and that consequently it is at least as certain that God, who is this Perfect Being, is, or exists, as any demonstration of Geometry can be.

But the reason which leads many to persuade themselves that there is a difficulty in knowing this truth, and even also in knowing what their mind really is, is that they never raise their thoughts above sensible objects, and are so accustomed to consider nothing except by way of imagination, which is a mode of thinking limited to material objects, that all that is not imaginable seems to them not intelligible. The truth of this is sufficiently manifest from the single circumstance, that the philosophers of the Schools accept as a maxim that there is nothing in the Understanding which was not previously in the Senses, in which however it is certain that the ideas of God and of the soul have never been; and it appears to me that they who make use of their imagination to comprehend these ideas do exactly the same thing as if, in order to hear sounds or smell odours, they strove to avail themselves of their eyes; unless indeed that there is this difference., that the sense of sight does not afford us an inferior assurance to those of smell or hearing; in place of which, neither our imagination nor our senses can give us assurance of anything unless our understanding intervene.

Finally, if there be, still persons who are not sufficiently persuaded of the existence of God and of the soul, by the reasons I have adduced, I am desirous that they should know that all the other propositions, of the truth of which they deem themselves perhaps more assured, as that we have a body, and that there exist stars and an earth, and such like, are less certain; for, although we have a moral assurance of these things, which is so strong that there is an appearance of extravagance in doubting of their existence, yet at the same time no one, unless his intellect is impaired, can deny, when the question relates to a metaphysical certitude, that there is sufficient reason to exclude entire assurance, in the observation that when asleep we can in the same way imagine ourselves possessed of another body and that we see other stars and another earth, when there is nothing of the kind. For how do we know that the thoughts which occur in dreaming are false rather than those other which we experience when awake, since the former are often not less vivid and distinct than the latter? And though men of the highest genius study this question as long as they please, I do not believe that they will be able to give any reason which can be sufficient to remove this doubt, unless they presuppose the existence of God. For, in the first place, even the principle which I have already taken as a rule, viz., that all the things which we clearly and distinctly conceive are true, is certain only because

God is or exists, and because he is a Perfect Being, and because all that we possess is derived from him: whence it follows that our ideas or notions, which to the extent of their clearness and distinctness are real, and proceed from God, must to that extent be true. Accordingly, whereas we not unfrequently have ideas or notions in which some falsity is contained, this can only be the case with such as are to some extent confused and obscure, and in this proceed from nothing, (participate of negation,) that is, exist in us thus confused because we are not wholly perfect. And it is evident that it is not less repugnant that falsity or imperfection, in so far as it is imperfection, should proceed from God, than that truth or perfection should proceed from nothing. But if we did not know that all which we possess of real and true proceeds from a Perfect and Infinite Being, however clear and distinct our ideas might be, we should have no ground on that account for the assurance that they possessed the perfection of being true.

But after the knowledge of God and of the soul has rendered us certain of this rule, we can easily understand that the truth of the thoughts we experience when awake, ought not in the slightest degree to be called in question on account of the illusions of our dreams. For if it happened that an individual, even when asleep, had some very distinct idea, as, for example, if a geometer should discover some new demonstration, the circumstance of his being asleep would not militate against its truth; and as for the most ordinary error of our dreams, which consists in their representing to us various objects in the same way as our external senses, this is not prejudicial, since it leads us very properly to suspect the truth of the ideas of sense; for we are not unfrequently deceived in the same manner when awake; as when persons in the jaundice see all objects yellow, or when the stars or bodies at a great distance appear to us much smaller than they are. For, in fine, whether awake or asleep, we ought never to allow ourselves to be persuaded of the truth of anything unless on the evidence of our Reason. And it must be noted that I say of our Reason, and not of our imagination or of our senses: thus, for example, although we very clearly see the sun, we ought not therefore to determine that it is only of the size which our sense of sight presents; and we may very distinctly imagine the head of a lion joined to the body of a goat, without being therefore shut up to the conclusion that a chimera exists; for it is not a dictate of Reason that what we thus see or imagine is in reality existent; but it plainly tells us that all our ideas or notions contain in them some truth; for otherwise it could not be that God, who is wholly perfect and veracious, should have placed them in us. And because our reasonings are never so clear or so complete during sleep as when we are awake, although sometimes the acts of our imagination are then as lively and distinct, if not more so than in our waking moments, Reason further dictates that, since all our thoughts cannot be true because of our partial imperfection, those possessing truth must infallibly be found in the experience of our waking moments rather than in that of our dreams.

PART V

I WOULD here willingly have proceeded to exhibit the whole chain of truths which I deduced from these primary; but as with a view to this it would have been necessary now to treat of many questions in dispute among the learned, with whom I do not wish to be embroiled, I believe that it will be better for me to refrain from this exposition, and only mention in general what these truths are, that the more judicious may be able to determine whether a more special account of them would conduce to the public advantage. I have ever remained firm in my original resolution to suppose no other principle than that of which I have recently availed myself in demonstrating the existence of God and of the soul, and to accept as true nothing that did not appear to me more clear and certain than the demonstrations of the geometers had formerly appeared; and yet I venture to state that not only have I found means to satisfy myself in a short time on all the principal difficulties which are usually treated of in Philosophy, but I have also observed certain laws established in nature by God in such a manner, and of which he has impressed on our minds such notions, that after we have reflected sufficiently upon these, we cannot doubt that they are accurately observed in all that exists or takes place in the world: and farther, by considering the concatenation of these laws, it appears to me that I have discovered many truths more useful and more important than all I had before learned, or even had expected to learn.

But because I have essayed to expound the chief of these discoveries in a Treatise which certain considerations prevent me from publishing, I cannot make the results known more conveniently than by here giving a summary of the contents of this Treatise. It was my design to comprise in it all that, before I set myself to write it, I thought I knew of the nature of material objects. But like the painters who, finding themselves unable to represent equally well on a plain surface all the different faces of a solid body, select one of the chief, on which alone they make the light fall, and throwing the rest into the shade, allow them to appear only in so far as they can be seen while looking at the principal one; so, fearing lest I should not be able to comprise in my discourse all that was in my mind, I resolved to expound singly, though at considerable length, my opinions regarding light; then to take the opportunity of adding something on the sun and the fixed stars, since light almost wholly proceeds from them; on the heavens since they transmit it; on the planets, comets, and earth, since they reflect it; and particularly on all the bodies that are upon the earth, since they are either coloured, or transparent, or luminous; and finally on man, since he is the spectator of these objects. Further, to enable me to cast this variety of subjects somewhat into the shade, and to express my judgment regarding them with greater freedom, without being necessitated to adopt or refute the opinions of the learned, I resolved to leave all the people here to their disputes, and to speak only of what would happen in a new world, if God were now to create somewhere in the imaginary spaces matter sufficient to compose one, and were to agitate variously and confusedly the different parts of this matter, so that there resulted a chaos as disordered as the poets ever feigned, and after that did nothing more than lend his ordinary concurrence to nature, and allow her to act in accordance with the laws which he had established. On this supposition, I, in the first place, described this matter, and essayed to represent it in such a manner that to my mind there can be nothing clearer and more intelligible, except what has been recently said regarding God and the soul; for I even expressly supposed that it possessed none of those forms or qualities which are so debated in the Schools, nor in general anything the knowledge of which is not so natural to our minds that no one can so much as imagine himself ignorant of it. Besides, I have pointed out what are the laws of nature; and, with no other principle upon which to found my reasonings except the infinite perfection of God, I endeavoured to demonstrate all those about which there could be any room for doubt, and to prove that they are such, that even if God had created more worlds, there could have been none in which these laws were not observed. Thereafter, I showed how the greatest part of the matter of this chaos must, in accordance with these laws, dispose and arrange itself in such a way as to present the appearance of heavens; how in the meantime some of its parts must compose an earth and some planets and comets, and others a sun and fixed stars. And, making a digression at this stage on the subject of light, I expounded at considerable length what the nature of that light must be which is found in the sun and the stars, and how thence in an instant of time it traverses the immense spaces of the heavens, and how from the planets and comets it is reflected towards the earth. To this I likewise added much respecting the substance, the situation, the motions, and all the different qualities of these heavens and stars; so that I thought I had said enough respecting them to show that there is nothing observable in the heavens or stars of our system that must not, or at least may not appear precisely alike in those of the system which I described. I came next to speak of the earth in particular, and to show how, even though I had expressly supposed that God had given no weight to the matter of which it is composed, this should not prevent all its parts from tending exactly to its centre; how with water and air on its surface, the disposition of the heavens and heavenly bodies, more especially of the moon, must cause a flow and ebb, like in all its circumstances to that observed in our seas, as also a certain current, both of water and air from east to west, such as is likewise observed between the tropics; how the mountains, seas, fountains, and rivers might naturally be formed in it, and the metals produced in the mines, and the plants grow in the fields; and in general, how all the bodies which are commonly denominated mixed or composite might be generated: and, among other things in the discoveries alluded to, inasmuch as besides the stars, I knew nothing except fire which produces light, I spared no pains to set forth all that pertains to its nature,—the manner of its production and support, and to explain how heat is sometimes found without light, and light without heat; to show how it can induce various colours upon different bodies and other diverse qualities; how it reduces some to a liquid state and hardens others; how it can consume almost all bodies, or convert them into ashes and smoke; and finally, how from these ashes, by the mere intensity of its action, it forms glass: for as this transmutation of ashes into glass appeared to me as wonderful as any other in nature, I took a special pleasure in describing it.

I was not, however, disposed, from these circumstances, to conclude that this world had been created in the manner I described; for it is much more likely that God made it at the first such as it was to be. But this is certain, and an opinion commonly received among theologians, that the action by which he now sustains it is the same with that by which he originally created it; so that even although he had from the beginning given it no other form than that of chaos, provided only he had established certain laws of nature, and had lent it his concurrence to enable it to act as it is wont to do, it may be believed, without discredit to the miracle of creation, that, in this way alone, things purely material might, in course of time, have become such as we observe them at present; and their nature is much more easily conceived when they are beheld coming in this manner gradually into existence, than when they are only considered as produced at once in a finished and perfect state.

From the description of inanimate bodies and plants, I passed to animals, and particularly to man. But since I had not as yet sufficient knowledge to enable me to treat of these in the same manner as of the rest, that is to say, by deducing effects from their causes, and by showing from what elements and in what manner Nature must produce them, I remained satisfied with the supposition that God formed the body of man wholly like to one of ours, as well in the external shape of the members as in the internal conformation of the organs, of the same matter with that I had described, and at first placed in it no Rational Soul, nor any other principle, in room of the Vegetative or Sensitive Soul, beyond kindling; in the heart one of those fires without light, such as I had already described, and which I thought was not different from the heat in hay that has been heaped together before it is dry, or that which causes fermentation in new wines before they are run clear of the fruit. For, when I examined the kind of functions which might, as consequences of this supposition, exist in this body, I found precisely all those which may exist in us independently of all power of thinking, and consequently without being in any measure owing to the soul; in other words, to that part of us which is distinct from the body, and of which it has been said above that the nature distinctively consists in thinking,—functions in which the animals void of Reason may be said wholly to resemble us; but among which I could not discover any of those that, as dependent on thought alone, belong to us as men, while, on the other hand, I did afterwards discover these as soon as I supposed God to have created a Rational Soul, and to have annexed it to this body in a particular manner which I described.

But, in order to show how I there handled this matter, I mean here to give the explication of the motion of the heart and arteries, which, as the first and most general motion observed in animals, will afford the means of readily determining what should be thought of all the rest. And that there may be less difficulty in understanding what I am about to say on this subject, I advise those who are not versed in Anatomy, before they commence the perusal of these observations, to take the trouble of getting dissected in their presence the heart of some large animal possessed of lungs, (for this is throughout sufficiently like the human,) and to have shewn to them its two ventricles or cavities: in the first place, that in the right side, with which correspond two very ample tubes, viz the hollow vein, *(vena cava,)* which is the principal receptacle of the blood, and the trunk of the tree, as it were, of which all the other veins in the body are branches; and the arterial vein, *(vena arteriosi,)* inappropriately so denominated, since it is in truth only an artery, which, taking its rise in the heart, is divided after passing out from it, into many branches which presently disperse themselves all over the lungs; in the second place, the cavity in the left side, with which correspond in the same manner two canals in size equal to or larger than the preceding, viz., the venous artery, (arteria venosa,) likewise inappropriately thus designated, because it is simply a vein which comes in from the lungs, where it is divided into many branches, interlaced with those of the arterial vein, and those of the tube called the windpipe, through which the air we breathe enters; and the great artery which, issuing from the heart, sends it branches all over the body. I should wish also that such persons were carefully shewn the eleven pellicles which, like so many small valves, open and shut the four orifices that are in these two cavities, viz., three at the entrance of the hollow vein, where they are disposed in such a manner as by no means to prevent the blood which it contains from flowing into the right ventricle of the heart, and yet exactly to prevent its flowing out; three at the entrance to the arterial vein, which, arranged in a manner exactly the opposite of the former, readily permit the blood contained in this cavity to pass into the lungs, but hinder that contained in the lungs from returning to this cavity; and, in like manner, two others at the mouth of the venous artery, which allow the blood from the lungs to flow into the left cavity of the heart, but preclude its return; and three at the mouth of the great artery, which suffer the blood to flow from the heart, but prevent its reflux. Nor do we need to seek any other reason for the number of these pellicles beyond this that the orifice of the venous artery being of an oval shape from the nature of its situation, can be adequately closed with two, whereas the others being round are more conveniently closed with three. Besides, I wish such persons to observe that the grand artery and the arterial vein are of much harder and firmer texture than the venous artery and the hollow vein; and that the two last expand before entering the heart, and there form, as it were, two pouches denominated the auricles of the heart, which are composed of a substance similar to that of the heart itself; and that there is always more warmth in the heart than in any other part of the body; and, finally, that this heat is capable of causing any drop of blood that passes into the cavities rapidly to expand and dilate, just as all liquors do when allowed to fall drop by drop into a highly heated vessel.

For, after these things, it is not necessary for me to say anything more with a view to explain the motion of the heart, except that when its cavities are not full of blood, into these the blood of necessity flows,—from the hollow vein into the right, and from the venous artery into the left; because these two vessels are always full of blood, and their orifices, which are turned towards the heart, cannot then be closed. But as soon as two drops of blood have thus passed, one into each of the cavities, these drops which can not but be very large, because the orifices through which they pass are wide, and the vessels from which they come full of blood, are immediately rarefied, and dilated by the heat they meet with. In this way they cause the whole heart to expand, and at the same time press home and shut the five small valves that are at the entrances of the two vessels from which they flow, and thus prevent any more blood from coming down into the heart, and becoming more and more rarefied, they push open the six small valves that are in the orifices of the other two vessels, through which they pass out, causing in this way all the branches of the arterial vein and of the grand artery to expand almost simultaneously with the heart—which immediately thereafter begins to contract, as do also the arteries, because the blood that has entered them has cooled, and the six small valves close, and the five of the hollow vein and of the venous artery open anew and allow a passage to other two drops of blood, which cause the heart and the arteries again to expand as before. And, because the blood which thus enters into the heart passes through these two pouches called auricles, it thence happens that their motion is the contrary of that of the heart, and that when it expands they contract. But lest those who are ignorant of the force of mathematical demonstrations, and who are not accustomed to distinguish true reasons from mere verisimilitudes, should venture, without examination, to deny what has been said, I wish it to be considered that the motion which I have now explained follows as necessarily from the very arrangement of the parts, which may be observed in the heart by the eye alone, and from the heat which may be felt with the fingers, and from the nature of the blood as learned from experience, as does the motion of a clock from the power, the situation, and shape of its counter-weights and wheels.

But if it be asked how it happens that the blood in the veins, flowing in this way continually into the heart, is not exhausted, and why the arteries do not become too full, since all the blood which passes through the heart flows into them, I need only mention in reply what has been written by a physician of England, who has the honour of having broken the ice on this subject, and of having been the first to teach that there are many small passages at the extremities of the arteries, through which the blood received by them from the heart passes into the small branches of the veins, whence it again returns to the heart; so that its course amounts precisely to a perpetual circulation. Of this we have abundant proof in the ordinary experience of surgeons, who, by binding the arm with a tie of moderate straitness above the part where they open the vein, cause the blood to flow more copiously than it would have done without any ligature; whereas quite the contrary would happen were they to bind it below; that is, between the hand and the opening, or were to make the ligature above the opening very tight. For it is manifest that the tie, moderately straitened, while adequate to hinder the blood already in the arm from returning towards the heart by the veins, cannot on that account prevent new blood from coming forward through the arteries, because these are situated below the veins, and their coverings, from their greater consistency, are more difficult to compress; and also that the blood which comes from the heart tends to pass through them to the hand with greater force than it does to return from the hand to the heart through the veins. And since the latter current escapes from the arm by the opening made in one of the veins, there must of necessity be certain passages below the ligature, that is, towards the extremities of the arm through which it can come thither from the arteries. This physician likewise abundantly establishes what he has advanced respecting the motion of the blood, from the existence of certain pellicles, so disposed in various places along the course of the veins, in the manner of small valves, as not to permit the blood to pass from the middle of the body towards the extremities, but only to return from the extremities to the heart; and farther, from experience which shows that all the blood which is in the body may flow out of it in a very short time through a single artery that has been cut, even although this had been closely tied in the immediate neighbourhood of the heart and cut between the heart and the ligature, so as to prevent the supposition that the blood flowing out of it could come from any other quarter than the heart.

But there are many other circumstances which evince that what I have alleged is the true cause of the motion of the blood: thus, in the first place, the difference that is observed between the blood which flows from the veins, and that from the arteries, can only arise from this, that being rarefied, and, as it were, distilled by passing through the heart, it is thinner, and more vivid, and warmer immediately after leaving the heart, in other words, when in the arteries, than it was a short time before passing into either, in other words, when it was in the veins; and if attention be given, it will be found that this difference is very marked only in the neighbourhood of the heart; and is not so evident in parts more remote from it. In the next place, the consistency of the coats of which the arterial vein and the great artery are composed, sufficiently shows that the blood is impelled against them with more force than against the veins. And why should the left cavity of the heart and the great artery be wider and larger than the right cavity and the arterial vein, were it not that the blood of the venous artery, having only been in the lungs after it has passed through the heart, is thinner, and rarefies more readily, and in a higher degree, than the blood which proceeds immediately from the hollow vein? And what can physicians conjecture from feeling the pulse unless they know that according as the blood changes its nature it can be rarefied by the warmth of the heart, in a higher or lower degree, and more or less quickly than before? And if it he inquired how this heat is communicated to the other members, must it not be admitted that this is effected by means of the blood, which, passing through the heart, is there heated anew, and thence diffused over all the body? Whence it happens, that if the blood be withdrawn from any part, the heat is likewise withdrawn by the same means; and although the heart were as hot as glowing iron, it would not be capable of warming the feet and hands as at present, unless it continually sent thither new blood. We likewise perceive from this, that the true use of respiration is to bring sufficient fresh air into the lungs, to cause the blood which flows into them from the right ventricle of the heart, where it has been rarefied and, as it were, changed into vapours, to become thick, and to convert it anew into blood, before it flows into the left cavity, without which process it would be unfit for the nourishment of the fire that is there. This receives confirmation from the circumstance, that it is observed of animals destitute of lungs that they have also but one cavity in the heart, and that in children who cannot use them while in the womb, there is a hole through which the blood flows from the hollow vein into the left cavity of the heart, and a tube through which it passes from the arterial vein into the grand artery without passing through the lung. In the next place, how could digestion be carried on in the stomach unless the heart communicated heat to it through the arteries, and along with this certain of the more fluid parts of the blood, which assists in the dissolution of the food that has been taken in? Is not also the operation which converts the juice of food into blood easily comprehended, when it is considered that it is distilled by passing and repassing through the heart perhaps more than one or two hundred times in a day? And what more need be adduced to explain nutrition, and the production of the different humours of the body, beyond saying, that the force with which the blood, in being rarefied, passes from the heart towards the extremities of the arteries, causes certain of its parts to remain in the members at which they arrive, and there occupy the place of some others expelled by them; and that according to the situation, shape, or smallness of the pores with which they meet, some rather than others flow into certain parts, in the same way that some sieves are observed to act, which, by being variously perforated, serve to separate different species of grain? And, in the last place, what above all is here worthy of observation, is the generation of the animal spirits, which are like a very subtle wind, or rather a very pure and vivid flame which, continually ascending in great abundance from the heart to the brain, thence penetrates through the nerves into the muscles, and gives motion to all the members; so that to account for other parts of the blood which, as most agitated and penetrating, are the fittest to compose these spirits, proceeding towards the brain, it is not necessary to suppose any other cause, than simply, that the arteries which carry them thither proceed from the heart in the most direct lines, and that, according to the rules of Mechanics, which are the same with those of Nature, when many objects tend at once to the same point where there is not sufficient room for all, (as is the case with the parts of the blood which flow forth from the left cavity of the heart and tend towards the brain,) the weaker and less agitated parts must necessarily be driven aside from that point by the stronger which alone in this way reach it.

I had expounded all these matters with sufficient minuteness in the Treatise which I formerly thought of publishing. And after these, I had shewn what must be the fabric of the nerves and muscles of the human body to give the animal spirits contained in it the power to move the members, as when we see heads shortly after they have been struck off still move and bite the earth, although no longer animated; what changes must take place in the brain to produce waking, sleep, and dreams; how light, sounds, odours, tastes, heat, and all the other qualities of external objects impress it with different ideas by means of the senses; how hunger, thirst, and the other internal affections can likewise impress upon it divers ideas; what must be understood by the common sense (*sensus communis)* in which these ideas are received, by the memory which retains them, by the fantasy which can change them in various ways, and out of them compose new ideas, and which, by the same means, distributing the animal spirits through the muscles, can cause the members of such a body to move in as many different ways, and in a manner as suited, whether to the objects that are presented to its senses or to its internal affections, as can take place in our own case apart from the guidance of the will. Nor will this appear at all strange to those who are acquainted with the variety of movements performed by the different automata, or moving machines fabricated by human industry, and that with help of but few pieces compared with the great multitude of bones, muscles, nerves, arteries, veins, and other parts that are found in the body of each animal. Such persons will look upon this body as a machine made by the hands of God, which is incomparably better arranged, and adequate to movements more admirable than is any machine of human invention. And here I specially stayed to show that, were there such machines exactly resembling in organs and outward form an ape or any other irrational animal, we could have no means of knowing that they were in any respect of a different nature from these animals; but if there were machines bearing the image of our bodies, and capable of imitating our actions as far as it is morally possible, there would still remain two most certain tests whereby to know that they were not therefore really men. Of these the first is that they could never use words or other signs arranged in such a manner as is competent to us in order to declare our thoughts to others: for we may easily conceive a machine to be so constructed that it emits vocables, and even that it emits some correspondent to the action upon it of external objects which cause a change in its organs; for example, if touched in a particular place it may demand what we wish to say to it; if in another it may cry out that it is hurt, and such like; but not that it should arrange them variously so as appositely to reply to what is said in its presence, as men of the lowest grade of intellect can do. The second test is, that although such machines might execute many things with equal or perhaps greater perfection than any of us, they would, without doubt, fail in certain others from which it could be discovered that they did not act from knowledge, but solely from the disposition of their organs: for while Reason is an universal instrument that is alike available on every occasion, these organs, on the contrary, need a particular arrangement for each particular action; whence it must be morally impossible that there should exist in any machine a diversity of organs sufficient to enable it to act in all the occurrences of life, in the way in which our reason enables us to act. Again, by means of these two tests we may likewise know the difference between men and brutes. For it is highly deserving of remark, that there are no men so dull and stupid, not even idiots, as to be incapable of joining together different words, and thereby constructing a declaration by which to make their thoughts understood; and that on the other hand, there is no other animal, however perfect or happily circumstanced which can do the like. Nor does this inability arise from want of organs: for we observe that magpies and parrots can utter words like ourselves, and are yet unable to speak as we do, that is, so as to show that they understand what they say; in place of which men born deaf and dumb, and thus not less, but rather more than the brutes, destitute of the organs which others use in speaking, are in the habit of spontaneously inventing certain signs by which they discover their thoughts to those who, being usually in their company, have leisure to learn their language. And this proves not only that the brutes have less Reason than man, but that they have none at all: for we see that very little is required to enable a person to speak; and since a certain inequality of capacity is observable among animals of the same species, as well as among men, and since some are more capable of being instructed than others, it is incredible that the most perfect ape or parrot of its species, should not in this be equal to the most stupid infant of its kind, or at least to one that was crackbrained, unless the soul of brutes were of a nature wholly different from ours. And we ought not to confound speech with the natural movements which indicate the passions and can be imitated by machines as well as manifested by animals; nor must it be thought with certain of the ancients, that the brutes speak, although we do not understand their language. For if such were the case, since they are endowed with many organs analogous to ours, they could as easily communicate their thoughts to us as to their fellows. It is also very worthy of remark, that, though there are many animals which manifest more industry than we in certain of their actions, the same animals are yet observed to show none at all in many others: so that the circumstance that they do better than we does not prove that they are endowed with mind, for it would thence follow that they possessed greater Reason than any of us, and could surpass us in all things; on the contrary, it rather proves that they are destitute of Reason, and that it is Nature which acts in them according to the disposition of their organs: thus it is seen, that a clock composed only of wheels and weights can number the hours and measure time more exactly than we with all our skill.

I had after this described the Reasonable Soul, and shewn that it could by no means be educed from the power of matter, as the other things of which I had spoken but that it must be expressly created; and that it is not sufficient that it be lodged in the human body exactly like a pilot in a ship, unless perhaps to move its members, but that it is necessary for it to be joined and united more closely to the body in order to have sensations and appetites similar to ours, and thus constitute a real man. I here entered, in conclusion, upon the subject of the soul at considerable length, because it is of the greatest moment: for after the error of those who deny the existence of God, an error which I think I have already sufficiently refuted, there is none that is more powerful in leading feeble minds astray from the straight path of virtue than the supposition that the soul of the brutes is of the same nature with our own; and consequently that after this life we have nothing to hope for or fear, more than flies and ants; in place of which, when we know how far they differ we much better comprehend the reasons which establish that the soul is of a nature wholly independent of the body, and that consequently it is not liable to die with the latter; and, finally, because no other causes are observed capable of destroying it, we are naturally led thence to judge that it is immortal.

PART VI

THREE years have now elapsed since I finished the Treatise containing all these matters; and I was beginning to revise it, with the view to put it into the hands of a printer, when I learned that persons to whom I greatly defer, and whose authority over my actions is hardly less influential than is my own Reason over my thoughts, had condemned a certain doctrine in Physics, published a short time previously by another individual [Galileo], to which I will not say that I adhered, but only that, previously to their censure, I had observed in it nothing which I could imagine to be prejudicial either to religion or to the state, and nothing therefore which would have prevented me from giving expression to it in writing, if Reason had persuaded me of its truth; and this led me to fear lest among my own doctrines likewise some one might be found in which I had departed from the truth, notwithstanding the great care I have always taken not to accord belief to new opinions of which I had not the most certain demonstrations, and not to give expression to aught that might tend to the hurt of any one. This has been sufficient to make me alter my purpose of publishing them; for although the reasons by which I had been induced to take this resolution were very strong, yet my inclination, which has always been hostile to writing books, enabled me immediately to discover other considerations sufficient to excuse me for not undertaking the task. And these reasons, on one side and the other, are such, that not only is it in some measure my interest here to state them, but that of the public, perhaps, to know them.

I have never made much account of what has proceeded from my own mind; and so long as I gathered no other advantage from the Method I employ beyond satisfying myself on some difficulties belonging to the speculative sciences, or endeavouring to regulate my actions according to the principles it taught me, I never thought myself bound to publish anything respecting it. For in what regards manners, every one is so full of his own wisdom, that there might be found as many reformers as heads, if any were allowed to take upon themselves the task of mending them, except those whom God has constituted the supreme rulers of his people, or to whom he has given sufficient grace and zeal to be prophets; and although my speculations greatly pleased myself, I believed that others had theirs, which perhaps pleased them still more. But as soon as I had acquired some general notions respecting Physics, and beginning to make trial of them in various particular difficulties, had observed how far they can carry us, and how much they differ from the principles that have been employed up to the present time I believed that I could not keep them concealed without sinning grievously against the law by which we are bound to promote, as far as in us lies, the general good of mankind. For by them I perceived it to be possible to arrive at knowledge highly useful in life; and in room of the Speculative Philosophy usually taught in the Schools, to discover a Practical, by means of which, knowing the force and action of fire, water, air, the stars, the heavens, and all the other bodies that surround us, as distinctly as we know the various crafts of our artisans, we might also apply them in the same way to all the uses to which they are adapted, and thus render ourselves the lords and possessors of nature. And this is a result to be desired, not only in order to the invention of an infinity of arts, by which we might be enabled to enjoy without any trouble the fruits of the earth, and all its comforts, but also and especially for the preservation of health, which is without doubt, of all the blessings of this life, the first and fundamental one; for the mind is so intimately dependent upon the condition and relation of the organs of the body. that if any means can ever be found to render men wiser and more ingenious than hitherto, I believe that it is in medicine they must be sought for. It is true that the science of Medicine, as it now exists, contains few things whose utility is very remarkable: but without any wish to depreciate it, I am confident that there is no one, even among those whose profession it is, who does not admit that all at present known in it is almost nothing in comparison of what remains to be discovered; and that we could free ourselves from an infinity of maladies of body as well as of mind, and perhaps also even from the debility of age, if we had sufficiently ample knowledge of their causes, and of all the remedies provided for us by Nature. But since I designed to employ my whole life in the search after so necessary a Science, and since I had fallen in with a path which seems to me such, that if any one follow it he must inevitably reach the end desired, unless he be hindered either by the shortness of life or the want of experiments, I judged that there could be no more effectual provision against these two impediments than if I were faithfully to communicate to the public all the little I might myself have found, and incite men of superior genius to strive to proceed farther, by contributing, each according to his inclination and ability, to the experiments which it would be necessary to make, and also by informing the public of all they might discover, so that, by the last beginning where those before them had left off, and thus connecting the lives and labours of many, we might collectively proceed much farther than each by himself could do.

I remarked, moreover, with respect to experiments, that they become always more necessary the more one is advanced in knowledge; for, at the commencement, it is better to make use only of what is spontaneously presented to our senses, and of which we cannot remain ignorant, provided we bestow on it any reflection, however slight, than to concern ourselves about more uncommon and recondite phenomena: the reason of which is, that the more uncommon often only mislead us so long as the causes of the more ordinary are still unknown; and the circumstances upon which they depend are almost always so special and minute as to be highly difficult to detect. But in this I have adopted the following order: first, I have essayed to find in general the principles, or first causes of all that is or can be in the world, without taking into consideration for this end anything but God himself who has created it, and without educing them from any other source than from certain germs of truths naturally existing in our minds. In the second place, I examined what were the first and most ordinary effects that could be deduced from these causes; and it appears to me that, in this way I have found heavens, stars, an earth, and even on the earth, water, air, fire, minerals, and some other things of this kind, which of all others are the most common and simple, and hence the easiest to know. Afterwards, when I wished to descend to the more particular, so many diverse objects presented themselves to me, that I believed it to be impossible for the human mind to distinguish the forms or species of bodies that are upon the earth, from an infinity of others which might have been, if it had pleased God to place them there, or consequently to apply them to our use, unless we rise to causes through their effects, and avail ourselves of many particular experiments. Thereupon, turning over in my mind all the objects that had ever been presented to my senses, I freely venture to state that I have never observed any which I could not satisfactorily explain by the principles I had discovered. But it is necessary also to confess that the power of nature is so ample and vast, and these principles so simple and general, that I have hardly observed a single particular effect which I cannot at once recognize as capable of being deduced in many different modes from the principles, and that my greatest difficulty usually is to discover in which of these modes the effect is dependent upon them; for out of this difficulty I cannot otherwise extricate myself than by again seeking certain experiments, which may be such that their result is not the same, if it is in the one of these modes that we must explain it, as it would be if it were to be explained in the other. As to what remains, I am now in a position to discern, as I think, with sufficient clearness what course must be taken to make the majority of those experiments which may conduce to this end: but I perceive likewise that they are such and so numerous, that neither my hands nor my income, though it were a thousand times larger than it is, would be sufficient for them all; so that, according as henceforward I shall have the means of making more or fewer experiments, I shall in the same proportion make greater or less progress in the knowledge of nature. This was what I had hoped to make known by the Treatise I had written, and so clearly to exhibit the advantage that would thence accrue to the public, as to induce all who have the common good of man at heart, that is, all who are virtuous in truth, and not merely in appearance, or according to opinion, as well to communicate to me the experiments they had already made, as to assist me in those that remain to be made.

But since that time other reasons have occurred to me, by which I have been led to change my opinion, and to think that I ought indeed to go on committing to writing all the results which I deemed of any moment, as soon as I should have tested their truth, and to bestow the same care upon them as I would have done had it been my design to publish them. This course commended itself to me, as well because I thus afforded myself more ample inducement to examine them thoroughly, for doubtless that is always more narrowly scrutinized which we believe will be read by many, than that which is written merely for our private use, (and frequently what has seemed to me true when I first conceived it, has appeared false when I have set about committing it to writing;) as because I thus lost no opportunity of advancing the interests of the public, as far as in me lay, and since thus likewise, if my writings possess any value, those into whose hands they may fall after my death may be able to put them to what us they deem proper. But I resolved by no means to consent to their publication during my lifetime, lest either the oppositions or the controversies to which they might give rise, or even the reputation, such as it might be, which they would acquire for me, should be any occasion of my losing the time that I had set apart for my own improvement. For though it be true that every one is bound to promote to the extent of his ability the good of others, and that to be useful to no one is really to be worthless, yet it is likewise true that our cares ought to extend beyond the present; and it is good to omit doing what might perhaps bring some profit to the living, when we have in view the accomplishment of other ends that will be of much greater advantage to posterity. And in truth, I am quite willing it should be known that the little I have hitherto learned is almost nothing in comparison with that of which I am ignorant, and to the knowledge of which I do not despair of being able to attain; for it is much the same with those who gradually discover truth in the Sciences, as with those who when growing rich find less difficulty in making great acquisitions, than they formerly experienced when poor in making acquisitions of much smaller amount. Or they may be compared to the commanders of armies, whose forces usually increase in proportion to their victories, and who need greater prudence to keep together the residue of their troops after a defeat than after a victory to take towns and provinces. For he truly engages in battle who endeavors to surmount all the difficulties and errors which prevent him from reaching the knowledge of truth, and he is overcome in fight who admits a false opinion touching a matter of any generality and importance, and he requires thereafter much more skill to recover his former position than to make great advances when once in possession of thoroughly ascertained principles. As for myself, if I have succeeded in discovering any truths in the Sciences, (and I trust that what is contained in this volume will show that I have found some,) I can declare that they are but the consequences and results of five or six principal difficulties which I have surmounted, and my encounters with which I reckoned as battles in which victory declared for me. I will not hesitate even to avow my belief that nothing further is wanting to enable me fully to realize my designs than to gain two or three similar victories; and that I am not so far advanced in years but that, according to the ordinary course of nature, I may still have sufficient leisure for this end. But I conceive myself the more bound to husband the time that remains, the greater my expectation of being able to employ it aright, and I should doubtless have much to rob me of it, were I to publish the principles of my Physics: for although they are almost all so evident that to assent to them no more is needed than simply to understand them, and although there is not one of them of which I do not expect to be able to give demonstration, yet, as it is impossible that they can be in accordance with all the diverse opinions of others, I foresee that I should frequently be turned aside from my grand design, on occasion of the opposition which they would be sure to awaken.

It may be said, that these oppositions would be useful both in making me aware of my errors, and, if my speculations contain anything of value, in bringing others to a fuller understanding of it; and still farther, as many can see better than one, in leading others who are now beginning to avail themselves of my principles, to assist me in turn with their discoveries. But though I recognise my extreme liability to error, and scarce ever trust to the first thoughts which occur to me, yet the experience I have had of possible objections to my views prevents me from anticipating any profit from them. For I have already had frequent proof of the judgments, as well of those I esteemed friends, as of some others to whom I thought I was an object of indifference, and even of some whose malignity and envy would, I knew, determine them to endeavour to discover what partiality concealed from the eyes of my friends. But it has rarely happened that anything has been objected to me which I had myself altogether overlooked, unless it were something far removed from the subject: so that I have never met with a single critic of my opinions who did not appear to me either less rigorous or less equitable than myself. And further, I have never observed that any truth before unknown has been brought to light by the disputations that are practised in the Schools; for while each strives for the victory, each is much more occupied in making the best of mere verisimilitude, than in weighing the reasons on both sides of the question; and those who have been long good advocates are not afterwards on that account the better judges.

As for the advantage that others would derive from the communication of my thoughts, it could not be very great; because I have not yet so far prosecuted them as that much does not remain to be added before they can be applied to practice. And I think I may say without vanity, that if there is any one who can carry them out that length, it must be myself rather than another: not that there may not be in the world many minds incomparably superior to mine, but because one cannot so well seize a thing and make it one's own, when it has been learned from another, as when one has himself discovered it. And so true is this of the present subject that, though I have often explained some of my opinions to persons of much acuteness, who, whilst I was speaking, appeared to understand them very distinctly, yet, when they repeated them, I have observed that they almost always changed them to such an extent that I could no longer acknowledge them as mine. I am glad, by the way, to take this opportunity of requesting posterity never to believe on hearsay that anything has proceeded from me which has not been published by myself; and I am not at all astonished at the extravagances attributed to those ancient philosophers whose own writings we do not possess; whose thoughts, however, I do not on that account suppose to have been really absurd, seeing they were among the ablest men of their times, but only that these have been falsely represented to us. It is observable, accordingly, that scarcely in a single instance has any one of their disciples surpassed them; and I am quite sure that the most devoted of the present followers of Aristotle would think themselves happy if they had as much knowledge of nature as he possessed, were it even under the condition that they should never afterwards attain to higher. In this respect they are like the ivy which never strives to rise above the tree that sustains it, and which frequently even returns downwards when it has reached the top; for it seems to me that they also sink, in other words, render themselves less wise than they would be if they gave up study, who, not contented with knowing all that is intelligibly explained in their author, desire in addition to find in him the solution of many difficulties of which he says not a word, and never perhaps so much as thought. Their fashion of philosophizing, however, is well suited to persons whose abilities fall below mediocrity; for the obscurity of the distinctions and principles of which they make use enables them to speak of all things with as much confidence as if they really knew them, and to defend all that they say on any subject against the most subtle and skilful, without its being possible for any one to convict them of error. In this they seem to me to be like a blind man, who, in order to fight on equal terms with a person that sees, should have made him descend to ·he bottom of an intensely dark cave: and I may say that such persons have an interest in my refraining from publishing the principles of the Philosophy of which I make use; for, since these are of a kind the simplest and most evident, I should, by publishing them, do much the same as if I were to throw open the windows, and allow the light of day to enter the cave into which the combatants had descended. But even superior men have no reason for any great anxiety to know these principles, for if what they desire is to be able to speak of all things, and to acquire a reputation for learning, they will gain their end more easily by remaining satisfied with the appearance of truth, which can be found without much difficulty in all sorts of matters, than by seeking the truth itself which unfolds itself but slowly and that only in some departments, while it obliges us, when we have to speak of others, freely to confess our ignorance. If, however, they prefer the knowledge of some few truths to the vanity of appearing ignorant of none, as such knowledge is undoubtedly much to be preferred, and, if they choose to follow a course similar to mine, they do not require for this that I should say anything more than I have already said in this Discourse. For if they are capable of making greater advancement than I have made, they will much more be able of themselves to discover all that I believe myself to have found; since as I have never examined aught except in order, it is certain that what yet remains to be discovered is in itself more difficult and recondite, than that which I have already been enabled to find, and the gratification would be much less in learning it from me than in discovering it for themselves. Besides this, the habit which they will acquire, by seeking first what is easy, and then passing onward slowly and step by step to the more difficult, will benefit them more than all my instructions. Thus, in my own case, I am persuaded that if I had been taught from my youth all the truths of which I have since sought out demonstrations, and had thus learned them without labour, I should never, perhaps, have known any beyond these; at least, I should never have acquired the habit and the facility which I think I possess in always discovering new truths in proportion as I give myself to the search. And, in a single word, if there is any work in the world which cannot be so well finished by another as by him who has commenced it, it is that at which I labour.

It is true, indeed, as regards the experiments which may conduce to this end, that one man is not equal to the task of making them all; but yet he can advantageously avail himself, in this world, of no hands besides his own, unless those of artisans, or parties of the same kind, whom he could pay, and whom the hope of gain (a means of great efficacy) might stimulate to accuracy in the performance of what was prescribed to them. For as to those who, through curiosity or a desire of learning, of their own accord, perhaps, offer him their services, besides that in general their promises exceed their performance, and that they sketch out fine designs of which not one is ever realized, they will, without doubt, expect to be compensated for their trouble by the explication of some difficulties, or, at least, by compliments and useless speeches, in which he cannot spend any portion of his time without loss to himself. And as for the experiments that others have already made, even although these parties should be willing of themselves to communicate them to him, (which is what those who esteem them secrets will never do,) the experiments are, for the most part, accompanied with so many circumstances and superfluous elements, as to make it exceedingly difficult to disentangle the truth from its adjuncts; besides, he will find almost all of them so ill described, or even so false, (because those who made them have wished to see in them only such facts as they deemed conformable to their principles,) that, if in the entire number there should be some of a nature suited to his purpose, still their value could not compensate for the time that would be necessary to make the selection. So that if there existed any one whom we assuredly knew to be capable of making discoveries of the highest kind, and of the greatest possible utility to the public; and if all other men were therefore eager by all means to assist him in successfully prosecuting his designs, I do not see that they could do aught else for him beyond contributing to defray the expenses of the experiments that might be necessary; and for the rest, prevent his being deprived of his leisure by the unseasonable interruptions of any one. But besides that I neither have so high an opinion of myself as to be willing to make promise of anything extraordinary, nor feed on imaginations so vain as to fancy that the public must he much interested in my designs. I do not, on the other hand, own a soul so mean as to be capable of accepting from any one a favour of which it could be supposed that I was unworthy.

These considerations taken together were the reason why, for the last three years, I have been unwilling to publish the Treatise I had on hand, and why I even resolved to give publicity during my life to no other that was so general, or by which the principles of my Physics might be understood. But since then, two other reasons have come into operation that have determined me here to subjoin some particular specimens, and give the public some account of my doings and designs. Of these considerations, the first is, that if I failed to do so, many who were cognizant of my previous intention to publish some writings, might have imagined that the reasons which induced me to refrain from so doing, were less to my credit than they really are; for although I am not immoderately desirous of glory, or even, if I may venture so to say, although I am averse from it in so far as I deem it hostile to repose which I hold in greater account than aught else, yet, at the same time, I have never sought to conceal my actions as if they were crimes, nor made use of many precautions that I might remain unknown; and this partly because I should have thought such a course of conduct a wrong against myself, and partly because it would have occasioned me some sort of uneasiness which would again have been contrary to the perfect mental tranquillity which I court. And for as much as, while thus indifferent to the thought alike of fame or of forgetfulness, I have yet been unable to prevent myself from acquiring some sort of reputation, I have thought it incumbent on me to do my best to save myself at least from being ill spoken of. The other reason that has determined me to commit to writing these specimens of philosophy is, that I am becoming daily more and more alive to the delay which my design of self-instruction suffers, for want of the infinity of experiments I require, and which it is impossible for me to make without the assistance of others; and, without flattering myself so much as to expect the public to take a large share in my interests, I am yet unwilling to be found so far wanting in the duty I owe to myself as to give occasion to those who shall survive me to make it matter of reproach against me some day, that I might have left them many things in a much more perfect state than I have done, had I not too much neglected to make them aware of the ways in which they could have promoted the accomplishment of my designs.

And I thought that it was easy for me to select some matters which should neither be obnoxious to much controversy, nor should compel me to expound more of my principles than I desired, and which should yet be sufficient clearly to exhibit what I can or cannot accomplish in the Sciences. Whether or not I have succeeded in this it is not for me to say; and I do not wish to forestall the judgments of others by speaking myself of my writings; but it will gratify me if they be examined, and, to afford the greater inducement to this, I request all who may have any objections to make to them, to take the trouble of forwarding these to my publisher, who will give me notice of them, that I may endeavour to subjoin at the same time my reply; and in this way readers seeing both at once will more easily determine where the truth lies; for I do not engage in any case to make prolix replies, but only with perfect frankness to avow my errors if I am convinced of them, or if I cannot perceive them, simply to state what I think is required for defence of the matters I have written, adding thereto no explication of any new matter that it may not be necessary to pass without end from one thing to another.

If some of the matters of which I have spoken in the beginning of the Dioptrics and Meteorics should offend at first sight, because I call them hypotheses and seem indifferent about giving proof of them, I request a patient and attentive reading of the whole, from which I hope those hesitating will derive satisfaction; for it appears to me that the reasonings are so mutually connected in these Treatises, that, as the last are demonstrated by the first which are their causes, the first are in their turn demonstrated by the last which are their effects. Nor must it be imagined that I here commit the fallacy which the logicians call a circle; for since experience renders the majority of these effects most certain, the causes from which I deduce them do not serve so much to establish their reality as to explain their existence; but on the contrary, the reality of the causes is established by the reality of the effects. Nor have I called them hypotheses with any other end in view except that it may be known that I think I am able to deduce them from those first truths which I have already expounded; and yet that I have expressly determined not to do so, to prevent a certain class of minds from thence taking occasion to build some extravagant Philosophy upon what they may take to be my principles, and my being blamed for it. I refer to those who imagine that they can master in a day all that another has taken twenty years to think out, as soon as he has spoken two or three words to them on the subject; or who are the more liable to error and the less capable of perceiving truth in very proportion as they are more subtle and lively. As to the opinions which are truly and wholly mine, I offer no apology for them as new,—persuaded as I am that if their reasons be well considered they will be found to be so simple and so conformed to common sense as to appear less extraordinary and less paradoxical than any others which can be held on the same subjects; nor do I even boast of being the earliest discoverer of any of them, but only of having adopted them, neither because they had nor because they had not been held by others, but solely because Reason has convinced me of their truth.

Though artisans may not be able at once to execute the invention which is explained in the Dioptrics, I do not think that any one on that account is entitled to condemn it; for since address and practice are required in order so to make and adjust the machines described by me as not to overlook the smallest particular, I should not be less astonished if they succeeded on the first attempt than if a person were in one day to become an accomplished performer on the guitar, by merely having excellent sheets of music set up before him. And if I write in French, which is the language of my country, in preference to Latin, which is that of my preceptors, it is because I expect that those who make use of their unprejudiced natural Reason will be better judges of my opinions than those who give heed to the writings of the ancients only; and as for those who unite good sense with habits of study, whom alone I desire for judges, they will not, I feel assured, be so partial to Latin as to refuse to listen to my reasonings merely because I expound them in the vulgar Tongue.

In conclusion, I am unwilling here to say anything very specific of the progress which I expect to make for the future in the Sciences, or to bind myself to the public by any promise which I am not certain of being able to fulfil; but this only will I say, that I have resolved to devote what time I may still have to live to no other occupation than that of endeavouring to acquire some knowledge of Nature, which shall be of such a kind as to enable us therefrom to deduce rules in Medicine of greater certainty than those at present in use; and that my inclination is so much opposed to all other pursuits, especially to such as cannot be useful to some without being hurtful to others, that if, by any circumstances, I had been constrained to engage in such, I do not believe that I should have been able to succeed. Of this I here make a public declaration, though well aware that it cannot serve to procure for me any consideration in the world, which, however, I do not in the least affect; and I shall always hold myself more obliged to those through whose favour I am permitted to enjoy my retirement without interruption than to any who might offer me the highest earthly preferments.

PRINCIPIA MATHEMATICAS
OPTICKS
ISAAC NEWTON

Isaac Newton (1642-1727) has been called "one of the greatest names in the history of human thought" and "generally regarded as the most original and influential theorist in the history of science." He built on the foundations laid by Copernicus, Kepler, Galileo, Descartes, and others to bring to a climax, in effect, the scientific revolution. During an eighteen month period from 1665 to 1667, while Cambridge University was closed on account of the plague and he remained at home, he developed the basis for his theory of universal gravitation and his laws of motion, for his theories of color and light and his invention of the calculus. But it would be a number of years before these discoveries had any impact on the scientific community, mainly with the publication of *Philosophie Naturalis Principia Mathematica* in 1686, the publication of *Opticks* in 1704, and the publication of *De analysi* in 1711.

Newton overcame the handicaps of an unhappy childhood and poor health to make his great contributions. Born prematurely on Christmas Day 1642 (January 4, 1643, New Style calendar), three months after the death of his father who was an illiterate yeoman, little Isaac at the age of two was turned over to his grandmother so that his mother could give her attention to a second family with a new husband whom Isaac learned to hate. Still he was able to attend grammar school in Grantham and then to enter Trinity College, Cambridge.

After the plague, Newton returned to Cambridge as a fellow of Trinity College, and in 1669 became professor of mathematics. In 1689 he was elected member of parliament for Cambridge. In 1699 he became a member of the Royal Society council and an associate of the French Academy. He left Cambridge for permanent residence in London in 1701. He had become warden of the mint in 1696 and then served as master of the mint for thirty years, until his death at the age of eighty. He was knighted by Queen Anne in 1705.

1. What fundamental assumption regarding the nature of the universe do Galileo and Newton share?
2. How does Newton describe time and space? Why do you suppose he found it necessary to be so careful regarding these definitions? How do Newton's definitions relate to your own concepts of time and space?
3. What are the implications of Newton's first rule of reasoning in philosophy for choosing between two theories which purport to explain the same set of phenomena?
4. How would Bacon have viewed Newton's "Rules of Reasoning in Philosophy"?
5. Why do you suppose that Galileo, Newton, and the ancients were so fascinated with the science of motion?
6. According to Newton, what is color? How did his view differ from the general belief of the time?
7. As colors go, what is the difference between white and black?
8. What was Newton's notion of the nature of light? How has that held up over the years?

PRINCIPIA MATHEMATICAS

ISAAC NEWTON

PREFACE TO THE FIRST EDITION

Since the ancients (as we are told by Pappus) esteemed the science of mechanics of greatest importance in the investigation of natural things, and the moderns, rejecting substantial forms and occult qualities, have endeavored to subject the phenomena of nature to the laws of mathematics, I have in this treatise cultivated mathematics as far as it relates to philosophy. The ancients considered mechanics in a twofold respect; as rational, which proceeds accurately by demonstration, and practical. To practical mechanics all the manual arts belong, from which mechanics took its name. But as artificers do not work with perfect accuracy, it comes to pass that mechanics is so distinguished from geometry that which is perfectly accurate is called geometrical; what is less so, is called mechanical. However, the errors are not in the art, but in the artificer. He that works with less accuracy is an imperfect mechanic; and if any could work with perfect accuracy, he would be the most perfect mechanic of all, for the description of right lines and circles, upon which geometry is founded, belongs to mechanics. Geometry does not teach us to draw these lines, but requires them to be drawn, for it requires that the learner should first be taught to describe these accurately before he enters upon geometry, then it shows how by these operations problems may be solved. To describe right lines and circles are problems, but not geometrical problems. The solution of these problems is required from mechanics, and by Geometry the use of them, when so solved, is shown; and it is the glory of Geometry that from those few principles, brought from without, it is able to produce so many things. Therefore geometry is founded in mechanical practice, and is nothing but that part of universal mechanics which accurately proposes and demonstrates the art of measuring. But since the manual arts are chiefly employed in the moving of bodies, it happens that Geometry is commonly referred to their magnitude, and mechanics to their motion. In this sense rational mechanics will be the science of motions resulting from any forces whatsoever, and of the forces required to produce any motions, accurately proposed and demonstrated. This part of mechanics, as far as it extended to the five powers which relate to manual arts, was cultivated by the ancients, who considered gravity (it not being a manual power) no otherwise than in moving weights by those powers. But I consider philosophy rather than arts and write not concerning manual but natural powers, and consider chiefly those things which relate to gravity, levity, elastic force the resistance of fluids. and the like forces, whether attractive or impulsive, and therefore I offer this work as the mathematical principles of philosophy, for the whole burden of philosophy deems to consist in this—from the phenomena of motions to investigate the forces of nature, and then from these forces to demonstrate the other phenomena; and to this end the general propositions in the first and second books are directed. In the third book I give an example of this in the explication of the system of the world, for by the propositions mathematically demonstrated in the former books in the third I derive from the celestial phenomena the forced of gravity with which bodies tend to the sun and the several planets. Then from these forces by other propositions which are also mathematical, I deduce the motions of the planets, the comets, the moon, and the sea. I wish we could derive the rest of the phenomena of nature by the same kind of reasoning from mechanical principles, for I am induced by many reasons to suspect that they may all depend upon certain forces by which the particles of bodies, by some causes hitherto unknown, are either mutually impelled towards one another, and cohere in regular figures, or are repelled and recede from one another. These forces being unknown, philosophers have hitherto attempted the search of Nature in vain, but I hope the principles here laid down will afford some light either to this or some truer method of philosophy.

In the publication of this work the most acute and universally learned Mr. Edmund Halley not only assisted me in correcting the errors of the press and preparing the geometrical figures, but it was through his solicitations that it came to be published for when he had obtained of me my demonstrations of the figure of the celestial orbits, he continually pressed me to communicate the same to the Royal Society, who afterwards, by their kind encouragement and entreaties, engaged me to think of publishing them. But after I had begun to consider the inequalities of the lunar motions, and had entered upon some other things relating to the laws and measures of gravity and other forces; and the figures that would be described by bodies attracted according to given laws, and the motion of several bodies moving among themselves, the motion of bodies in resisting mediums; the forces, densities, and motions, of mediums; the orbits of the comets, and such Like, I deferred that publication till I had made a search into those matters, and could put forth the whole together. What relates to the lunar motions (being imperfect), I have put all together in the corollaries of Prop. 66, to avoid being obliged to propose and distinctly demonstrate the several things there contained in a method more prolix than the subject deserved and interrupt the series of the other propositions. Some things, found out after the rest, I chose to insert in places less suitable, rather than change the number of the propositions and the citations. I heartily beg that what I have here done may be read with forbearance; and that my labors in a subject so difficult may be examined, not so much with the view to censure as to remedy their defects.

Is. NEWTON

Cambridge, Trinity College, *May* 8. 1686

SCHOLIUM

Hitherto I have laid down the definitions of such words as are less known and explained the sense in which I would have them to be understood in the following discourse. I do not define time, space. place, and motion, as being well known to all. Only I must observe, that the common people conceive those quantities under no other notions but from the relation they bear to sensible objects. And thence arise certain prejudices, for the removing of which it will be convenient to distinguish them into absolute and relative, true and apparent, mathematical and common.

1. Absolute, true, and mathematical time, of itself, and from its own nature, flows equably without relation to anything external, and by another name is called duration: relative, apparent, and common time, is some sensible and external (whether accurate or unequable) measure of duration by the means of motion, which is commonly used instead of true time; such as an hour, a day, a month, a year.

II. Absolute space, in its own nature, without relation to anything external, remains always similar and immovable. Relative space is some movable dimension or measure of the absolute spaces; which our senses determine by its position to bodies; and which is commonly taken for immovable space; such is the dimension of a subterraneous, an aerial, or celestial space, determined by its position in respect of the earth. Absolute and relative space are the same in figure And magnitude; but they do not remain always numerically the same. For if the earth, for instance, moves, a space of our air, which relatively and in respect of the earth remains always the same, will at one time be one part of the absolute space into which the air passes, at another time it will be another part of the same, and so, absolutely understood, it will be continually changed.
III. Place is a part of space which a body takes up. and is according to the space, either absolute or relative. I say, a part of space; not the situation, nor the external surface of the body. For the places of equal solids are always equal but their surfaces, by reason of their dissimilar figures, are often unequal. Positions properly have no quantity, nor are they so much the places themselves, as the properties of places. The motion of the whole is the same with the sum of the motions of the parts, that is, the translation of the whole, out of its place, is the same thing with the sum of the translations of the parts out of their places, and therefore the place of the whole is the same as the sum of the places of the parts, and for that reason, it is internal, and in the whole body.
IV. Absolute motion is the translation of a body from one absolute place into another; and relative motion, the translation from one relative place into another. Thus in a ship under sail, the relative place of a body is that part of the ship which the body possesses; or that part of the cavity which the body fills, and which therefore moves together with the ship: and relative rest is the continuance of the body in the same part of the ship, or of its cavity. But real absolute rest, is the continuance of the body in the same part of that immovable space, in which the ship itself, its cavity, and all that it contains, is moved. Wherefore, if the earth is really at rest, the body, which relatively rests in the ship, will really and absolutely move with the same velocity which the ship has on the earth. But if the earth also moves, the true and absolute motion of the body will arise, partly from the true motion of the earth, in immovable space partly from the relative motion of the ship on the earth; and if the body moves also relatively in the ship, its true motion will arise, partly from the true motion of the earth, in immovable space, and partly from the relative motions as well of the ship on the earth, as of the body in the ship; and from these relative motions will arise the relative motion of the body on the earth. As if that part of the earth, where the ship is, was truly moved towards the east, with a velocity of 10,010 parts, while the ship itself, with a fresh gale, and full sails, is carried towards the west, with a velocity expressed by 10 of those parts; but a sailor walks in the ship towards the east, with 1 part of the said velocity, then the sailor will be moved truly in immovable space towards the east, with a velocity of 10,001 parts, and relatively on the earth towards the west, with a velocity of 9 of those parts.

Absolute time, in astronomy, is distinguished from relative, by the equation or correction of the apparent time. For the natural days are truly unequal, though they are commonly considered as equal, and used for a measure of time; astronomers correct this inequality that they may measure the celestial motions by a more accurate time. It may be, that there is no such thing as an equable motion, whereby time may be accurately measured. All motions may be accelerated and retarded, but the flowing of absolute time is not liable to any change. The duration of perseverance of the existence of things remains the same, Whether the motions are swift or slow, or none at all: and therefore this duration ought to be distinguished from what are only sensible measures thereof; and from which we deduce it, by means of the astronomical equation. The necessity of this equation, for determining the times of a phenomenon, is evinced as well from the experiments of the pendulum clock, as by eclipses of the satellites of Jupiter.

As the order of the parts of time is immutable, so also is the order of the parts of space. Suppose those parts to be moved out of their places, and they will be moved (if the expression may be allowed) out of themselves. For times and spaces are, as it were, the places as well of themselves as of all other things. All things are placed in time as to order of succession; and in space as to order of situation. It is from their essence or nature that they are places; and that the primary places of things should be movable, is absurd. These are therefore the absolute places; and translations out of those places, are the only absolute motions.

But because the parts of space cannot be seen, or distinguished from one another by our senses, therefore in their stead we use sensible measures of them. For from the positions and distances of things from any body considered as immovable, we define all places; and then with respect to such places, we estimate all motions, considering bodies as transferred from some of those places into others. And so, instead of absolute places and motions, we use relative ones; and that without any inconvenience in common affairs; but in philosophical disquisitions, we ought to abstract from our senses, and consider things themselves, distinct from what are only sensible measures of them. For it may be that there is no body really at rest, to which the places and motions of others may be referred.

But we may distinguish rest and motion, absolute and relative, one from the other by their properties, causes, and effects. It is a property of rest, that bodies really at rest do rest in respect to one another. And therefore as it is possible, that in the remote regions of the fixed stars, or perhaps far beyond them, there may be some body absolutely at rest; but impossible to know, from the position of bodies to one another in our regions, whether any of these do keep the same position to that remote body, it follows that absolute rest cannot be determined from the position of bodies m our regions.

It is a property of motion, that the parts, which retain given positions to their wholes, do partake of the motions of those wholes. For all the parts of revolving bodies endeavor to recede from the axis of motion; and the impetus of bodies moving forwards arises from the joint impetus of all the parts. Therefore, if surrounding bodies are moved, those that are relatively at rest within them will partake of their motion. Upon which account, the true and absolute motion of a body cannot be determined by the translation of it from those which only seem to rest; for the external bodies ought not only to appear at rest, but to be really at rest. For otherwise, all included bodies, besides their translation from near the surrounding ones, partake likewise of their true motions; and though that translation were not made, they would not be really at rest, but only seem to be so. For the surrounding bodies stand in the like relation to the surrounded as the exterior part of a whole does to the interior, or as the shell does to the kernel; but if the shell moves, the kernel will also move, as being part of the whole, without any removal from near the shell.

A property, near akin to the preceding, is this, that if a place is moved, whatever is placed therein moves along with it; and therefore a body, which is moved from a place in motion, partakes also of the motion of its place. Upon which account, all motions, from places in motion, are no other than parts of entire and absolute motions; and every entire motion is composed of the motion of the body out of its first place, and the motion of this place out of its place, and so on, until we come to some immovable place, as in the before-mentioned example of the sailor. Wherefore, entire and absolute motions can be no otherwise determined than by immovable places, and for that reason I did before refer those absolute motions to immovable places, but relative ones to movable places. Now no other places are immovable but those that, from infinity to infinity, do all retain the same given position one to another; and upon this account must ever remain unmoved; and do thereby constitute immovable space.

The causes by which true and relative motions are distinguished, one from the other, are the forces impressed upon bodies to generate motion. True motion is neither generated nor altered, but by some force impressed upon the body moved; but relative motion may be generated or altered without any force impressed upon the body. For it is sufficient only to impress some force on other bodies with which the former is compared, that by their giving way, that relation may be changed, in which the relative rest or motion of this other body did consist. Again, true motion suffers always some change from any force impressed upon the moving body; but relative motion dues not necessarily undergo any change by such forces. For if the same forces are likewise impressed on those other bodies with which the comparison is made, that the relative position may be preserved, then that condition will be preserved in which the relative motion consists. And therefore any relative motion may be changed when the true motion remains unaltered, and the relative may be preserved when the true suffers some change. Thus, true motion by no means consists in such relations.

The effects which distinguish absolute from relative motion are, the forces of receding from the axis of circular motion. For there are no such forces in a circular motion purely relative, but in a true and absolute circular motion, they are greater or less, according to the quantity of the motion. If a vessel, hung by a long cord, is so often turned about that the cord is strongly twisted, then filled with water, and held at rest together with the water; thereupon, by the sudden action of another force, it is whirled about the contrary way, and while the cord is untwisting itself, the vessel continues for some time in this motion; the surface of the water will at first be plain, as before the vessel began to move, but after that, the vessel, by gradually communicating its motion to the water, sill make it begin sensibly to revolve, and recede by little and little from the middle, and ascend to the sides of the vessel, forming itself into a concave figure (as I have experienced). and the swifter the motion becomes, the higher will the water rise, till at last, performing its revolutions in the same times with the vessel it becomes relatively at rest in it. This ascent of the water shows its endeavor to recede from the axis of its motion; and the true and absolute circular motion of the water, which is here directly contrary to the relative, becomes known, and may be measured by this endeavor. At first, when the relative motion of the water in the vessel was greatest. it produced no endeavor to recede from the axis; the water showed no tendency to the circumference, nor any ascent towards the sides of the vessel, bur remained of a plain surface, and therefore its true circular motion had not yet begun. But afterwards, when the relative motion of the water had decreased, the ascent thereof towards the sides of the vessel proved its endeavor to recede from the axis; and this endeavor showed the real circular motion of the water continually increasing, till it had acquired its greatest quantity, when the water rested relatively in the vessel. And therefore this endeavor does not depend upon any translation of the water in respect of the ambient bodies, nor can true circular motion be defined by such translation. There is only one real circular motion of any one revolving body, corresponding to only one power of endeavoring to recede from its axis of motion, as its proper and adequate effect; but relative motions, in one and the same body, are innumerable, according to the various relations it bears to external bodies, and, like other relations, are altogether destitute of any real effect, any otherwise than they may perhaps partake of that one only true motion. And therefore in their system who suppose that our heavens, revolving below the sphere of the fixed stars, carry the planets along with them; the several parts of those heavens, and the planets, which are indeed relatively at rest in their heavens, do yet really move. For they change their position one to another (which never happens to bodies truly at rest), and being carried together with their heavens, partake of their motions, and as parts of revolving wholes, endeavor to recede from the axis of their motions.

Wherefore relative quantities are not the quantities themselves, whose names they bear, but those sensible measures of them (either accurate or inaccurate), which are commonly used instead of the measured quantities themselves. And if the meaning of words is to be determined by their use, then by the names time, space, place, and motion, their [sensible] measures are properly to be understood; and the expression will be unusual, and purely mathematical, if the measured quantities themselves are meant. On this account, those violate the accuracy of language, which ought to be kept precise, who interpret these words for the measured quantities. Nor do those less defile the purity of mathematical and philosophical truths, who confound real quantities with their relations and sensible measures.

It is indeed a matter of great difficulty to discover, and effectually to distinguish. the true motions of particular bodies from the apparent; because the parts of that immovable space, in which those motions are performed, do by no means come under the observation of our senses. Yet the thing is not altogether desperate; for we have some arguments to guide us, partly from the apparent motions, which are the differences of the true motions; partly from the forces, which are the causes and effects of the true motions. For instance, if two globes, kept at a given distance one from the other by means of a cord that connects them, were revolved about their common centre of gravity, we might, from the tension of the cord, discover the endeavor of the globes to recede from the axis of their motion, and from thence we might compute the quantity of their circular motions. And then if any equal forces should be impressed at once on the alternate faces of the globes to augment or diminish their circular motions, from the increase or decrease of the tension of the cord, we might infer the increment or decrement of their motions; and thence would be found on what faces those forces ought to be impressed. that the motions of the globes might be most augmented; that is, we might discover their hindmost faces, or those which, in the circular motion, do follow. But the faces which follow being known, and consequently the opposite ones that precede, we should likewise know the determination of their motions. And thus we might find both the quantity and the determination of this circular motion, even in an immense vacuum, where there was nothing external or sensible with which the globes could be compared. But now, if in that space some remote bodies were placed that kept always a given position one to another, as the fixed stars do in our regions, we could not indeed determine from the relative translation of the globes among those bodies, whether the motion did belong to the globes or to the bodies. But if we observed the cord and found that its tension was that very tension which the motions of the globes required, we might conclude the motion to be in the globes. and the bodies to be at rest; and then, lastly, from the translation of the globes among the bodies, we should find the determination of their motions. But how we are to obtain the true motions from their causes, effects, and apparent differences, and the converse, shall be explained more at large in the following treatise. For to this end it was that I composed it.

* * * * *

AXIOMS, OR LAWS OF MOTION

LAW I

Every body continues in its state of rest, or of uniform motion in a right line, unless it is compelled to change that state by forces impressed upon it.

Projectiles continue in their motions, so far as they are not retarded by the resistance of the air, or impelled downwards by the force of gravity. A top, whose parts by their cohesion are continually drawn aside from rectilinear motions, does not cease its rotation, otherwise than as it is retarded by the air. The greater bodies of the planets and comets, meeting with less resistance in freer spaces, preserve their motions both progressive and circular for a much longer time.

LAW II

The change of motion is *proportional to the motive force impressed; and is made in the direction of the right line in which that force is impressed.*

If any force generates a motion, a double force will generate double the motion, a triple force triple the motion, whether that force be impressed altogether and at once, or gradually and successively. And this motion (being always directed the same way with the generating force), if the body moved before, is added to or subtracted from the former motion, according as they directly conspire with or are directly contrary to each other; or obliquely joined, when they are oblique, so as to produce a new motion compounded from the determination of both.

LAW III

To every action there is always opposed an equal reaction: or, the mutual actions of two bodies upon each other are always equal, and directed to contrary parts.

Whatever draws or presses another is as much drawn or pressed by the other. If you press a stone with your finger, the finger is also pressed by the stone. If a horse draws a stone tied to a rope, the horse (if I may so say) will be equally drawn back towards the stone; for the distended rope, by the same endeavor to relax or unbend itself, will draw the horse as much towards the stone as it does the stone towards the horse, and will obstruct the progress of the one as much as it advances that of the other. If a body impinge upon another, and by its force change the motion of the other, that body also (because of the equality of the mutual pressure) will undergo an equal change, in its own motion, towards the contrary part. The changes made by these actions are equal, not in the velocities but in the motions of bodies; that is to say;, if the bodies are not hindered by any other impediments. For, because the motions are equally changed, the changes of the velocities made towards contrary parts are inversely proportional to the bodies. This law takes place also in attractions, as will be proved in the next Scholium.

* * * * *

RULES OF REASONING IN PHILOSOPHY

RULE I

We are to admit no more causes of natural things than such as are both true and sufficient to explain their appearances.

To this purpose the philosophers say that Nature does nothing in vain, and more is in vain when less will serve; for Nature is pleased with simplicity, and affects not the pomp of superfluous causes.

RULE II

Therefore to the same natural effects we must, as far as possible, assign the same causes.

As to respiration in a man and in a beast; the descent of stones in Europe and in America, the light of our culinary fire and of the sun, the reflection of light in the earth, and in the planets.

RULE III

The qualities of bodies, which admit neither intensification nor remission of degrees, and which are found to belong to all bodies within the reach of our experiments, are to be esteemed the universal qualities of all bodies whatsoever.

For since the qualities of bodies are only known to us by experiments, we are to hold for universal all such as universally agree with experiments; and such as are not liable to diminution can never be quite taken away. We are certainly not to relinquish the evidence of experiments for the sake of dreams and vain fictions of our own devising, nor are we to recede from the analogy of Nature, which is wont to be simple, and always consonant to itself. We no other way know the extension of bodies than by our senses, nor do these reach it in all bodies; but because we perceive extension in all that are sensible, therefore describe it universally to all others also. That abundance of bodies are hard, we learn by experience; and because the hardness of the whole arises from the hardness of the parts, we therefore justly infer the hardness of the undivided particles not only of the bodies we feel but of all others. That all bodies are impenetrable, we gather not from reason, but from sensation. The bodies which we handle we find impenetrable, and thence conclude impenetrability to be an universal property of all bodies whatsoever. That all bodies are movable, and endowed with certain powers (which we call the inertia) of persevering in their motion, or in their rest, we only infer from the like properties observed in the bodies which we have seen. The extension, hardness. impenetrability, mobility, and inertia of the whole, result from the extension, hardness impenetrability, mobility, and inertia of the parts; and hence we conclude the least particles of all bodies to be also all extended and hard and impenetrable, and movable, and endowed with their proper inertia. And this is the foundation of all philosophy. Moreover, that the divided but contiguous particles of bodies may be separated from one another. is matter of observation, and, in the particles that remain undivided. our minds are able to distinguish yet lesser parts, as is mathematically demonstrated. But whether the parts so distinguished, and not yet divided, may by the powers of Nature, be actually divided and separated from one another, we cannot certainly determine. Yet, had we the proof of but one experiment that any undivided particle, in breaking a hard and solid body, suffered a division, we might by virtue of this rule conclude that the undivided as well as the divided particles may be divided and actually separated to infinity.

Lastly, if it universally appears, by experiments and astronomical observations, that all bodies about the earth gravitate towards the earth, and that in proportion to the quantity of matter which they severally contain; that the moon likewise, according to the quantity of its matter, gravitates towards the earth; that, on the other band, our sea gravitates towards the moon; and all the planets one towards another, and the comets in like manner towards the sun; we must, in consequence of this rule, universally allow that all bodies whatsoever are endowed with a principle of mutual gravitation. For the argument from the appearances concludes with more force for the universal gravitation of all bodies than for their impenetrability; of which, among those in the celestial regions. we have no experiments, nor any manner of observation. Not that I affirm gravity to be essential to bodies: by their *vis insita I* mean nothing but their inertia. This is immutable. Their gravity is diminished as they recede from the earth.

RULE IV

In experimental philosophy we are to look upon propositions inferred by general inductions from phenomena as accurately or very nearly true, notwithstanding any contrary hypotheses that may be imagined, till such time as other phenomena occur, by which they may either be made more accurate, or liable to exceptions.

This rule we must follow, that the argument of induction may not be evaded by hypotheses.

* * * * *

GENERAL SCHOLIUM

The hypothesis of vortices is pressed with many difficulties. That every planet by a radius drawn to the sun may describe areas proportional to the times of description, the periodic times of the several parts of the vortices should observe the square of their distances from the sun; but that the periodic times of the planets may obtain the (3/2)th power of their distances from the sun, the periodic times of the parts of the vortex ought to be as the (3/2)th power of their distances. That the smaller vortices may maintain their lesser revolutions about Saturn, Jupiter, and other planets, and swim quietly and undisturbed in the greater vortex of the sun, the periodic times of the parts of the sun's vortex should be equal, but the rotation of the sun and planets about their axes, which ought to correspond with the motions of their vortices, recede far from all these proportions. The motions of the comets are exceedingly regular, are governed by the same laws with the motions of the planets, and can by no means be accounted for by the hypothesis of vortices, for comets are carried with very eccentric motions through all parts of the heavens indifferently, with a freedom that is incompatible with the notion of a vortex.

Bodies projected in our air suffer no resistance but from the air. Withdraw the air, as is done in Mr. Boyle's vacuum, and the resistance ceases; for in this void a bit of fine down and a piece of solid gold descend with equal velocity. And the same argument must apply to the celestial spaces above the earth's atmosphere, in these spaces, where there is no air to resist their motions, all bodies will move with the greatest freedom; and the planets and comets win constantly pursue their revolutions in orbits given in kind and position, according to the laws above explained; but though these bodies may, indeed, continue in their orbits by the mere laws of gravity, yet they could by no means have at first derived the regular position of the orbits themselves from those laws.

The six primary planets are revolved about the sun in circles concentric with the sun, and with motions directed towards the same parts, and almost in the same plane. Ten moons are revolved about the earth, Jupiter, and Saturn, in circles concentric with them, with the same direction of motion, and nearly in the planes of the orbits of those planets; but it is not to be conceived that mere mechanical causes could give birth to so many regular motions, since the comets range over all parts of the heavens in very eccentric orbits for by that kind of motion they pass easily through the orbs of the planets and with great rapidity; and in their aphelions, where they move the slowest, and are detained the longest, they recede to the greatest distances from each other, and hence suffer the least disturbance from their mutual attractions. This most beautiful system of the sun, planets, and comets, could only proceed from the counsel and dominion of an intelligent and powerful Being. And if the fixed stars are the centres of other like systems, these, being formed by the like wise counsel, must be all subject to the dominion of One; especially since the light of the fixed stars is of the same nature with the light of the sun, and from every system light passes into all the other systems: and lest the systems of the fixed stars should, by their gravity, fall on each other, he hath placed those systems at immense distances from one another.

This Being governs all things, not as the soul of the world, but as Lord over all, and on account of his dominion he is wont to be called Lord God παντοκράτωρ, or *Universal Ruler*, for God is a relative word, and has a respect to servants, and Deity is the dominion of God not over his own body, as those imagine who fancy God to be the soul of the world, but over servants. The Supreme God is a Being eternal, infinite, absolutely perfect, but a being, however perfect, without dominion, cannot be said to be Lord God; for we say, my God, your God, the God of Israel, the God of Gods, and Lord of Lords; but we do not say, my Eternal, your Eternal, the Eternal of Israel, the Eternal of Gods; we do not say, my Infinite, or my Perfect: these are titles which have no respect to servants. The word God usually signifies *Lord;* but every lord is not a God. It is the dominion of a spiritual being which constitutes a God: a true, supreme, or imaginary dominion makes a true, supreme, or imaginary God. And from his true dominion it follows that the true God is a living, intelligent, and powerful Being, and, from his other perfections, that he is supreme, or most perfect. He is eternal and infinite, omnipotent and omniscient; that is, his duration reaches from eternity to eternity; his presence from infinity to infinity, he governs all things, and knows all things that are or can be done. He is not eternity and infinity, but eternal and infinite; he is not duration or space, but he endures and is present. He endures forever, and is everywhere present; and, by existing always and everywhere, he constitutes duration and space. Since every particle of space is always, and every indivisible moment of duration *is everywhere,* certainly the Maker and Lord of all things cannot be *never* and *nowhere.* Every soul that has perception is, though in different times and in different organs of sense and motion, still the same indivisible person. There are given successive parts in duration, coexistent parts in space, but neither the one nor the other in the person of a man, or his thinking principle; and much less can they be found in the thinking substance of God. Every man, so far as he is a thing that has perception, is one and the same man during his whole life, in all and each of his organs of sense. God is the same God, always and everywhere. He is omnipresent not *virtually* only, but also *substantially;* for virtue cannot subsist without substance. In him are all things contained and moved; yet neither affects the other: God suffers nothing from the motion of bodies, bodies find no resistance from the omnipresence of God.

* * * * *

OPTICKS

DISPERSION OF LIGHT

In the year 1666, (at which time I applied myself to the grinding of optick glasses of other figures than spherical) I procured me a triangular glass prism, to try therewith the celebrated phaenomena of colours. And in order thereto, having darkened my chamber, and made a small hole in my window-shuts, to let in a convenient quantity of the sun's light, I placed my prism at its entrance, that it might be thereby refracted to the opposite wall. It was at first a very pleasing divertissement, to view the vivid and intense colours produced thereby; but after a while applying my self to consider them more circumspectly, I became surprised, to see them in an oblong form; which, according to the received laws of refraction, I expected should have been circular. They were terminated at the sides with straight lines, but at the ends, the decay of light was so gradual that it was difficult to determine justly, what was their figure; yet they seemed semicircular.

Comparing the length of this colour'd Spectrum with its breadth, I found it about five times greater, a disproportion so extravagant, that it excited me to a more than ordinary curiosity to examining from whence it might proceed. I could scarce think, that the various thicknesses of the glass, or the termination with shadow or darkness, could have any influence on light to produce such an effect; yet I thought it not amiss, first to examine those circumstances, and so try'd what would happen by transmitting light through parts of the glass of divers thicknesses, or through holes in the window of diverse bignesses, or by setting the prism without, so that the light might pass through it, and be refracted, before it was terminated by the hole: But I found none of those circumstances material. The fashion of the colours was in all these cases the same.

Then I suspected, whether by any unevenness in the glass or other contingent irregularity, these colours might be thus dilated. And to try this, I took another prism like the former, and so placed it, that the light passing through them both might be refracted contrary ways, and so by the latter returned into that course from which the former had diverted it. For by this means I thought the regular effects of the first prism would be destroyed by the second prism, but the irregular ones more augmented, by the multiplicity of refractions. The event was, that the light, which by the first prism was diffused into an oblong form was by the second reduced into an orbicular one, with as much regularity as when it did not at all pass through them.

Then I began to suspect, whether the rays, after their trajection through the prism, did not move in curve lines, and according to their more or less curvity tend to divers parts of the wall. And it increased my suspicion, when I remembered that I had often seen a tennis ball struck with an oblique racket, describe such a curve line. For, a circular as well as a progressive motion being communicated to it by that stroke, its parts on that side where the motions conspire, must press and beat the contiguous air more violently than on the other, and there excite a reluctancy and reaction of the air proportionably greater. And for the same reason, if the rays of light should possibly be globular bodies, and by their oblique passage out of one medium into another, acquire a circulating motion, they ought to feel the greater resistance from the ambient ether, on that side, where the motions conspire, and thence be continually bowed to the other. But notwithstanding this plausible ground of suspicion, which I came to examine it, I could observe no such curvity in them. And besides (which was enough for my purpose) I observed, that the difference 'twixt the length of the image, and the diameter of the hole, through which the light was transmitted, was proportionable to their distance.

The gradual removal of these suspicions at length led me to the *Experimentum Crucis,* which was this: I took two boards, and placed one of them close behind the prism at the window, so that the light might pass through a small hole, made in it for the purpose, and fall on the other board, which I placed at about 12 feet distance, having first made a small hole in it also, for some of the incident light to pass through. Then I placed another prism behind this second board, so that the light trajected through both the boards might pass through that also, and be again refracted before it arrived at the wall. This done, I took the first prism in my hand, and turned it to and fro slowly about its axis, so much as to make the several parts of the image cast, on the second board, successively pass through the hole in it, that I might observe to what places on the wall the second prism would refract them. And I saw by the variation of those places, that the light, tending to that end of the image, towards which the refraction of the first prism was made, did in the second prism suffer a refraction considerably greater than the light tending to the other end. And so the true cause of the length of that image was detected to be no other, than that light is not similar or homogenial, but consists of *Difform Rays, some of which are more Refrangible than others; so* that without any difference in their incidence on the same medium, some shall be more Refracted than others; and therefore that, according to their *particular Degrees of Refrangibility,* they were transmitted through the prism to divers parts of the opposite wall.

Now I shall proceed to acquaint you with another more notable *Deformity* in its rays, wherein the origin of colours is unfolded: concerning which I shall lay down the doctrine first, and then for its examination give you an instance or two of the experiments, as a specimen of the rest.

The doctrine you will find comprehended and illustrated in the following propositions:

1. As the rays of light differ in degrees of refrangibility so they also differ in their disposition to exhibit, this or that particular colour. Colours are not qualifications of light, derived from refractions, or reflections of natural bodies (as 'tis generally believed) but original and connate properties, which in divers rays are divers. Some rays are disposed to exhibit a red colour and no other; some a yellow and no other, some a green and no other, and so of the rest. Nor are there only rays proper and particular to the more eminent colours, but even to all their intermediate gradations.

2. To the same degree of refrangibility ever belongs the same colour, and to the same colour ever belongs the same degree of refrangibility. The least refrangible rays are all disposed to exhibit a red colour, and contrarily those rays which are disposed to exhibit a red colour, are all the least refrangible: so the most refrangible rays are all disposed to exhibit a deep violet colour, and contrarily those which are apt to exhibit such a violet colour are all the most refrangible. And so to all the intermediate colours in a continued series belong intermediate degrees of refrangibility. And this Analogy 'twixt colours and refrangibility is very precise and strict; the rays always either exactly agreeing in both, or proportionally disagreeing in both.

3. The species of colour, and degree of refrangibility proper to any particular sort of rays, is not mutable by refraction, nor by reflection from natural bodies, nor by any other cause that I could yet observe. When any one sort of rays hath been well parted from those of other kinds, it hath afterwards obstinately retained its colour, notwithstanding my utmost endeavors to change it. I have refracted it with prisms, and reflected it with bodies, which in daylight were of other colours; I have intercepted it with the coloured film of air, interceded two compressed plates of glass; transmitted it through coloured mediums, and through mediums irradiated with other sorts of rays, and diversely terminated it; and yet could never produce any new colour out of it. It would by contracting or dilating become more brisk, or faint, and by the loss of many rays, in some cases very obscure and dark; but I could never see it changed in specie.

4. Yet seeming transmutations of colours may be made, where there is any mixture of divers sorts of rays. For in such mixtures, the component colours appear not, but, by their mutual allaying each other, constitute a midling colour.

And therefore, if by refraction, or any other of the aforesaid causes, the difform rays, latent in such a mixture, be separated, there shall emerge colours different from the colour of the composition. Which colours are not new generated, but only made apparent by being parted; for if they be again entirely mixt and blended together, they will again compose that colour, which they did before separation. And for the same reason, transmutations made by the convening of divers colours are not real; for when the difform rays are again severed, they will exhibit the very same colours which they did before they entered the composition; as you see blue and yellow powders, when finely mixed, appear to the naked eye, green, and yet the colours of the component corpuscles are not thereby really transmuted, but only blended. For when viewed with a good microscope they still appear blue and yellow interspersedly.

5. There are therefore two sorts of colours. The one original and simple, and the other compounded of these. The original or primary colours are red, yellow, green, blue, and a violet-purple, together with orange, indigo, and an indefinite variety of intermediate gradations.

6. The same colours in specie with these primary ones, may be also produced by composition. For a mixture of yellow and blue makes green; of red and yellow makes orange; of orange and yellowish green makes yellow. And in general, if any two colours be mixed, which in the series of those generated by the prism are not too far distant one from another, they by their mutual alloy compound that colour, which in the said series appeareth in the midway between them. But those which are situated at too great a distance, do not so. Orange and indigo produce not the intermediate green, nor scarlet and green the intermediate yellow.

7. But the most surprising, and wonderful composition was that of whiteness. There is no one sort of rays which alone can exhibit this. 'Tis ever compounded, and to its composition, are requisite all the aforesaid primary colours, mixed in a due proportion. I have often with admiration beheld that all the colours of the prism being made to converge, and thereby to be again mixed, as they were in the light before it was incident upon the prism, reproduced light, entirely and perfectly white, and not at all sensibly differing from a direct light of the sun, unless when the glasses, I used, were not sufficiently clear; for then they would a little incline it to their colour.

8. Hence therefore it comes to pass, that whiteness is the usual colour of light; for light is a confused aggregate of rays indued with all sorts of colours, as they were promiscuously darted from the various parts of luminous bodies. And of such a confused aggregate, as I said, is generated whiteness, if there be a due proportion of the ingredients; but if any one predominate, the light must incline to that colour; as it happens in the blue flame of brimstone; the yellow flame of a candle; and the various colours of the fixed stars.

9. These things considered, the manner how colours are produced by the prism is evident. For, of the rays, constituting the incident light, since those which differ in colour proportionally differ in refrangibility, they by their unequal refractions must be severed and dispersed into an oblong form in an orderly succession, from the least refracted scarlet, to the most refracted violet. And for the same reason it is, that objects when looked upon through a prism, appear coloured. For the difform rays, by their unequal refractions, are made to diverge towards several parts of the Retina, and these express the images of things coloured, as in the former case they did the sun's image upon a wall. And by

this inequality of refractions, they become not only coloured, but also very confused and indistinct.

10. Why the colours of the rainbow appear in falling drops of rain, is also from hence evident. For those drops which refract the rays, disposed to appear purple, in greatest quantity to the spectator's eye, refract the rays of other sorts so much less, as to make them pass beside it; and such are the drops on the inside of the primary bow, and on the outside of the secondary or exterior one. So those drops, which refract in greatest plenty the rays, apt to appear red, toward the spectator's eye, refract those of other sorts so much more, as to make them pass beside it; and such are the drops on the exterior part of the primary, and interior part of the secondary bow.

11. The odd phaenomena of an infusion of *Lignum Nepbriticum,* leaf-gold, fragments of coloured glass, and some other transparently coloured bodies, appearing in one position of one colour, and of another in another, are on these grounds no longer riddles. For those are substances apt to reflect one sort of light, and transmit another; as may be seen in a dark room, by illuminating them with familiar or uncompounded light. For then they appear of that colour only, with which they are illuminated, but yet in one position more vivid and luminous that in another, accordingly as they are disposed more or less to reflect or transmit the incident colour.

12. From hence also is manifest the reason of an unexpected experiment which Mr. *Hook,* somewhere in *Micrography* relates to have made with two wedge-like transparent vessels, filled the one with a red, the other with a blue liquor: namely, that though they were severally transparent enough, yet both together became opake; for if one transmitted only red, and the other only blue, no rays could pass through both.

13. I might add more instances of this nature, but I shall conclude with this general one. That the colours of all natural bodies have no other origin than is, that they are variously qualified, to reflect one sort of light in greater plenty than another. And this I have experimented in a dark room, by illuminating those bodies with uncompounded light of divers colours. For by that means any body may be made to appear of any colour. They have there no appropriate colour, but ever appear of the colour of the light cast upon them, but yet with this difference, that they are most brisk and vivid in the light of their own daylight colour. Minium appeareth there of any colour indifferently, with which it is illustrated, but yet most luminous in red, and so bise appeareth indifferently of any colour, but yet most luminous in blue. And therefore minium reflecteth rays of any colour, but most copiously those endowed with red, and consequently when illustrated with daylight; that is, with all sorts of rays promiscuously blended, those qualified with red shall abound most in the reflected light, and by their prevalence cause it to appear of that colour. And for the same reason bise, reflecting blue most copiously, shall appear blue by the excess of those rays in its reflected light; and the like of other bodies. And that this is the entire and adequate cause of their colours, is manifest, because they have no power to change or alter the colours of any sort of rays incident apart, but put on all colours indifferently, with which they are enlightened.

These things being so, it can be no longer disputed, whether there be colours in the dark, or whether they be the qualities of the objects we see, no nor perhaps, whether light be a body. For, since colours are the qualities of light, having its rays for their entire and immediate subject, how can we think those rays qualities also, unless one quality may be the subject of, and sustain another; which in effect is to call it substance. We should not know bodies for substances; were it not for their sensible qualities, and the principal of those being now found due to something else, we have as good reason to believe that to be a substance also.

Besides, who ever thought any quality to be a heterogeneous aggregate, such as light is discovered to be? But to determine more absolutely what light is, after what manner refracted, and by what modes or actions it produceth in our minds the phantasms of colours, is not so easie; and I shall not mingle conjectures with certainties.

The rest of the paper contains suggestions of various experiments by which the views presented can be tested.

The Nature of Light

Qu. 28. Are not all Hypotheses erroneous, in which Light is supposed to consist in Pression or Motion, propagated through a fluid Medium? For in all these Hypotheses the Phaenomena of Light have been hitherto explained by supposing that they arise from new Modifications of the Rays; which is an erroneous Supposition.

If Light consisted only in Pression propagated without actual Motion, it would not be able to agitate and heat the Bodies which refract and reflect it. If it consisted in Motion propagated to all distances in an instant, it would require an infinite force every moment, in every shining Particle, to generate that Motion. And if it consisted in Pression or Motion, propagated either in an instant or in time, it would bend into the Shadow. For Pression or Motion cannot be propagated in a Fluid in right Lines, beyond an Obstacle which stops part of the Motion, but will bend and spread every way into the quiescent Medium which lies beyond the Obstacle. Gravity tends downward, but the Pressure of Water arising from Gravity tends every way with equal Force, and is propagated as readily, and with as much force sideways as downwards, and through crooked passages as through strait ones. The Waves on the surface of stagnating Water, passing by the sides of a broad Obstacle which stops part of them, bend afterwards and dilate themselves gradually into the quiet Water behind the Obstacle. The Waves, Pulses or Vibrations of the Air, wherein Sounds consist, bend manifestly, though not so much as the Waves of Water. For a Bell or a Cannon may be heard beyond a Hill which intercepts the sight of the sounding Body, and Sounds are propagated as readily through crooked Pipes as through streight ones. But Light is never known to follow crooked Passages nor to bend into the Shadow. For the fix'd Stars by the Interposition of any of the Planets cease to be seen. And so do the Parts of the Sun by the Interposition of the Moon, *Mercury* or *Venus*. The Rays which pass very near to thee edges of any Body, are bent a little by the action of the Body, as we showed above; but this bending is not towards but from the Shadow, and is perform'd only in the passage of the Ray by the Body, and at a very small distance from it. So soon as the Ray is past the Body, it goes right on.

The rest of this Query is omitted.

Vous avez confirmé dans ces lieux pleins d'ennui
Ce que Newton connut sans sortir de chez lui.
You have confirmed in these tedious places what Newton knew without leaving his home.

François Marie Arouet [Voltaire]

As we read the *Principia* we feel as when we are in an ancient armoury where the weapons are of gigantic size; and as we look at them we marvel what manner of man was he who could use as a weapon what we can scarcely lift as a burden.

William Whewell

When Newton saw an apple fall, he found. . .
A mode of proving that the earth turn'd round
In a most natural whirl, called gravitation,
And thus is the sole mortal who could grapple
Since Adam, with a fall or an apple.

George Gordon, Lord Byron, *Don Juan*

If I have been able to see farther than others, it is because I stood on the shoulders of giants.

Isaac Newton

Every experiment is like a weapon which must be used in its particular way—a spear to thrust, a club to strike. Experimenting requires a man who knows when to thrust and when to strike, each according to need and fashion.

Philippus Aureolus Paracelsus, *Surgeon's Book*, 1605

THE NEW ORGANON

FRANCIS BACON

Francis Bacon (1561-1626), one of the first to challenge the method and reasoning of Aristotle, was one of the pioneers of the scientific revolution.

He was born in London at York House, the residence of his father as keeper of the Great Seal under Queen Elizabeth. His mother, Lady Ann, was a scholarly woman whose sister was the lady whose husband would become Elizabeth's prime minister. Bacon spent his first twelve years at York House and then entered Trinity College, Cambridge.

In 1576 he joined the staff of the English Ambassador in Paris. While on that assignment, he traveled widely around Europe. After his father's death in 1579 he returned to London and was admitted to Gray's Inn for studies in law. He became a barrister in 1582 and two years later was elected to Parliament. In 1603 he was knighted. In 1605 he published *Advancement of Learning*, dedicated to the King, and the following summer he married Alice Barnham, daughter of a wealthy London alderman.

Bacon was made Baron Verulam in 1618 and appointed lord chancellor of England. Two years later he was created Viscount St. Alban. It was during the period 1608-1620 that he wrote his greatest work, *Novum Organum*. He prepared at least twelve drafts of the book before it was published.

1. What was Bacon's criticism of ancient science and philosophy?

2. How does Bacon define the scientific method? What is his system of reasoning?

3. What does Bacon mean by "Idols of the Tribe," "Idols of the Cave," "Idols of the Market Place," and "Idols of the Theater"? What is the significance of each?

4. What are the greatest hindrances to the advancement of science?

5. How does Bacon distinguish between "experiments of light" and "experiments of fruit"? What is the relationship between the two?

THE NEW ORGANON

FRANCIS BACON

AUTHOR'S PREFACE

Those who have taken upon them to lay down the law of nature as a thing already searched out and understood, whether they have spoken in simple assurance or professional affectation, have therein done philosophy and the sciences great injury. For as they have been successful in inducing belief, so they have been effective in quenching and stopping inquiry; and have done more harm by spoiling and putting an end to other men's efforts than good by their own. Those on the other hand who have taken a contrary course, and asserted that absolutely nothing can be known—whether it were from hatred of the ancient sophists, or from uncertainty and fluctuation of mind, or even from a kind of fullness of learning, that they fell upon this opinion—have certainly advanced reasons for it that are not to be despised; but yet they have neither started from true principles nor rested in the just conclusion, zeal and affectation having carried them much too far. The more ancient *of* the Greeks (whose writings are lost) took up with better judgment a position between these two extremes—between the presumption of pronouncing on everything, and the despair of comprehending anything; and though frequently and bitterly complaining of the difficulty of inquiry and the obscurity of things, and like impatient horses champing at the bit, they did not the less follow up their object and engage with nature, thinking (it seems) that this very question—viz., whether or not anything can be known—was to be settled not by arguing, but by trying. And yet they too, trusting entirely to the force of their understanding, applied no rule, but made everything turn upon hard thinking and perpetual working and exercise of the mind.

Now my method, though hard to practice, is easy to explain; and it is this. I propose to establish progressive stages of certainty. The evidence of the sense, helped and guarded by a certain process of correction, I retain. But the mental operation which follows the act of sense I for the most part reject; and instead of it I open and lay out a new and certain path for the mind to proceed in, starting directly from the simple sensuous perception. The necessity of this was felt, no doubt, by those who attributed so much importance to logic, showing thereby that they were in search of helps for the understanding, and had no confidence in the native and spontaneous process of the mind. But this remedy comes too late to do any good, when the mind is already, through the daily intercourse and conversation of life, occupied with unsound doctrines and beset on all sides by vain imaginations. And therefore that art of logic, coming (as I said) too late to the rescue, and no way able to set matters right again, has had the effect of fixing errors rather than disclosing truth. There remains but one course for the recovery of a sound and healthy condition—namely, that the entire work of the understanding be commenced afresh, and the mind itself be from the very outset not left to take its own course, but guided at every step; and the business be done as if by machinery. Certainly if in things mechanical men had set to work with their naked hands, without help or force of instruments, just as in things intellectual they have set to work with little else than the naked forces of the understanding, very small would the matters have been which, even with their best efforts applied in conjunction, they could have attempted or accomplished. Now (to pause a while upon this example and look in it as in a glass) let us suppose that some vast obelisk were (for the decoration of a triumph or some such magnificence) to be removed from its place, and that men should set to work upon it with their naked hands, would not any sober spectator think them mad? And if they should then send for more people, thinking that in that way they might manage it, would he not think them all the madder? And if they then proceeded to make a selection, putting away the weaker hands, and using only the strong and vigorous, would he not think them madder than ever? And if lastly, not content with this, they resolved to call in aid the art of athletics, and required all their men to come with hands, arms, and sinews well anointed and medicated according to the rules of the art, would he not cry out that they were only taking pains to show a kind of method and discretion in their madness? Yet just so it is that men proceed in matters intellectual—with just the same kind of mad effort and useless combination of forces—when they hope great things either from the number and cooperation or from the excellency and acuteness of individual wits; yea, and when they endeavor by logic (which may be considered as a kind of athletic art) to strengthen the sinews of the understanding, and yet with all this study and endeavor it is apparent to any true judgment that they are but applying the naked intellect all the time; whereas in every great work to be done by the hand of man it is manifestly impossible, without instruments and machinery, either for the strength of each to be exerted or the strength of to be united.

Upon these premises two things occur to me of which, that they may not be overlooked, I would have men reminded. First, it falls out fortunately as I think for the allaying of contradictions and heartburnings, that the honor and reverence due to the ancients remains untouched and undiminished, while I may carry out my designs and at the same time reap the fruit of my modesty. For if I should profess that I, going the same road as the ancients, have something better to produce, there must needs have been some comparison or rivalry between us (not to be avoided by any art of words) in respect of excellency or ability of wit; and though in this there would be nothing unlawful or new (for if there be anything misapprehended by them, or falsely laid down, why may not I, using a liberty common to all, take exception to it?) yet the contest, however just and allowable, would have been an unequal one perhaps, in respect of the measure of my own powers. As it is, however (my object being to open a new way for the understanding, a way by them untried and unknown), the case is altered: party zeal and emulation are at an end, and I appear merely as a guide to point out the road—an office of small authority, and depending more upon a kind of luck than upon any ability or excellency. And thus much relates to the persons only. The other point of which I would have men reminded relates to the matter itself.

Be it remembered then that I am far from wishing to interfere with the philosophy which now flourishes, or with any other philosophy more correct and complete than this which has been or may hereafter be propounded. For I do not object to the use of this received philosophy, or others like it, for supplying matter for disputations or ornaments for discourse—for the professor's lecture and for the business of life. Nay, more, I declare openly that for these uses the philosophy which I bring forward will not be much available. It does not lie in the way. It cannot be caught up in passage. It does not flatter the understanding by conformity with preconceived notions. Nor will it come down to the apprehension of the vulgar except by its utility and effects.

Let there be therefore (and may it be for the benefit of both) two streams and two dispensations of knowledge, and in like manner two tribes or kindreds of students in philosophy—tribes not hostile or alien to each other, but bound together by mutual services, let there in short be one method for the cultivation, another for the invention, of knowledge.

And for those who prefer the former, either from hurry or from considerations of business or for want of mental power to take in and embrace the other (which must needs be most men's case), I wish that they may succeed to their desire in what they are about, and obtain what they are pursuing. But if there be any man who, not content to rest in and use the knowledge which has already been discovered, aspires to penetrate further; to overcome, not an adversary in argument, but nature in action; to seek, not pretty and probable conjectures, but certain and demonstrable knowledge—I invite all such to join themselves, as true sons of knowledge, with me, that passing by the outer courts of nature, which numbers have trodden, we may find a way at length into her inner chambers. And to make my meaning clearer and to familiarize the thing by giving it a name, I have chosen to call one of these methods or ways *Anticipation of the Mind*, the other *Interpretation of Nature.*

Moreover, I have one request to make. I have on my own part made it my care and study that the things which I shall propound should not only be true, but should also be presented to men's minds, how strangely soever preoccupied and obstructed, in a manner not harsh or unpleasant. It is but reasonable, however (especially in so great a restoration of learning and knowledge), that I should claim *of* men one favor in return, which is this: if anyone would form an opinion or judgment either out of his own observation, or out of the crowd of authorities, or out of the forms of demonstration (which have now acquired a sanction like that of judicial laws), concerning these speculations of mine, let him not hope that he can do it in passage or by the by; but let him examine the thing thoroughly; let him make some little trial for himself of the way which I describe and lay out; let him familiarize his thoughts with that subtlety of nature to which experience bears witness; let him correct by seasonable patience and due delay the depraved and deep-rooted habits of his mind; and when all this is done and he has begun to be his own master, let him (if he will) use his own judgment.

APHORISMS

[BOOK ONE]

I

Man, being the servant and interpreter of Nature, can do and understand so much and so much only as he has observed in fact or in thought of the course of nature. Beyond this he neither knows anything nor can do anything.

II

Neither the naked hand nor the understanding left to itself can effect much. It is by instruments and helps that the work is done, which are as much wanted for the understanding as for the hand. And as the instruments of the hand either give motion or guide it, so the instruments of the mind supply either suggestions for the understanding or cautions.

III

Human knowledge and human power meet in one; for where the cause is not known the effect cannot be produced. Nature to be commanded must be obeyed; and that which in contemplation is as the cause is in operation as the rule.

IV

Toward the effecting of works, all that man can do is to put together or put asunder natural bodies. The rest is done by nature working within.

V

The study of nature with a view to works is engaged in by the mechanic, the mathematician, the physician, the alchemist, and the magician; but by all (as things now are) with slight endeavor and scanty success.

VI

It would be an unsound fancy and self-contradictory to expect that things which have never yet been done can be done except by means which have never yet been tried.

VII

The productions of the mind and hand seem very numerous in books and manufactures. But all this variety lies in an exquisite subtlety and derivations from a few things already known, not in the number of axioms.

VIII

Moreover, the works already known are due to chance and experiment rather than to sciences; for the sciences we now possess are merely systems for the nice ordering and setting forth of things already invented, not methods of invention or directions for new works.

IX

The cause and root of nearly all evils in the sciences is this —that while we falsely admire and extol the powers of the human mind we neglect to seek for its true helps.

X

The subtlety of nature is greater many times over than the subtlety of the senses and understanding; so that all those specious meditations, speculations, and glosses in which men indulge are quite from the purpose, only there is no one by to observe it.

XI

As the sciences which we now have do not help us in finding out new works, so neither does the logic which we now have help us in finding out new sciences.

XII

The logic now in use serves rather to fix and give stability to the errors which have their foundation in commonly received notions than to help the search after truth. So it does more harm than good.

XIII

The syllogism is not applied to the first principles of sciences, and is applied in vain to intermediate axioms, being no match for the subtlety of nature. It commands assent therefore to the proposition, but does not take hold of the thing.

XIV

The syllogism consists of propositions, propositions consist of words, words are symbols of notions. Therefore if the notions themselves (which is the root of the matter) are confused and overhastily abstracted from the facts, there can be no firmness in the superstructure. Our only hope therefore lies in a true induction.

XV

There is no soundness in our notions, whether logical or physical. Substance, Quality, Action, Passion, Essence itself, are not sound notions; much less are Heavy, Light, Dense, Rare, Moist, Dry, Generation, Corruption, Attraction, Repulsion, Element, Matter, Form, and the like; but all are fantastical and ill defined.

XVI

Our notions of less general species, as Man, Dog, Dove, and of the immediate perceptions of the sense, as Hot, Cold, Black, White, do not materially mislead us; yet even these are sometimes confused by the flux and alteration of matter and the mixing of one thing with another. All the others which men have hitherto adopted are but wanderings, not being abstracted and formed from things by proper methods.

XVII

Nor is there less of willfulness and wandering in the construction of axioms than in the formation of notions, not excepting even those very principles which are obtained by common induction; but much more in the axioms and lower propositions educed by the syllogism.

XVIII

The discoveries which have hitherto been made in the sciences are such as lie close to vulgar notions, scarcely beneath the surface. In order to penetrate into the inner and further recesses of nature, it is necessary that both notions and axioms be derived from things by a more sure and guarded way, and that a method of intellectual operation be introduced altogether better and more certain.

XIX

There are and can be only two ways of searching into and discovering truth. The one flies from the senses and particulars to the most general axioms, and from these principles, the truth of which it takes for settled and immovable, proceeds to judgment and to the discovery of middle axioms. And this way is now in fashion. The other derives axioms from the senses and particulars, rising by a gradual and unbroken ascent, so that it arrives at the most general axioms last of all. This is the true way, but as yet untried.

XX

The understanding left to itself takes the same course (namely, the former) which it takes in accordance with logical order. For the mind longs to spring up to positions of higher generality, that it may find rest there, and so after a little while wearies of experiment. But this evil is increased by logic, because of the order and solemnity of its disputations.

XXI

The understanding left to itself, in a sober, patient, and grave mind, especially if it be not hindered by received doctrines, tries a little that other way, which is the right one, but with little progress, since the understanding, unless directed and assisted, is a thing unequal, and quite unfit to contend with the obscurity of things.

XXII

Both ways set out from the senses and particulars and rest in the highest generalities; but the difference between them is infinite. For the one just glances at experiment and particulars in passing, the other dwells duly and orderly among them. The one, again, begins at once by establishing certain abstract and useless generalities, the other rises by gradual steps to that which is prior and better known in the order of nature.

XXIII

There is a great difference between the Idols of the human mind and the Ideas of the divine. That is to say, between certain empty dogmas, and the true signatures and marks set upon the works of creation as they are found in nature.

XXIV

It cannot be that axioms established by argumentation should avail for the discovery of new works, since the subtlety of nature is greater many times over than the subtlety of argument. But axioms duly and orderly formed from particulars easily discover the way to new particulars, and thus render sciences active.

XXV

The axioms now in use, having been suggested by a scanty and manipular experience and a few particulars of most general occurrence, are made for the most part just large enough to fit and take these in; and therefore it is no wonder if they do not lead to new particulars. And if some opposite instance, not observed or not known before, chance to come in the way, the axiom is rescued and preserved by some frivolous distinction; whereas the truer course would be to correct the axiom itself.

XXVI

The conclusions of human reason as ordinarily applied in matters of nature, I call for the sake of distinction *Anticipations of Nature* (as a thing rash or premature). That reason which is elicited from facts by a just and methodical process, I call *Interpretation of Nature.*

XXVII

Anticipations are a ground sufficiently firm for consent, for even if men went mad all after the same fashion, they might agree one with another well enough.

XXVIII

For the winning of assent, indeed, anticipations are far more powerful than interpretations, because being collected from a few instances, and those for the most part of familiar occurrence. they straightway touch the understanding and fill the imagination: whereas interpretations, on the other hand, being gathered here and there from very various and widely dispersed facts, cannot suddenly strike the understanding; and therefore they must needs, in respect of the opinions of the time, seem harsh and out of tune, much as the mysteries of faith do.

XXIX

In sciences founded on opinions and dogmas, the use of anticipations and logic is good; for in them the object is to command assent to the proposition, not to master the thing.

XXX

Though all the wits of all the ages should meet together and combine and transmit their labors, yet will no great progress ever be made in science by means of anticipations; because radical errors in the first concoction of the mind are not to be cured by the excellence of functions and subsequent remedies.

XXXI

It is idle to expect any great advancement in science from the superinducing and engrafting of new things upon old. We must begin anew from the very foundations, unless we would revolve forever in a circle with mean and contemptible progress.

XXXII

The honor of the ancient authors, and indeed of all, remains untouched, since the comparison I challenge is not of wits or faculties, but of ways and methods, and the part I take upon myself is not that of a judge, but of a guide.

XXXIII

This must be plainly avowed: no judgment can be rightly formed either of my method or of the discoveries to which it leads, by means of anticipations (that is to say, of the reasoning which is now in use); since I cannot be called on to abide by the sentence of a tribunal which is itself on trial.

XXXIV

Even to deliver and explain what I bring forward is no easy matter, for things in themselves new will yet be apprehended with reference to what is old.

XXXV

It was said by Borgia of the expedition of the French into Italy, that they came with chalk in their hands to mark out their lodgings, not with arms to force their way in. I in like manner would have my doctrine enter quietly into the minds that are fit and capable of receiving it; for confutations cannot be employed when the difference is upon first principles and very notions, and even upon forms of demonstration.

XXXVI

One method of delivery alone remains to us which is simply this: we must lead men to the particulars themselves, and their series and order; while men on their side must force themselves for a while to lay their notions by and begin to familiarize themselves with facts.

XXXVII

The doctrine of those who have denied that certainty could be attained at all has some agreement with my way of proceeding at the first setting out; but they end in being infinitely separated and opposed. For the holders of that doctrine assert simply that nothing can be known. I also assert that not much can be known in nature by the way which is now in use. But then they go on to destroy the authority of the senses and understanding; whereas I proceed to devise and supply helps for the same.

XXXVIII

The idols and false notions which are now in possession of the human understanding, and have taken deep root therein, not only so beset men's minds that truth can hardly find entrance, but even after entrance is obtained, they will again in the very instauration of the sciences meet and trouble us, unless men being forewarned of the danger fortify themselves as far as may be against their assaults.

XXXIX

There are four classes of Idols which beset men's minds. To these for distinction's sake I have assigned names, calling the first class *Idols of the Tribe;* the second, *Idols of the Cave;* the third, *Idols of the Market Place;* the fourth, *Idols of the Theater.*

XL

The formation of ideas and axioms by true induction is no doubt the proper remedy to be applied for the keeping off and clearing away of idols. To point them out, however, is of great use; for the doctrine of Idols is to the interpretation of nature what the doctrine of the refutation of sophisms is to common logic.

XLI

The Idols of the Tribe have their foundation in human nature itself, and in the tribe or race of men. For it is a false assertion that the sense of man is the measure of things. On the contrary, all perceptions as well of the sense as of the mind are according to the measure of the individual and not according to the measure of the universe. And the human understanding is like a false mirror, which, receiving rays irregularly, distorts and discolors the nature of things by mingling its own nature with it.

XLII

The Idols of the Cave are the idols of the individual man. For everyone (besides the errors common to human nature in general) has a cave or den of his own, which refracts and discolors the light of nature, owing either to his own proper and peculiar nature; or to his education and conversation with others; or to the reading of books, and the authority of those whom he esteems and admires; or to the differences of impressions, accordingly as they take place in a mind preoccupied and predisposed or in a mind indifferent and settled; or the like. So that the spirit of man (according as it is meted out to different individuals) is in fact a thing variable and full of perturbation, and governed as it were by chance. Whence it was well observed by Heraclitus that men look for sciences in their own lesser worlds, and not in the greater or common world.

XLIII

There are also Idols formed by the intercourse and association of men with other, which I call Idols of the Market Place, on account of the commerce and consort of men there. For it is by discourse that men associate, and words are imposed according to the apprehension of the vulgar. And therefore the ill and unfit choice of words wonderfully obstructs the understanding. Nor do the definitions or explanations wherewith in some things learned men are wont to guard and defend themselves, by any means set the matter right. But words plainly force and overrule the understanding, and throw all into confusion, and lead men away into numberless empty controversies and idle fancies.

XLIV

Lastly, there are Idols which have immigrated into men's minds from the various dogmas of philosophies, and also from wrong laws of demonstration. These I call Idols of the Theater, because in my judgment all the received systems are but so many stage plays, representing worlds of their own creation after an unreal and scenic fashion. Nor is it only of the systems now in vogue, or only of the ancient sects and philosophies, that I speak; for many more plays of the same kind may yet be composed and in like artificial manner set forth; seeing that errors the most widely different have nevertheless causes for the most part alike. Neither again do I mean this only of entire systems, but also of many principles and axioms in science, which by tradition, credulity, and negligence have come to be received. But of these several kinds of Idols I must speak more largely and exactly, that the understanding may be duly cautioned.

XLV

The human understanding is of its own nature prone to suppose the existence of more order and regularity in the world than it finds. And though there be many things in nature which are singular and unmatched, yet it devises for them parallels and conjugates and relatives which do not exist. Hence the fiction that all celestial bodies move in perfect circles, spirals and dragons being (except in name) utterly rejected. Hence too the element of fire with its orb is brought in, to make up the square with the other three which the sense perceives. Hence also the ratio of density of the so-called elements is arbitrarily fixed at ten to one. And so on of other dreams. And these fancies affect not dogmas only, but simple notions also.

XLVI

The human understanding when it has once adopted an opinion (either as being the received opinion or as being agreeable to itself) draws all things else to support and agree with it. And though there be a greater number and weight of instances to be found on the other side, yet these it either neglects and despises, or else by some distinction sets aside and rejects, in order that by this great and pernicious predetermination the authority of its former conclusions may remain inviolate. And therefore it was a good answer that was made by one who, when they showed him hanging in a temple a picture of those who had paid their vows as having escaped shipwreck, and would have him say whether he did not now acknowledge the power of the gods—"Aye," asked he again, "but where are they painted that were drowned after their vows?" And such is the way of all superstition, whether in astrology, dreams, omens, divine judgments, or the like; wherein men, having a delight in such vanities, mark the events where they are fulfilled, but where they fail, though this happen much oftener, neglect and pass them by. But with far more subtlety does this mischief insinuate itself into philosophy and the sciences; in which the first conclusion colors and brings into conformity with itself all that come after, though far sounder and better. Besides, independently of that delight and vanity which I have described, it is the peculiar and perpetual error of the human intellect to be more moved and excited by affirmatives than by negatives; whereas it ought properly to hold itself indifferently disposed toward both alike. Indeed, in the establishment of any true axiom, the negative instance is the more forcible of the two.

XLVII

The human understanding is moved by those things most which strike and enter the mind simultaneously and suddenly, and so fill the imagination; and then it feigns and supposes all other things to be somehow, though it cannot see how, similar to those few things by which it is surrounded. But for that going to and fro to remote and heterogeneous instances by which axioms are tried as in the fire, the intellect is altogether slow and unfit, unless it be forced thereto by severe laws and overruling authority.

XLVIII

The human understanding, is unquiet; it cannot stop or rest, and still presses onward, but in vain. Therefore it is that we cannot conceive of any end or limit to the world, but always as of necessity it occurs to us that there is something beyond. Neither, again, can it be conceived how eternity has flowed down to the present day, for that distinction which is commonly received of infinity in time past and in time to come can by no means hold; for it would thence follow that one infinity is greater than another, and that infinity is wasting away and tending to become finite. The like subtlety arises touching the infinite divisibility of lines, from the same inability of thought to stop. But this inability interferes more mischievously in the discovery of causes; for although the most general principles in nature ought to be held merely positive, as they are discovered, and cannot with truth be referred to a cause, nevertheless the human understanding being unable to rest still seeks something prior in the order of nature. And then it is that in struggling toward that which is further off it falls back upon that which is nearer at hand, namely, on final causes, which have relation clearly to the nature of man rather than to the nature of the universe; and from this source have strangely defiled philosophy. But he is no less an unskilled and shallow philosopher who seeks causes of that which is most general, than he who in things subordinate and subaltern omits to do so.

XLIX

The human understanding is no dry light, but receives an infusion from the will and affections; whence proceed sciences which may be called "sciences as one would." For what a man had rather were true he more readily believes. Therefore he rejects difficult things from impatience of research; sober things, because they narrow hope; the deeper things of nature, from superstition; the light of experience, from arrogance and pride, lest his mind should seem to be occupied with things mean and transitory; things not commonly believed, out of deference to the opinion of the vulgar. Numberless, in short, are the ways, and sometimes imperceptible, in which the affections color and infect the understanding.

L

But by far the greatest hindrance and aberration of the human understanding proceeds from the dullness, incompetency, and deceptions of the senses; in that things which strike the sense outweigh things which do not immediately strike it, though they be more important. Hence it is that speculation commonly ceases where sight ceases; insomuch that of things invisible there is little or no observation. Hence all the working of the spirits enclosed in tangible bodies lies hid and unobserved of men. So also all the more subtle changes of form in the parts of coarser substances (which they commonly call alteration, though it is in truth local motion through exceedingly small spaces) is in like manner unobserved. And yet unless these two things just mentioned be searched out and brought to light, nothing great can be achieved in nature, as far as the production of works is concerned. So again the essential nature of our common air, and of all bodies less dense than air (which are very many), is almost unknown. For the sense by itself is a thing infirm and erring; neither can instruments for enlarging or sharpening the senses do much; but all the truer kind of interpretation of nature is effected by instances and experiments fit and apposite; wherein the sense decides touching the experiment only, and the experiment touching the point in nature and the thing itself.

LI

The human understanding is of its own nature prone to abstractions and gives a substance and reality to things which are fleeting. But to resolve nature into abstractions is less to our purpose than to dissect her into parts; as did the school of Democritus, which went further into nature than the rest. Matter rather than forms should be the object of our attention, its configurations and changes of configuration, and simple action, and law of action or motion; for forms are figments of the human mind, unless you will call those laws of action forms.

LII

Such then are the idols which I call *Idols of the Tribe,* and which take their rise either from the homogeneity of the substance of the human spirit, or from its preoccupation, or from its narrowness, or from its restless motion, or from an infusion of the affections, or from the incompetency of the senses, or from the mode of impression.

LIII

The *Idols of the Cave* take their rise in the peculiar constitution, mental or bodily, of each individual; and also in education, habit, and accident. Of this kind there is a great number and variety. But I will instance those the pointing out of which contains the most important caution, and which have most effect in disturbing the clearness of the understanding.

LIV

Men become attached to certain particular sciences and speculations, either because they fancy themselves the authors and inventors thereof, or because they have bestowed the greatest pains upon them and become most habituated to them. But men of this kind, if they betake themselves to philosophy and contemplation of a general character, distort and color them in obedience to their former fancies; a thing especially to be noticed in Aristotle, who made his natural philosophy a mere bond servant to his logic, thereby rendering it contentious and well-nigh useless. The race of chemists, again out of a few experiments of the furnace, have built up a fantastic philosophy, framed with reference to a few things; and Gilbert also, after he had employed himself most laboriously in the study and observation of the loadstone, proceeded at once to construct an entire system in accordance with his favorite subject.

LV

There is one principal and as it were radical distinction between different minds, in respect of philosophy and the sciences, which is this: that some minds are stronger and apter to mark the differences of things, others to mark their resemblances. The steady and acute mind can fix its contemplations and dwell and fasten on the subtlest distinctions; the lofty and discursive mind recognizes and puts together the finest and most general resemblances. Both kinds, however, easily err in excess, by catching the one at gradations, the other at shadows.

LVI

There are found some minds given to an extreme admiration of antiquity. others to an extreme love and appetite for novelty; but few so duly tempered that they can hold the mean, neither carping at what has been well laid down by the ancients; nor despising what is well introduced by the moderns. This, however, turns to the great injury of the sciences and philosophy, since these affectations of antiquity and novelty are the humors of partisans rather than judgments; and truth is to be sought for not in the felicity of any age, which is an unstable thing, but in the light of nature and experience, which is eternal. These factions therefore must be abjured, and care must be taken that the intellect be not hurried by them into assent.

LVII

Contemplations of nature and of bodies in their simple form break up and distract the understanding, while contemplations of nature and bodies in their composition and configuration overpower and dissolve the understanding, a distinction well seen in the school of Leucippus and Democritus as compared with the other philosophies. For that school is so busied with the particles that it hardly attends to the structure, while the others are so lost in admiration of the structure that they do not penetrate to the simplicity of nature. These kinds of contemplation should therefore be alternated and taken by turns, so that the understanding may be rendered at once penetrating and comprehensive, and the inconveniences above mentioned, with the idols which proceed from them, may be avoided.

LVIII

Let such then be our provision and contemplative prudence for keeping off and dislodging the *Idols of the Cave,* which grow for the most part either out of the predominance of a favorite subject, or out of an excessive tendency to compare or to distinguish, or out of partiality for particular ages, or out of the largeness or minuteness of the objects contemplated. And generally let every student of nature take this as a rule: that whatever his mind seizes and dwells upon with peculiar satisfaction is to be held in suspicion, and that so much the more care is to be taken in dealing with such questions to keep the understanding even and clear.

LIX

But the *Idols of the Market Place* are the most troublesome of all—idols which have crept into the understanding through the alliances of words and names. For men believe that their reason governs words; but it is also true that words react on the understanding; and this it is that has rendered philosophy and the sciences sophistical and inactive. Now words, being commonly framed and applied according to the capacity of the vulgar, follow those lines of division which are most obvious to the vulgar understanding. And whenever an understanding of greater acuteness or a more diligent observation would alter those lines to suit the true divisions of nature, words stand in the way and resist the change. Whence it comes to pass that the high and formal discussions of learned men end oftentimes in disputes about words and names; with which (according to the use and wisdom of the mathematicians) it would be more prudent to begin, and so by means of definitions reduce them to order. Yet even definitions cannot cure this evil in dealing with natural and material things, since the definitions themselves consist of words, and those words beget others. So that it is necessary to recur to individual instances, and those in due series and order, as I shall say presently when I come to the method and scheme for the formation of notions and axioms.

LX

The idols imposed by words on the understanding are of two kinds. They are either names of things which do not exist (for as there are things left unnamed through lack of observation, so likewise are there names which result from fantastic suppositions and to which nothing in reality corresponds), or they are names of things which exist, but yet confused and ill-defined, and hastily and irregularly derived from realities. Of the former kind are Fortune, the Prime

Mover, Planetary Orbits, Element of Fire, and like fictions which owe their origin to false and idle theories. And this class of idols is more easily expelled, because to get rid of them it is only necessary that all theories should be steadily rejected and dismissed as obsolete.

But the other class, which springs out of a faulty and unskillful abstraction, is intricate and deeply rooted. Let us take for example such a word as *humid* and see how far the several things which the word is used to signify agree with each other, and we shall find the word *humid* to be nothing else than a mark loosely and confusedly applied to denote a variety of actions which will not bear to be reduced to any constant meaning. For it both signifies that which easily spreads itself round any other body; and that which in itself is indeterminate and cannot solidize; and that which readily yields in every direction; and that which easily divides and scatters itself; and that which easily unites and collects itself; and that which readily flows and is put in motion; and that which readily clings to another body and wets it; and that which is easily reduced to a liquid, or being solid easily melts. Accordingly, when you come to apply the word, if you take it in one sense, flame is humid; if in another, air is not humid; if in another, fine dust is humid; if in another, glass is humid. So that it is easy to see that the notion is taken by abstraction only from water and common and ordinary liquids, without any due verification.

There are, however, in words certain degrees of distortion and error. One of the least faulty kinds is that of names of substances, especially of lowest species and well-deduced (for the notion of *chalk* and of *mud is* good, of *earth* bad); a more faulty kind is that of actions, as to *generate, to corrupt, to alter;* the most faulty is of qualities (except such as are the immediate objects of the sense) as *heavy, light, rare, dense,* and the like. Yet in all these cases some notions are of necessity a little better than others, in proportion to the greater variety of subjects that fall within the range of the human sense.

LXI

But the *Idols of the Theater* are not innate, nor do they steal into the understanding secretly, but are impressed and received into the mind from the playbooks of philosophical systems and the perverted rules of demonstration. To attempt refutations in this case would be merely inconsistent with what I have already said, for since we agree neither upon principles nor upon demonstrations there is no place for argument. And this is so far well, inasmuch as it leaves the honor of the ancients untouched. For they are no wise disparaged—the question between them and me being only as to the way. For as the saying is, the lame man who keeps the right road outstrips the runner who takes a wrong one. Nay, it is obvious that when a man runs the wrong way, the more active and swift he is, the further he will go astray.

But the course I propose for the discovery of sciences is such as leaves but little to the acuteness and strength of wits, but places all wits and understandings nearly on a level. For as in the drawing of a straight line or a perfect circle, much depends on the steadiness and practice of the hand, if it be done by aim of hand only, but if with the aid of rule or compass, little or nothing; so is it exactly with my plan. But though particular confutations would be of no avail, yet touching the sects and general divisions of such systems I must say something; something also touching the external signs which show that they are unsound; and finally something touching the causes of such great infelicity and of such lasting and general agreement in error; that so the access to truth may be made less difficult, and the human understanding may the more willingly submit to its purgation and dismiss its idols.

LXII

Idols of the Theater, or of Systems, are many, and there can be and perhaps will be yet many more. For were it not that now for many ages men's minds have been busied with religion and theology; and were it not that civil governments, especially monarchies, have been averse to such novelties, even in matters speculative; so that men labor therein to the peril and harming of their fortunes—not only unrewarded, but exposed also to contempt and envy—doubtless there would have arisen many other philosophical sects like those which in great variety flourished once among the Greeks. For as on the phenomena of the heavens many hypotheses may be constructed, so likewise (and more also) many various dogmas may be set up and established on the phenomena of philosophy. And in the plays of this philosophical theater you may observe the same thing which is found in the theater of the poets, that stories invented for the stage are more compact and elegant, and more as one would wish them to be, than true stories out of history.

In general, however, there is taken for the material of philosophy either a great deal out of a few things, or a very little out of many things; so that on both sides philosophy is based on too narrow a foundation of experiment and natural history, and decides on the authority of too few cases. For the Rational School of philosophers snatches from experience a variety of common instances, neither duly ascertained nor diligently examined and weighed, and leaves all the rest to meditation and agitation of wit.

There is also another class of philosophers who, having bestowed much diligent and careful labor on a few experiments, have thence made bold to educe and construct systems, wresting all other facts in a strange fashion to conformity therewith.

And there is yet a third class, consisting of those who out of faith and veneration mix their philosophy with theology and traditions; among whom the vanity of some has gone so far aside as to seek the origin of sciences among spirits and genii. So that this parent stock of errors—this false philosophy—is of three kinds: the Sophistical, the Empirical, and the Superstitious.

LXIII

The most conspicuous example of the first class was Aristotle, who corrupted natural philosophy by his logic: fashioning the world out of categories; assigning to the human soul, the noblest of substances, a genus from words of the second intention; doing the business of density and rarity (which is to make bodies of greater or less dimensions, that is, occupy greater or less spaces), by the frigid distinction of act and power; asserting that single bodies have each a single and proper motion, and that if they participate in any other, then this results from an external cause; and imposing countless other arbitrary restrictions on the nature of things; being always more solicitous to provide an answer to the question and affirm something positive in words, than about the inner truth of things; a failing best shown when his philosophy is compared with other systems of note among the Greeks. For the *homoeomera* of Anaxagoras; the Atoms of Leucippus and Democritus; the Heaven and Earth of Parmenides; the Strife and Friendship of Empedocles; Heraclitus' doctrine how bodies are resolved into the indifferent nature of fire, and remolded into solids, have all of them some taste of the natural philosopher—some savor of the nature of things, and experience, and bodies; whereas in the physics of Aristotle you hear hardly anything but the words of logic, which in his metaphysics also, under a more imposing name, and more forsooth as a realist than a nominalist, he has handled over again. Nor let any weight be given to the fact that in his books on animals and his problems, and other of his treatises, there is frequent dealing with experiments. For he had come to his conclusion before; he did not consult experience, as he should have done, for the purpose of framing his decisions and axioms, but having first determined the question according to his will, he then resorts to experience, and bending her into conformity with his placets, leads her about like a captive in a procession. So that even on this count he is more guilty than his modern followers, the schoolmen, who have abandoned experience altogether.

LXIV

But the Empirical school of philosophy gives birth to dogmas more deformed and monstrous than the Sophistical or Rational school. For it has its foundations not in the light of common notions (which though it be a faint and superficial light, is yet in a manner universal, and has reference to many things), but in the narrowness and darkness of a few experiments. To those therefore who are daily busied with these experiments and have infected their imagination with them, such a philosophy seems probable and all but certain; to all men else incredible and vain. Of this there is a notable instance in the alchemists and their dogmas, though it is hardly to be found elsewhere in these times, except perhaps in the philosophy of Gilbert. Nevertheless, with regard to philosophies of this kind there is one caution not to be omitted; for I foresee that if ever men are roused by my admonitions to betake themselves seriously to experiment and bid farewell to sophistical doctrines, then indeed through the premature hurry of the understanding to leap or fly to universals and principles of things, great danger may be apprehended from philosophies of this kind, against which evil we ought even now to prepare.

LXV

But the corruption of philosophy by superstition and an admixture of theology is far more widely spread, and does the greatest harm, whether to entire systems or to their parts. For the human understanding is obnoxious to the influence of the imagination no less than to the influence of common notions. For the contentious and sophistical kind of philosophy ensnares the understanding; but this kind, being fanciful and tumid and half poetical, misleads it more by flattery. For there is in man an ambition of the understanding, no less than of the will, especially in high and lofty spirits.

Of this kind we have among the Greeks a striking example in Pythagoras, though he united with it a coarser and more cumbrous superstition; another in Plato and his school, more dangerous and subtle. It shows itself likewise in parts of other philosophies, in the introduction of abstract forms and final causes and first causes, with the omission in most cases of causes intermediate, and the like. Upon this point the greatest caution should be used. For nothing is so mischievous as the apotheosis of error; and it is a very plague of the understanding for vanity to become the object of veneration. Yet in this vanity some of the moderns have with extreme levity indulged so far as to attempt to found a system of natural philosophy on the first chapter of Genesis, on the book of Job, and other parts of the sacred writings, seeking for the dead among the living; which also makes the inhibition and repression of it the more important, because from this unwholesome mixture of things human and divine there arises not only a fantastic philosophy but also a heretical religion. Very meet it is therefore that we be sober-minded, and give to faith that only which is faith's.

LXVI

So much, then, for the mischievous authorities of systems, which are founded either on common notions, or on a few experiments, or on superstition. It remains to speak of the faulty subject matter of contemplations, especially in natural philosophy. Now the human understanding is infected by the sight of what takes place in the mechanical arts, in which the alteration of bodies proceeds chiefly by composition or separation, and so imagines that something similar goes on in the universal nature of things. From this source has flowed the fiction of elements, and of their concourse for the formation of natural bodies. Again, when man contemplates nature working freely, he meets with different species of things, of animals, of plants, of minerals; whence he readily passes into the opinion that there are in nature certain primary forms which nature intends to educe, and that the remaining variety proceeds from hindrances and aberrations of nature in the fulfillment of her work, or from the collision of different species and the transplanting of one into another. To the first of these speculations we owe our primary qualities of the elements; to the other our occult properties and specific virtues; and both of them belong to those empty compendia of thought wherein the mind rests, and whereby it is diverted from more solid pursuits. It is to better purpose that the physicians bestow their labor on the secondary qualities of matter, and the operations of attraction, repulsion, attenuation, conspissation, dilatation, astriction, dissipation, maturation, and the like; and were it not that by those two compendia which I have mentioned (elementary qualities, to wit, and specific virtues) they corrupted their correct observations in these other matters—either reducing them to first qualities and their subtle and incommensurable mixtures, or not following them out with greater and more diligent observations to third and fourth qualities, but breaking off the scrutiny prematurely—they would have made much greater progress. Nor are powers of this kind (I do not say the same, but similar) to be sought for only in the medicines of the human body, but also in the changes of all other bodies.

But it is a far greater evil that they make the quiescent principles, *wherefrom,* and not the moving principles, *whereby,* things are produced, the object of their contemplation and inquiry. For the former tend to discourse, the latter to works. Nor is there any value in those vulgar distinctions of motion which are observed in the received system of natural philosophy, as generation, corruption, augmentation, diminution, alteration, and local motion. What they mean no doubt is this: if a body in other respects not changed be moved from its place, *this is local motion;* if without change of place or essence, it be changed in quality, this is *alteration;* if by reason of the change the mass and quantity of the body do not remain the same, this is *augmentation* or *diminution;* if they be changed to such a degree that they change their very essence and substance and turn to something else, this is *generation* and *corruption.* But all this is merely popular, and does not at all go deep into nature; for these are only measures and limits, not kinds of motion. What they intimate is *how far,* not *by what means,* or *from what source.* For they do not suggest anything with regard either to the desires of bodies or to the development of their parts. It is only when that motion presents the thing grossly and palpably to the sense as different from what it was that they begin to mark the division. Even when they wish to suggest something with regard to the causes of motion, and to establish a division with reference to them, they introduce with the greatest negligence a distinction between motion natural and violent, a distinction which is itself drawn entirely from a vulgar notion, since all violent motion is also in fact natural; the external efficient simply setting nature working otherwise than it was before. But if, leaving all this, anyone shall observe (for instance) that there is in bodies a desire of mutual contact, so as not to suffer the unity of nature to be quite separated or broken and a vacuum thus made; or if anyone say that there is in bodies a desire of resuming their natural dimensions or tension, so that if compressed within or extended beyond them, they immediately strive to recover themselves, and fall back to their old volume and extent; or if anyone say that there is in bodies a desire of congregating toward masses of kindred nature—of dense bodies, for instance, toward the globe of the earth, of thin and rare bodies toward the compass of the sky; all these and the like are truly physical kinds of motion—but those others are entirely logical and scholastic, as is abundantly manifest from this comparison.

Nor again is it a lesser evil that in their philosophies and contemplations their labor is spent in investigating and handling the first principles of things and the highest generalities of nature; whereas utility and the means of working result entirely from things intermediate. Hence it is that men cease not from abstracting nature till they come to potential and uninformed matter, nor on the other hand from dissecting nature till they reach the atom; things which, even if true, can do but little for the welfare of mankind.

LXVII

A caution must also be given to the understanding against the intemperance which systems of philosophy manifest in giving or withholding assent, because intemperance of this kind seems to establish idols and in some sort to perpetuate them, leaving no way open to reach and dislodge them.

This excess is of two kinds: the first being manifest in those who are ready in deciding, and render sciences dogmatic and magisterial; the other in those who deny that we can know anything, and so introduce a wandering kind of inquiry that leads to nothing; of which kinds the former subdues, the latter weakens the understanding. For the philosophy of Aristotle, after having by hostile confutations destroyed all the rest (as the Ottomans serve their brothers), has laid down the law on all points; which done, he proceeds himself to raise new questions of his own suggestion, and dispose of them likewise, so that nothing may remain that is not certain and decided; a practice which holds and is in use among his successors.

The school of Plato, on the other hand, introduced *Acatalepsia,* at first in jest and irony, and in disdain of the older sophists, Protagoras, Hippias, and the rest, who were of nothing else so much ashamed as of seeming to doubt about anything. But the New Academy made a dogma of it, and held it as a tenet. And though theirs is a fairer seeming way than arbitrary decisions, since they say that they by no means destroy all investigation, like Pyrrho and his Refrainers, but allow of some things to be followed as probable, though of none to be maintained as true; yet still when the human mind has once despaired of finding truth, its interest in all things grows fainter, and the result is that men turn aside to pleasant disputations and discourses and roam as it were from object to object, rather than keep on a course of severe inquisition. But, as I said at the beginning and am ever urging, the human senses and understanding, weak as they are, are not to be deprived of their authority, but to be supplied with helps.

LXVIII

So much concerning the several classes of Idols and their equipage; all of which must be renounced and put away with a fixed and solemn determination, and the understanding thoroughly freed and cleansed; the entrance into the kingdom of man, founded on the sciences, being not much other than the entrance into the kingdom of heaven, whereinto none may enter except as a little child.

LXIX

But vicious demonstrations are as the strongholds and defenses of idols; and those we have in logic do little else than make the world the bondslave of human thought, and human thought the bondslave of words. Demonstrations truly are in effect the philosophies themselves and the sciences. For such as *they* are, well or ill established, such are the systems of philosophy and the contemplations which follow. Now in the whole of the process which leads from the sense and objects to axioms and conclusions, the demonstrations which we use are deceptive and incompetent. This process consists of four parts, and has as many faults. In the first place, the impressions of the sense itself are faulty; for the sense both fails us and deceives us. But its shortcomings are to be supplied, and its deceptions to be corrected. Secondly, notions are ill-drawn from the impressions of the senses, and are indefinite and confused, whereas they should be definite and distinctly bounded. Thirdly, the induction is amiss which infers the principles of sciences by simple enumeration, and does not, as it ought, employ exclusions and solutions (or separations) of nature. Lastly, that method of discovery and proof according to which the most general principles are first established, and then intermediate axioms are tried and proved by them, is the parent of error and the curse of all science. Of these things, however, which now I do but touch upon, I will speak more largely when, having performed these expiations and purgings of the mind, I come to set forth the true way for the interpretation of nature.

LXX

But the best demonstration by far is experience, if it go not beyond the actual experiment. For if it be transferred to other cases which are deemed similar, unless such transfer be made by a just and orderly process, it is a fallacious thing. But the manner of making experiments which men now use is blind and stupid. And therefore, wandering and straying as they do with no settled course, and taking counsel only from things as they fall out, they fetch a wide circuit and meet with many matters, but make little progress; and sometimes are full of hope, sometimes are distracted; and always find that there is something beyond to be sought. For it generally happens that men make their trials carelessly, and as it were in play; slightly varying experiments already known, and, if the thing does not answer, growing weary and abandoning the attempt. And even if they apply themselves to experiments more seriously and earnestly and laboriously, still they spend their labor in working out some one experiment, as Gilbert with the magnet, and the chemists with gold; a course of proceeding not less unskillful in the design than small in the attempt. For no one successfully investigates the nature of a thing in the thing itself; the inquiry must be enlarged so as to become more general.

And even when they seek to educe some science or theory from their experiments, they nevertheless almost always turn aside with overhasty and unseasonable eagerness to practice; not only for the sake of the uses and fruits of the practice, but from impatience to obtain in the shape of some new work an assurance for themselves that it is worth their while to go on; and also to show themselves off to the world, and so raise the credit of the business in which they are engaged. Thus, like Atalanta, they go aside to pick up the golden apple, but meanwhile they interrupt their course, and let the victory escape them. But in the true course of experience, and in carrying it on to the effecting of new works, the divine wisdom and order must be our pattern. Now God on the first day of creation created light only, giving to that work an entire day, in which no material substance was created. So must we likewise from experience of every kind first endeavor to discover true causes and axioms; and seek for experiments of Light, not for experiments of fruit. For axioms rightly discovered and established supply practice with its instruments, not one by one, but in clusters, and draw after them trains and troops of works. Of the paths, however, of experience, which no less than the paths of judgment are impeded and beset, I will speak hereafter; here I have only mentioned ordinary experimental research as a bad kind of demonstration. But now the order of the matter in hand leads me to add something both as to those *signs* which I lately mentioned (signs that the systems of philosophy and contemplation in use are in a bad condition), and also as to the *causes* of what seems at first so strange and incredible. For a knowledge of the signs prepares assent; an explanation of the causes removes the marvel—which two things will do much to render the extirpation of idols from the understanding more easy and gentle.

* * * * *

LXXXIX

Neither is it to be forgotten that in every age natural philosophy has had a troublesome and hard to deal with adversary—namely, superstition, and the blind and immoderate zeal of religion. For we see among the Greeks that those who first proposed to men's then uninitiated ears the natural causes for thunder and for storms were thereupon found guilty of impiety. Nor was much more forbearance shown by some of the ancient fathers of the Christian church to those who on most convincing grounds (such as no one in his senses would now think of contradicting) maintained that the earth was round, and of consequence asserted the existence of the antipodes.

Moreover, as things now are, to discourse of nature is made harder and more perilous by the summaries and systems of the schoolmen who, having reduced theology into regular order as well as they were able, and fashioned it into the shape of an art, ended in incorporating the contentious and thorny philosophy of Aristotle, more than was fit, with the body of religion.

To the same result, though in a different way, tend the speculations of those who have taken upon them to deduce the truth of the Christian religion from the principles of philosophers, and to confirm it by their authority, pompously solemnizing this union of the sense and faith as a lawful marriage, and entertaining men's minds with a pleasing variety of matter, but all the while disparaging things divine by mingling them with things human. Now in such mixtures of theology with philosophy only the received doctrines of philosophy are included; while new ones, albeit changes for the better, are all but expelled and exterminated.

Lastly, you will find that by the simpleness of certain divines, access to any philosophy, however pure, is well-nigh closed. Some are weakly afraid lest a deeper search into nature should transgress the permitted limits of sober-mindedness, wrongfully wresting and transferring what is said in Holy Writ against those who pry into sacred mysteries, to the hidden things of nature, which are barred by no prohibition. Others with more subtlety surmise and reflect that if second causes are unknown everything can more readily be referred to the divine hand and rod, a point in which they think religion greatly concerned—which is in fact nothing else but to seek to gratify God with a lie. Others fear from past example that movements and changes in philosophy will end in assaults on religion. And others again appear apprehensive that in the investigation of nature something may be found to subvert or at least shake the authority of religion, especially with the unlearned. But these two last fears seem to me to savor utterly of carnal wisdom; as if men in the recesses and secret thought of their hearts doubted and distrusted the strength of religion and the empire of faith over the sense, and therefore feared that the investigation of truth in nature might be dangerous to them. But if the matter be truly considered, natural philosophy is, after the word of God, at once the surest medicine against superstition and the most approved nourishment for faith, and therefore she is rightly given to religion as her most faithful handmaid, since the one displays the will of God, the other his power. For he did not err who said, "Ye err in that ye know not the Scriptures and the power of God," thus coupling and blending in an indissoluble bond information concerning his will and meditation concerning his power. Meanwhile it is not surprising if the growth of natural philosophy is checked when religion, the thing which has most power over men's minds, has by the simpleness and incautious zeal of certain persons been drawn to take part against her.

XC

Again, in the customs and institutions of schools, academies, colleges and similar bodies destined for the abode of learned men and the cultivation of learning, everything is found adverse to the progress of science. For the lectures and exercises there are so ordered that to think or speculate on anything out of the common way can hardly occur to any man. And if one or two have the boldness to use any liberty of judgment, they must undertake the task all by themselves; they can have no advantage from the company of others. And if they can endure this also, they will find their industry and largeness of mind no slight hindrance to their fortune. For the studies of men in these places are confined and as it were imprisoned in the writings of certain authors, from whom if any man dissent he is straightway arraigned as a turbulent person and an innovator. But surely there is a great distinction between matters of state and the arts; for the danger from new motion and from new light is not the same. In matters of state a change even for the better is distrusted, because it unsettles what is established; these things resting on authority, consent, fame and opinion, not on demonstration. But arts and sciences should be like mines, where the noise of new works and further advances is heard on every side. But though the matter be so according to right reason, it is not so acted on in practice; and the points above mentioned in the administration and government of learning put a severe restraint upon the advancement of the sciences.

XCI

Nay, even if that jealousy were to cease, still it is enough to check the growth of science that efforts and labors in this field go unrewarded. For it does not rest with the same persons to cultivate sciences and to reward them. The growth of them comes from great wits; the prizes and rewards of them are in the hands of the people, or of great persons, who are but in very few cases even moderately learned. Moreover, this kind of progress is not only unrewarded with prizes and substantial benefits; it has not even the advantage of popular applause. For it is a greater matter than the generality of men can take in, and is apt to be overwhelmed and extinguished by the gales of popular opinions. And it is nothing strange if a thing not held in honor does not prosper.

XCII

But by far the greatest obstacle to the progress of science and to the undertaking of new tasks and provinces therein is found in this—that men despair and think things impossible. For wise and serious men are wont in these matters to be altogether distrustful, considering with themselves the obscurity of nature, the shortness of life, the deceitfulness of the senses, the weakness of the judgment, the difficulty of experiment, and the like; and so supposing that in the revolution of time and of the ages of the world the sciences have their ebbs and flows; that at one season they grow and flourish, at another wither and decay, yet in such sort that when they have reached a certain point and condition they can advance no further. If therefore anyone believes or promises more, they think this comes of an ungoverned and unripened mind, and that such attempts have prosperous beginnings, become difficult as they go on, and end in confusion. Now since these are thoughts which naturally present themselves to men grave and of great judgment, we must take good heed that we be not led away by our love for a most fair and excellent object to relax or diminish the severity of our judgment. We must observe diligently what encouragement dawns upon us and from what quarter, and, putting aside the lighter breezes of hope, we must thoroughly sift and examine those which promise greater steadiness and constancy. Nay, and we must take state prudence too into our counsels, whose rule is to distrust, and to take the less favorable view of human affairs. I am now therefore to speak touching hope, especially as I am not a dealer in promises, and wish neither to force nor to ensnare men's judgments, but to lead them by the hand with their good will. And though the strongest means of inspiring hope will be to bring men to particulars, especially to particulars digested and arranged in my Tables of Discovery (the subject partly of the second, but much more of the fourth part of my Instauration), since this is not merely the promise of the thing but the thing itself; nevertheless, that everything may be done with gentleness, I will proceed with my plan of preparing men's minds, of which preparation to give hope is no unimportant part. For without it the rest tends rather to make men sad (by giving them a worse and meaner opinion of things as they are than they now have, and making them more fully to feel and know the unhappiness of their own condition) than to induce any alacrity or to whet their industry in making trial. And therefore it is fit that I publish and set forth those conjectures of mine which make hope in this matter reasonable, just as Columbus did, before that wonderful voyage of his across the Atlantic, when he gave the reasons for his conviction that new lands and continents might be discovered besides those which were known before; which reasons, though rejected at first, were afterwards made good by experience, and were the causes and beginnings of great events.

* * * * *

XCIX

Again, even in the great plenty of mechanical experiments, there is yet a great scarcity of those which are of most use for the information of the understanding. For the mechanic, not troubling himself with the investigation of truth, confines his attention to those things which bear upon his particular work, and will not either raise his mind or stretch out his hand for anything else. But then only will there be good ground of hope for the further advance of knowledge when there shall be received and gathered together into natural history a variety of experiments which are of no use in themselves but simply serve to discover causes and axioms, which I call Experimenta lucifera, experiments of light, to distinguish them from those which I call fructifera, experiments of fruit.

Now experiments of this kind have one admirable property and condition: they never miss or fail. For since they are applied, not for the purpose of producing any particular effect, but only of discovering the natural cause of some effect, they answer the end equally well whichever way they turn out; for they settle the question.

C

But not only is a greater abundance of experiments to be sought for and procured, and that too of a different kind from those hitherto tried; an entirely different method, order, and process for carrying on and advancing experience must also be introduced. For experience, when it wanders in its own track, is, as I have already remarked, mere groping in the dark, and confounds men rather than instructs them. But when it shall proceed in accordance with a fixed law, in regular order, and without interruption, then may better things be hoped of knowledge.

CI

But even after such a store of natural history and experience as is required for the work of the understanding, or of philosophy, shall be ready at hand, still the understanding is by no means competent to deal with it offhand and by memory alone; no more than if a man should hope by force of memory to retain and make himself master of the computation of an ephemeris. And yet hitherto more has been done in matter of invention by thinking than by writing; and experience has not yet learned her letters. Now no course of invention can be satisfactory unless it be carried on in writing. But when this is brought into use, and experience has been taught to read and write, better things may be hoped.

CII

Moreover, since there is so great a number and army of particulars, and that army so scattered and dispersed as to distract and confound the understanding, little is to be hoped for from the skirmishings and slight attacks and desultory movements of the intellect, unless all the particulars which pertain to the subject of inquiry shall, by means of Tables of Discovery, apt, well arranged, and, as it were, animate, be drawn up and marshaled; and the mind be set to work upon the helps duly prepared and digested which these tables supply.

CIII

But after this store of particulars has been set out duly and in order before our eyes, we are not to pass at once to the investigation and discovery of new particulars or works; or at any rate if we do so we must not stop there. For although I do not deny that when all the experiments of all the arts shall have been collected and digested, and brought within one man's knowledge and judgment, the mere transferring of the experiments of one art to others may lead, by means of that experience which I term literate, to the discovery of many new things of service to the life and state of man, yet it is no great matter that can be hoped from that; but from the new light of axioms, which having been educed from those particulars by a certain method and rule, shall in their turn point out the way again to new particulars, greater things may be looked for. For our road does not lie on a level, but ascends and descends; first ascending to axioms, then descending to works.

CIV

The understanding must not, however, be allowed to jump and fly from particulars to axioms remote and of almost the highest generality (such as the first principles, as they are called, of arts and things), and taking stand upon them as truths that cannot be shaken, proceed to prove and frame the middle axioms by reference to them; which has been the practice hitherto, the understanding being not only carried that way by a natural impulse, but also by the use of syllogistic demonstration trained and inured to it. But then, and then only, may we hope well of the sciences when in a just scale of ascent, and by successive steps not interrupted or broken, we rise from particulars to lesser axioms; and then to middle axioms, one above the other; and last of all to the most general. For the lowest axioms differ but slightly from bare experience, while the highest and most general (which we now have) are notional and abstract and without solidity. But the middle are the true and solid and living axioms, on which depend the affairs and fortunes of men; and above them again, last of all, those which are indeed the most general; such, I mean, as are not abstract, but of which those intermediate axioms are really limitations.

The understanding must not therefore be supplied with wings, but rather hung with weights, to keep it from leaping and flying. Now this has never yet been done; when it is done, we may entertain better hopes of the sciences.

CV

In establishing axioms, another form of induction must be devised than has hitherto been employed, and it must be used for proving and discovering not first principles (as they are called) only, but also the lesser axioms, and the middle, and indeed all. For the induction which proceeds by simple enumeration is childish; its conclusions are precarious and exposed to peril from a contradictory instance; and it generally decides on too small a number of facts, and on those only which are at hand. But the induction which is to be available for the discovery and demonstration of sciences and arts, must analyze nature by proper rejections and exclusions; and then, after a sufficient number of negatives, come to a conclusion on the affirmative instances—which has not yet been done or even attempted, save only by Plato, who does indeed employ this form of induction to a certain extent for the purpose of discussing definitions and ideas. But in order to furnish this induction or demonstration well and duly for its work, very many things are to be provided which no mortal has yet thought of; insomuch that greater labor will have to be spent in it than has hitherto been spent on the syllogism. And this induction must be used not only to discover axioms, but also in the formation of notions. And it is in this induction that our chief hope lies.

CVI

But in establishing axioms by this kind of induction, we must also examine and try whether the axiom so established be framed to the measure of those particulars only from which it is derived, or whether it be larger and wider. And if it be larger and wider, we must observe whether by indicating to us new particulars it confirm that wideness and largeness as by a collateral security, that we may not either stick fast in things already known, or loosely grasp at shadows and abstract forms, not at things solid and realized in matter. And when this process shall have come into use, then at last shall we see the dawn of a solid hope.

* * * * *

THE GROWTH AND MATURATION OF SCIENTIFIC KNOWLEDGE

Darwin

Agassiz

Clausius

Pasteur

Faraday

Einstein

Pierre and Marie Curie

Eve Curie

Heisenberg

Jeans

Cedering

When I heard the learn'd astronomer,
When the proofs, the figures, were ranged in columns before me,
When I was shown the charts and diagrams, to add, divide, and measure them,
When I sitting heard the astronomer where he lectured with much applause in the lecture-room,
How soon uncomfortable I became tired and sick,
Till rising and gliding out I wander'd off by myself,
In the mystical moist night air, and from time to time,
Look'd up in perfect silence at the Stars.

Walt Whitman

From the point of view of a physicist, a theory of matter is a policy rather than a creed; its object is to connect or coordinate apparently diverse phenomena, and above all to suggest, stimulate and direct experiment.

Sir J. J. Thomson, *The Corpuscular Theory of Matter*, 1907

Nature and Nature's laws lay hid in night.
Then God said, "Let Newton be," and all was light

Alexander Pope

It did not last. The devil howling, "Ho!
Let Einstein be!" restored the status quo.

Sir John Collings Squire

Charles Darwin, the Abraham of scientific men—a searcher as obedient to the command of truth as the patriarch was to the command of God.

John Tyndall

THE ORIGIN OF SPECIES

CHARLES DARWIN

The publication in 1859 of Darwin's *On the Origin of Species by Means of Natural Selection, or The Preservation of Favoured Races in the Struggle for Life* initiated a revolution in biological science and set off a controversy that has continued with greater or lesser intensity ever since. Others had proposed similar theories earlier, but they had not amassed the data and logical study that made Darwin's work so convincing.

Charles Darwin (1809-1882), the son of a wealthy physician of Shrewsburg, was educated at the University of Edinburgh and at Cambridge. Soon after graduation he joined the company of H.M.S. *Beagle* as a naturalist on a five-year voyage along the coast of South America and various islands, particularly the Galapagos, of the Pacific. He was deeply impressed by the fossils and the current species of animals that he found.

Back in London, Darwin began to study his data and he studied the domestic breeding of animals, seeking answers to questions about how the species had come about. A reading of the essay *On the Principle of Population* (1798) by Thomas R. Malthus (see Volume VI of the Series for Malthus' essay) led Darwin to the idea of a struggle for existence on the part of all creatures which progressed according to the advantages of variations.

In January 1839 he married Emma Wedgwood. In June 1842 he wrote a brief thirty-five page sketch that in effect was an outline of *The Origin of Species*. In September of that year, he moved with his family to Down, in Kent, to continue his study and research and writing. This went on over the next sixteen years until it appeared that he would have at least a three-volume work.

Thus far Darwin had kept his ideas secret, except for a few close friends. He was afraid that his theory was so controversial that it would be rejected by fellow scientists before he was prepared to defend it adequately.

Then, in June 1858, he received a paper from an address in the Celebes Islands of the East Indies. It was a paper that nearly duplicated the one Darwin had written in 1842. It was by a younger man, Alfred Russell Wallace, who, having visited South America as well as the Indies and also having read Malthus' essay *On the Principle of Population*, had arrived independently at the same conclusions as Darwin.

Two of Darwin's friends, Charles Lyell and Joseph Hooker, had been urging Darwin to proceed with publication lest he be preempted. Now Darwin undoubtedly was pleased to see this kind of support for his ideas, yet he naturally dreaded to see all his years of research and writing dimmed by competition.

Lyell and Hooker obtained a brief abstract of his theory from Darwin, and then they arranged for this and Wallace's paper both to be read at a meeting of the Linnean Society of London on July 1, 1858. This went well.

Then Lyell and Hooker urged Darwin to prepare his work for early publication. After another year of editing and consolidating and rewriting, Darwin's manuscript was ready. It was published on November 24, 1859, though only in 1,250 copies. A second edition of 3,000 copies was published in the next year, a third edition in 1861, a fourth in 1866, a fifth in 1869, and a sixth in 1872. Darwin lived another ten years but felt no need for any further modifications of the text. For his part, Wallace always modestly deferred to Darwin as the senior scholar in this whole area.

While finding certain errors in the work of Darwin, and especially on the part of his interpreters, Ashley Montagu wrote, "Next to the Bible no work has been quite as influential, in virtually every aspect of human thought, as *The Origin of Species.*" And George Gaylord Simpson declared, "Darwin was one of history's towering geniuses and ranks with the greatest heroes of man's intellectual progress."

1. What is a species? How do species relate to other biological classifications?

2. What is the theory of evolution?

3. What does Darwin mean by "natural selection"? How does he answer criticisms of the use of that term? Does he consider that the only factor in the development of a new species?

4. How does "sexual selection" relate to "natural selection"?

5. Are acquired characteristics passed on by heredity? How do you account for the appearance of variations?

6. Are plants subject to the same laws of natural selection as animals? Explain.

7. How does Darwin explain the formation of new species and the extinction of others?

8. What is the great "Tree of Life" that Darwin describes?

9. If species become extinct as a result of natural selection and survival of the fittest, what is the function of our "endangered species act"? Is that a wise policy?

10. How do you account for the strident opposition to Darwin's theories in some quarters?

11. How have recent scientific developments affected Darwinism?

THE ORIGIN OF SPECIES

CHARLES DARWIN

CHAPTER IV

NATURAL SELECTION; OR THE SURVIVAL OF THE FITTEST

Natural Selection—its power compared with man's selection—its power on characters of trifling importance—its power at all ages and on both sexes—Sexual selection—On the generality of intercrosses between individuals of the same species—Circumstances favourable and unfavourable to the results of Natural Selection, namely, intercrossing, isolation, number of individuals—Slow action—Extinction caused by Natural Selection—Divergence of Character, related to the diversity of inhabitants of any small area, and to naturalisation—Action of Natural Selection, through divergence of Character and Extinction, on the descendants from a common parent—Explains the grouping of all organic beings—Advance in organisation—Low forms preserved—Convergence of Character—Indefinite multiplication of species—Summary.

HOW will the struggle for existence, briefly discussed in the last chapter, act in regard to variation? Can the principle of selection, which we have seen is so potent in the hands of man, apply under nature? I think we shall see that it can act most efficiently. Let the endless number of slight variations and individual differences occurring in our domestic productions, and, in a lesser degree, in those under nature, be borne in mind; as well as the strength of the hereditary tendency. Under domestication, it may be truly said that the whole organisation becomes in some degree plastic. But the variability, which we almost universally meet with in our domestic productions, is not directly produced, as Hooker and Asa Gray have well remarked, by man; he can neither originate varieties, nor prevent their occurrence; he can only preserve and accumulate such as do occur. Unintentionally he exposes organic beings to new and changing conditions of life, and variability ensues; but similar changes of conditions might and do occur under nature. Let it also be borne in mind how infinitely complex and close-fitting are the mutual relations of all organic beings to each other and to their physical conditions of life; and consequently what infinitely varied diversities of structure might be of use to each being under changing conditions of life. Can it, then, be thought improbable, seeing that variations useful to man have undoubtedly occurred, that other variations useful in some way to each being in the great and complex battle of life, should occur in the course of many successive generations? If such do occur, can we doubt (remembering that many more individuals are born than can possibly survive) that individuals having any advantage, however slight, over others, would have the best chance of surviving and of procreating their kind? On the other hand, we may feel sure that any variation in the least degree injurious would be rigidly destroyed. This preservation of favourable individual differences and variations, and the destruction of those which are injurious, I have called Natural Selection, or the Survival of the Fittest. Variations neither useful nor injurious would not be affected by natural selection, and would be left either a fluctuating element, as perhaps we see in certain polymorphic species, or would ultimately become fixed, owing to the nature of the organism and the nature of the conditions.

Several writers have misapprehended or objected to the term Natural Selection. Some have even imagined that natural selection induces variability, whereas it implies only the preservation of such variations as arise and are beneficial to the being under its conditions of life. No one objects to agriculturists speaking of the potent effects of man's selection; and in this case the individual differences given by nature, which man for some object selects, must of necessity first occur. Others have objected that the term selection implies conscious choice in the animals which become modified; and it has even been urged that, as plants have no volition, natural selection is not applicable to them! In the literal sense of the word, no doubt, natural selection is a false term; but who ever objected to chemists speaking of the elective affinities of the various elements?—and yet an acid cannot strictly be said to elect the base with which it in preference combines. It has been said that I speak of natural selection as an active power or deity; but who objects to an author speaking of the attraction of gravity as ruling the movements of the planets? Every one knows what is meant and is implied by such metaphorical expressions; and they are almost necessary for brevity. So again it is difficult to avoid personifying the word Nature; but I mean by Nature, only the aggregate action and product of many natural laws, and by laws the sequence of events as ascertained by us. With a little familiarity such superficial objections will be forgotten.

We shall best understand the probable course of natural selection by taking the case of a country undergoing some slight physical change, for instance, of climate. The proportional numbers of its inhabitants will almost immediately undergo a change, and some species will probably become extinct. We may conclude, from what we have seen of the intimate and complex manner in which the inhabitants of each country are bound together, that any change in the numerical proportions of the inhabitants, independently of the change of climate itself, would seriously affect the others. If the country were open on its borders, new forms would certainly immigrate, and this would likewise seriously disturb the relations of some of the former inhabitants. Let it be remembered how powerful the influence of a single introduced tree or mammal has been shown to be. But in the case of an island, or of a country partly surrounded by barriers, into which new and better adapted forms could not freely enter, we should then have places in the economy of nature which would assuredly be better filled up, if some of the original inhabitants were in some manner modified; for, had the area been open to immigration, these same places would have been seized on by intruders. In such cases, slight modifications, which in any way favoured the individuals of any species, by better adapting them to their altered conditions, would tend to be preserved; and natural selection would have free scope for the work of improvement.

We have good reason to believe, as shown in the first chapter, that changes in the conditions of life give a tendency to increased variability; and in the foregoing cases the conditions have changed, and this would manifestly be favourable to natural selection, by affording a better chance of the occurrence of profitable variations. Unless such occur, natural selection can do nothing. Under the term of "variations," it must never be forgotten that mere individual differences are included. As man can produce a great result with his domestic animals and plants by adding up in any given direction individual differences, so could natural selection, but far more easily from having incomparably longer time for action nor do I believe that any great physical change, as of climate, or any unusual degree of isolation, to check immigration, is necessary in order that new and unoccupied places should be left, for natural selection to fill up by improving some of the varying inhabitants. For as all the inhabitants of each country are struggling together with nicely balanced forces, extremely slight modifications in the structure or habits of one species would often give it an advantage over others; and still further modifications of the same kind would often still further increase the advantage, as long as the species continued under the same conditions of life and profited by similar means of subsistence and defence. No country can be named in which all the native inhabitants are now so perfectly adapted to each other and to the physical conditions under which they live, that none of them could be still better adapted or improved; for in all countries, the natives have been so far conquered by naturalised productions, that they have allowed some foreigners to take firm possession of the land. And as foreigners have thus in every country beaten some of the natives, we may safely conclude that the natives might have been modified with advantage, so as to have better resisted the intruders.

As man can produce, and certainly has produced, a great result by his methodical and unconscious means of selection, what may not natural selection effect? Man can act only on external and visible characters. Nature, if I may be allowed to personify the natural preservation or survival of the fittest, cares nothing for appearances, except in so far as they are useful to any being. She can act on every internal organ, on every shade of constitutional difference, on the whole machinery of life. Man selects only for his own good; Nature only for that of the being which she tends. Every selected character is fully exercised by her, as is implied by the fact of their selection. Man keeps the natives of many climates in the same country; he seldom exercises each selected character in some peculiar and fitting manner; he feeds a long and a short beaked pigeon on the same food; he does not exercise a long-backed or long-legged quadruped in any peculiar manner; he exposes sheep with long and short wool to the same climate. He does not allow the most vigorous males to struggle for the females. He does not rigidly destroy all inferior animals, but protects during each varying season, as far as lies in his power, all his productions. He often begins his selection by some half-monstrous form; or at least by some modification prominent enough to catch the eye or to be plainly useful to him. Under nature, the slightest differences of structure or constitution may well turn the nicely balanced scale in the struggle for life, and so be preserved. How fleeting are the wishes and efforts of man! how short his time! and consequently how poor will be his results, compared with those accumulated by Nature during whole geological periods? Can we wonder, then, that Nature's productions should be far "truer" in character than man's productions; that they should be infinitely better adapted to the most complex conditions of life, and should plainly bear the stamp of far higher workmanship?

It may metaphorically be said that natural selection is daily and hourly scrutinising, throughout the world, the slightest variations; rejecting those that are bad, preserving and adding up all that are good; silently and insensibly working, *whenever and wherever opportunity offers*, at the improvement of each organic being in relation to its organic and inorganic conditions of life. We see nothing of these slow changes in progress, until the hand of time has marked the lapse of ages, and then so imperfect is our view into long-past geological ages, that we see only that the forms of life are now different from what they formerly were.

In order that any great amount of modification should be effected in a species, a variety, when once formed, must again, perhaps after a long interval of time, vary or present individual differences of the same favourable nature as before; and these must be again preserved, and so onwards step by step. Seeing that individual differences of the same kind perpetually recur, this can hardly be considered as an unwarrantable assumption. But whether it is true, we can judge only by seeing how far the hypothesis accords with and explains the general phenomena of nature. On the other hand, the ordinary belief that the amount of possible variation is a strictly limited quantity is likewise a simple assumption.

Although natural selection can act only through and for the good of each being, yet characters and structures, which we are apt to consider as of very trifling importance, may thus be acted on. When we see leaf-eating insects green, and bark-feeders mottled-grey; the alpine ptarmigan white in winter, the red grouse the colour of heather, we must believe that these tints are of service to these birds and insects in preserving them from danger. Grouse, if not destroyed at some period of their lives, would increase in countless numbers; they are known to suffer largely from birds of prey; and hawks are guided by eyesight to their prey—so much so, that on parts of the Continent persons are warned not to keep white pigeons, as being the most liable to destruction. Hence natural selection might be effective in giving the proper colour to each kind of grouse, and in keeping that colour, when once acquired, true and constant. Nor ought we to think that the occasional destruction of an animal of any particular colour would produce little effect: we should remember how essential it is in a flock of white sheep to destroy a lamb with the faintest trace of black. We have seen how the colour of the hogs, which feed on the "paint root" in Virginia, determines whether they shall live or die. In plants, the down on the fruit and the colour of the flesh are considered by botanists as characters of the most trifling importance, yet we hear from an excellent horticulturist, Downing, that in the United States smooth-skinned fruits suffer far more from a beetle, a Curculio, than those with down; that purple plums suffer far more from a certain disease than yellow plums, whereas another disease attacks yellow-fleshed peaches far more than those with other coloured flesh. If, with all the aids of art, these slight differences make a great difference in cultivating the several varieties, assuredly, in a state of nature, where the trees would have to struggle with other trees and with a host of enemies, such differences would effectually settle which variety, whether a smooth or downy, a yellow or purple fleshed fruit, should succeed.

In looking at many small points of difference between species, which, as far as our ignorance permits us to judge, seem quite unimportant, we must not forget that climate, food, etc., have no doubt produced some direct effect. It is also necessary to bear in mind that, owing to the law of correlation, when one part varies, and the variations are accumulated through natural selection, other modifications, often of the most unexpected nature, will ensue.

As we see that those variations which, under domestication, appear at any particular period of life, tend to reappear in the offspring at the same period; for instance, in the shape, size, and flavour of the seeds of the many varieties of our culinary and agricultural plants; in the caterpillar and cocoon stages of the varieties of the silkworm; in the eggs of poultry, and in the colour of the down of their chickens; in the horns of our sheep and cattle when nearly adult; so in a state of nature natural selection will be enabled to act on and modify organic beings at any age, by the accumulation of variations profitable at that age, and by their inheritance at a corresponding age. If it profit a plant to have its seeds more and more widely disseminated by the wind, I can see no greater difficulty in this being effected through natural selection, than in the cotton planter increasing and improving by selection the down in the pods on his cotton trees. Natural selection may modify and adapt the larva of an insect to a score of contingencies, wholly different from those which concern the mature insect; and these modifications may effect, through correlation, the structure of the adult. So, conversely, modifications in the adult may affect the structure of the larva; but in all cases natural selection will ensure that they shall not be injurious: for if they were so, the species would become extinct.

Natural selection will modify the structure of the young in relation to the parent, and of the parent in relation to the young. In social animals it will adapt the structure of each individual for the benefit of the whole community; if the community profits by the selected change. What natural selection cannot do, is to modify the structure of one species, without giving it any advantage, for the good of another species; and though statements to this effect may be found in works of natural history, I cannot find one case which will bear investigation. A structure used only once in an animal's life, if of high importance to it, might be modified to any extent by natural selection; for instance, the great jaws possessed by certain insects, used exclusively for opening the cocoon—or the hard tip to the beak of unhatched birds, used for breaking the egg. It has been asserted, that of the best short-beaked tumbler pigeons a greater number perish in the egg than are able to get out of it; so that fanciers assist in the act of hatching. Now, if nature had to make the beak of a full-grown pigeon very short for the bird's own advantage, the process of modification would be very slow, and there would be simultaneously the most rigorous selection of all the young birds within the egg, which had the most powerful and hardest beaks, for all with weak beaks would inevitably perish; or, more delicate and more easily broken shells might be selected, the thickness of the shell being known to vary like every other structure.

It may be well here to remark that with all beings there must be much fortuitous destruction, which can have little or no influence on the course of natural selection. For instance a vast number of eggs or seeds are annually devoured, and these could be modified through natural selection only if they varied in some manner which protected them from their enemies. Yet many of these eggs or seeds would perhaps, if not destroyed, have yielded individuals better adapted to their conditions of life than any of those which happened to survive. So again a vast number of mature animals and plants, whether or not they be the best adapted to their conditions, must be annually destroyed by accidental causes, which would not be in the least degree mitigated by certain changes of structure or constitution which would in other ways be beneficial to the species. But let the destruction of the adults be ever so heavy, if the number which can exist in any district be not wholly kept down by such causes, or again let the destruction of eggs or seeds be so great that only a hundredth or a thousandth part are developed,—yet of those which do survive, the best adapted individuals, supposing that there is any variability in a favourable direction, will tend to propagate their kind in larger numbers than the less well adapted. If the numbers be wholly kept down by the causes just indicated, as will often have been the case, natural selection will be powerless in certain beneficial directions;

but this is no valid objection to its efficiency at other times and in other ways; for we are far from having any reason to suppose that many species ever undergo modification and improvement at the same time in the same area.

SEXUAL SELECTION

Inasmuch as peculiarities often appear under domestication in one sex and become hereditarily attached to that sex, so no doubt it will be under nature. Thus it is rendered possible for the two sexes to be modified through natural selection in relation to different habits of life, as is sometimes the case; or for one sex to be modified in relation to the other sex, as commonly occurs. This leads me to say a few words on what I have called Sexual Selection. This form of selection depends, not on a struggle for existence in relation to other organic beings or to external conditions, but on a struggle between the individuals of one sex, generally the males, for the possession of the other sex. The result is not death to the unsuccessful competitor, but few or no offspring. Sexual selection is, therefore, less rigorous than natural selection. Generally, the most vigorous males, those which are best fitted for their places in nature, will leave most progeny. But in many cases, victory depends not so much on general vigour, as on having special weapons, confined to the male sex. A hornless stag or spurless cock would have a poor chance of leaving numerous offspring. Sexual selection, by always allowing the victor to breed, might surely give indomitable courage, length to the spur, and strength to the wing to strike in the spurred leg, in nearly the same manner as does the brutal cockfighter by the careful selection of his best cocks. How low in the scale of nature the law of battle descends, I know not; male alligators have been described as fighting, bellowing, and whirling round, like Indians in a war dance, for the possession of the females; male salmons have been observed fighting all day long; male stag beetles sometimes bear wounds from the huge mandibles of other males; the males of certain hymenopterous insects have been frequently seen by that inimitable observer M. Fabre, fighting for a particular female who sits by, an apparently unconcerned beholder of the struggle, and then retires with the conqueror. The war is, perhaps, severest between the males of polygamous animals, and these seem oftenest provided with special weapons. The males of carnivorous animals are already well armed; though to them and to others, special means of defence may be given through means of sexual selection, as the mane of the lion, and the hooked jaw to the male salmon; for the shield may be as important for victory, as the sword or spear.

Amongst birds, the contest is often of a more peaceful character. All those who have attended to the subject, believe that there is the severest rivalry between the males of many species to attract, by singing, the females. The rock thrush of Guiana, birds of paradise, and some others, congregate; and successive males display with the most elaborate care, and show off in the best manner, their gorgeous plumage; they likewise perform strange antics before the females, which, standing by as spectators, at last choose the most attractive partner. Those who have closely attended to birds in confinement well know that they often take individual preferences and dislikes; thus Sir R. Heron has described how a pied peacock was eminently attractive to all his hen birds. I cannot here enter on the necessary details; but if man can in a short time give beauty and an elegant carriage to his bantams, according to his standard of beauty, I can see no good reason to doubt that female birds, by selecting, during thousands of generations, the most melodious or beautiful males, according to their standard of beauty, might produce a marked effect. Some well-known laws, with respect to the plumage of male and female birds, in comparison with the plumage of the young, can partly be explained through the action of sexual selection on variations occurring at different ages, and transmitted to the males alone or to both sexes at corresponding ages; but I have not space here to enter on this subject.

Thus it is, as I believe, that when the males and females of any animal have the same general habits of life, but differ in structure, colour, or ornament, such differences have been mainly caused by sexual selection: that is, by individual males having had, in successive generations, some slight advantage over other males, in their weapons, means of defence, or charms, which they have transmitted to their male offspring alone. Yet, I would not wish to attribute all sexual differences to this agency for we see in our domestic animals peculiarities arising and becoming attached to the male sex, which apparently have not been augmented through selection by man. The tuft of hair on the breast of the wild turkey cock cannot be of any use, and it is doubtful whether it can be ornamental in the eyes of the female bird; indeed, had the tuft appeared under domestication, it would have been called a monstrosity.

ILLUSTRATIONS OF THE ACTION OF NATURAL SELECTION, OR THE SURVIVAL OF THE FITTEST

In order to make it clear how, as I believe, natural selection acts, I must beg permission to give one or two imaginary illustrations. Let us take the case of a wolf, which preys on various animals, securing some by craft, some by strength, and some by fleetness; and let us suppose that the fleetest prey, a deer for instance, had from any change in the country increased in numbers, or that other prey had decreased in numbers, during that season of the year when the wolf was hardest pressed for food. Under such circumstances the swiftest and slimmest wolves would have the best chance of surviving and so be preserved or selected,—provided always that they retained strength to master their prey at this or some other period of the year, when they were compelled to prey on other animals. I can see no more reason to doubt that this would be the result, than that man should be able to improve the fleetness of his greyhounds by careful and methodical selection, or by that kind of unconscious selection which follows from each man trying to keep the best dogs without any thought of modifying the breed. I may add, that, according to Mr. Pierce, there are two varieties of the wolf inhabiting the Catskill Mountains, in the United States, one with a light greyhound-like form, which pursues deer, and the other more bulky, with shorter legs, which more frequently attacks the shepherd's flocks.

It should be observed that, in the above illustration, I speak of the slimmest individual wolves, and not of any single strongly-marked variation having been preserved. In former editions of this work I sometimes spoke as if this latter alternative had frequently occurred. I saw the great importance of individual differences, and this led me fully to discuss the results of unconscious selection by man, which depends on the preservation of all the more or less valuable individuals, and on the destruction of the worst. I saw, also, that the preservation in a state of nature of any occasional deviation of structure, such as a monstrosity, would be a rare event; and that, if at first preserved, it would generally be lost by subsequent intercrossing with ordinary individuals. Nevertheless, until reading an able and valuable article in the 'North British Review' (1867), I did not appreciate how rarely single variations, whether slight or strongly-marked, could be perpetuated. The author takes the case of a pair of animals, producing during their lifetime two hundred offspring, of which, from various causes of destruction, only two on an average survive to procreate their kind. This is rather an extreme estimate for most of the higher animals, but by no means so for many of the lower organisms. He then shows that if a single individual were born, which varied in some manner, giving it twice as good a chance of life as that of the other individuals, yet the chances would be strongly against its survival. Supposing it to survive and to breed, and that half its young inherited the favourable variation; still, as the reviewer goes on to show, the young would have only a slightly better chance of surviving and breeding; and this chance would go on decreasing in the succeeding generations. The justice of these remarks cannot, I think, be disputed. If, for instance, a bird of some kind could procure its food more easily by having its beak curved, and if one were born with its beak strongly curved, and which consequently flourished, nevertheless there would be a very poor chance of this one individual perpetuating its kind to the exclusion of the common form; but there can hardly be a doubt, judging by what we see taking place under domestication, that this result would follow from the preservation during many generations of a large number of individuals with more or less strongly curved beaks, and from the destruction of a still larger number with the straightest beaks.

It should not, however, be overlooked that certain rather strongly marked variations, which no one would rank as mere individual differences, frequently recur owing to a similar organisation being similarly acted on—of which fact numerous instances could be given with our domestic productions. In such cases, if the varying individual did not actually transmit to its offspring its newly acquired character, it would undoubtedly transmit to them, as long as the existing conditions remained the same, a still stronger tendency to vary in the same manner. There can also be little doubt that the tendency to vary in the same manner has often been so strong that all the individuals of the same species have been similarly modified without the aid of any form of selection. Or only a third, fifth, or tenth part of the individuals may have been thus affected, of which fact several instances could be given. Thus Graba estimates that about one-fifth of the guillemots in the Faroe Islands consist of a variety so well marked, that it was formerly ranked as a distinct species under the name of Uria lacrymans. In cases of this kind, if the variation were of a beneficial nature, the original form would soon be supplanted by the modified form, through the survival of the fittest.

To the effects of intercrossing in eliminating variations of all kinds, I shall have to recur; but it may be here remarked that most animals and plants keep to their proper homes, and do not needlessly wander about; we see this even with migratory birds, which almost always return to the same spot. Consequently each newly formed variety would generally be at first local, as seems to be the common rule with varieties in a state of nature; so that similarly modified individuals would soon exist in a small body together, and would often breed together. If the new variety were successful in its battle for life, it would slowly spread from a central district, competing with and conquering the unchanged individuals on the margins of an ever-increasing circle.

It may be worthwhile to give another and more complex illustration of the action of natural selection. Certain plants excrete sweet juice, apparently for the sake of eliminating something injurious from the sap: this is effected, for instance, by glands at the base of the stipules in some Leguminosae, and at the backs of the leaves of the common laurel. This juice, though small in quantity, is greedily sought by insects; but their visits do not in any way benefit the plant. Now, let us suppose that the juice or nectar was excreted from the inside of the flowers of a certain number of plants of any species. Insects in seeking the nectar would get dusted with pollen, and would often transport it from one flower to another. The flowers of two distinct individuals of the same species would thus get crossed; and the act of crossing, as can be fully proved, gives rise to vigorous seedlings, which consequently would have the best chance of flourishing and surviving. The plants which produced flowers with the largest glands or nectaries, excreting most nectar, would oftenest be visited by insects, and would oftenest be crossed; and so in the long run would gain the upper hand and form a local variety. The flowers, also, which had their stamens and pistils placed, in relation to the size and habits of the particular insect which visited them, so as to favour in any degree the transportal of the pollen, would likewise be favoured. We might have taken the case of insects visiting flowers for the sake of collecting pollen instead of nectar; and as pollen is formed for the sole purpose of fertilisation, its destruction appears to be a simple loss to the plant; yet if a little pollen were carried, at first occasionally and then habitually, by the pollen-devouring insects from flower to flower, and a cross thus effected, although nine-tenths of the pollen were destroyed, it might still be a great gain to the plant to be thus robbed; and the individuals which produced more and more pollen, and had larger anthers, would be selected.

When our plant, by the above process long continued, had been rendered highly attractive to insects, they would, unintentionally on their part, regularly carry pollen from flower to flower; and that they do this effectually, I could easily show by many striking facts. I will give only one, as likewise illustrating one step in the separation of the sexes of plants. Some holly trees bear only male flowers, which have four stamens producing a rather small quantity of pollen, and a rudimentary pistil; other holly trees bear only female flowers; these have a full-sized pistil, and four stamens with shrivelled anthers, in which not a grain of pollen can be detected. Having found a female tree exactly sixty yards from a male tree, I put the stigmas of twenty flowers, taken from different branches, under the microscope, and on all, without exception, there were a few pollen grains, and on some a profusion. As the wind had set for several days from the female to the male tree, the pollen could not thus have been carried. The weather had been cold and boisterous, and therefore not favourable to bees, nevertheless every female flower which I examined had been effectually fertilised by the bees, which had flown from tree to tree in search of nectar. But to return to our imaginary case: as soon as the plant had been rendered so highly attractive to insects that pollen was regularly carried from flower to flower, another process might commence. No naturalist doubts the advantage of what has been called the "physiological division of labour"; hence we may believe that it would be advantageous to a plant to produce stamens alone in one flower or on one whole plant, and pistils alone in another flower or on another plant In plants under culture and placed under new conditions of life, sometimes the male organs and sometimes the female organs become more or less impotent; now if we suppose this to occur in ever so slight a degree under nature, then, as pollen is already carried regularly from flower to flower, and as a more complete separation of the sexes of our plant would be advantageous on the principle of the division of labour, individuals with this tendency more and more increased, would be continually favoured or selected, until at last a complete separation of the sexes might be effected. It would take up too much space to show the various steps, through dimorphism and other means, by which the separation of the sexes in plants of various kinds is apparently now in progress; but I may add that some of the species of holly in North America, are, according to Asa Gray, in an exactly intermediate condition, or, as he expresses it, are more or less dioeciously polygamous.

Let us now turn to the nectar-feeding insects; we may suppose the plant, of which we have been slowly increasing the nectar by continued selection, to be a common plant; and that certain insects depended in main part on its nectar for food I could give many facts showing how anxious bees are to save time for instance, their habit of cutting holes and sucking the nectar at the bases of certain flowers, which with a very little more trouble, they can enter by the mouth. Bearing such facts in mind, it may be believed that under certain circumstances individual differences in the curvature or length of the proboscis, etc, too slight to be appreciated by us, might profit a bee or other insect, so that certain individuals would be able to obtain their food more quickly than others; and thus the communities to which they belonged would flourish and throw off many swarms inheriting the same peculiarities. The tubes of the corolla of the common red and incarnate clovers (Trifolium pratense and incarnatum) do not on a hasty glance appear to differ in length; yet the hive bee can easily suck the nectar out of the incarnate clover, but not out of the common red clover, which is visited by humble bees alone; so that whole fields of the red clover offer in vain an abundant supply of precious nectar to the hive bee. That this nectar is much liked by the hive bee is certain; for I have repeatedly seen, but only in the autumn, many hive bees sucking the flowers through holes bitten in the base of the tube by humble bees. The difference in the length of the corolla in the two kinds of clover, which determines the visits of the hive bee, must be very trifling; for I have been assured that when red clover has been mown, the flowers of the second crop are somewhat smaller, and that these are visited by many hive bees. I do not know whether this statement is accurate; nor whether another published statement can be trusted, namely, that the Ligurian bee, which is generally considered a mere variety of the common hive bee, and which freely crosses with it, is able to reach and suck the nectar of the red clover. Thus, in a country where this kind of clover abounded, it might be a great advantage to the hive bee to have a slightly longer or differently constructed proboscis. On the other hand, as the fertility of this clover absolutely depends on bees visiting the flowers, if humble bees were to become rare in any country, it might be a great advantage to the plant to have a shorter or more deeply divided corolla, so that the hive bees should be enabled to suck its flowers. Thus I can understand how a flower and a bee might slowly become, either simultaneously or one after the other, modified and adapted to each other in the most perfect manner, by the continued preservation of all the individuals which presented slight deviations of structure mutually favourable to each other

I am well aware that this doctrine of natural selection, exemplified in the above imaginary instances, is open to the same objections which were first urged against Sir Charles Lyell's noble views on "the modern changes of the earth, as illustrative of geology"; but we now seldom hear the agencies which we see still at work, spoken of as trifling or insignificant, when used in explaining the excavation of the deepest valleys or the formation of long lines of inland cliffs. Natural selection acts only by the preservation and accumulation of small inherited modifications, each profitable to the preserved being; and as modern geology has almost banished such views as the excavation of a great valley by a single diluvial wave, so will natural selection banish the belief of the continued creation of new organic beings, or of any great and sudden modification in their structure.

ON THE INTERCROSSING OF INDIVIDUALS

I must here introduce a short digression. In the case of animals and plants with separated sexes, it is of course obvious that two individuals must always (with the exception of the curious and not well understood cases of parthenogenesis) unite for each birth; but in the case of hermaphrodites this is far from obvious. Nevertheless there is reason to believe that with all hermaphrodites two individuals, either occasionally or habitually, concur for the reproduction of their kind. This view was long ago doubtfully suggested by Sprengel, Knight, and Kölreuter. We shall presently see its importance; but I must here treat the subject with extreme brevity, though I have the materials prepared for an ample discussion. All vertebrate animals, all insects, and some other large groups of animals, pair for each birth. Modern research has much diminished the number of supposed hermaphrodites, and of real hermaphrodites a large number pair; that is, two individuals regularly unite for reproduction, which is all that concerns us. But still there are many hermaphrodite animals which certainly do not habitually pair, and a vast majority of plants are hermaphrodites. What reason, it may be asked, is there for supposing in these cases that two individuals ever concur in reproduction? As it is impossible here to enter on details, I must trust to some general considerations alone.

In the first place, I have collected so large a body of facts, and made so many experiments, showing, in accordance with the almost universal belief of breeders, that with animals and plants a cross between different varieties, or between individuals of the same variety but of another strain, gives vigour and fertility to the offspring; and on the other hand, that *close* interbreeding diminishes vigour and fertility; that these facts alone incline me to believe that it is a general law of nature that no organic being fertilises itself for a perpetuity of generations; but that a cross with another individual is occasionally—perhaps at long intervals of time—indispensable.

On the belief that this is a law of nature, we can, I think, understand several large classes of facts, such as the following, which on any other view are inexplicable. Every hybridizer knows how unfavourable exposure to wet is to the fertilisation of a flower, yet what a multitude of flowers have their anthers and stigmas fully exposed to the weather! If an occasional cross be indispensable, notwithstanding that the plant's own anthers and pistil stand so near each other as almost to insure self-fertilisation, the fullest freedom for the entrance of pollen from another individual will explain the above state of exposure of the organs. Many flowers, on the other hand, have their organs of fructification closely enclosed, as in the great papilionaceous or pea-family; but these almost invariably present beautiful and curious adaptations in relation to the visits of insects. So necessary are the visits of bees to many papilionaceous flowers, that their fertility is greatly diminished if these visits be prevented.

Now, it is scarcely possible for insects to fly from flower to flower, and not to carry pollen from one to the other, to the great good of the plant. Insects act like a camel-hair pencil, and it is sufficient, to ensure fertilisation, just to touch with the same brush the anthers of one flower and then the stigma of another; but it must not be supposed that bees would thus produce a multitude of hybrids between distinct species; for if a plant's own pollen and that from another species are placed on the same stigma, the former is so prepotent that it invariably and completely destroys, as has been shown by Gärtner, the influence of the foreign pollen.

When the stamens of a flower suddenly spring towards the pistil, or slowly move one after the other towards it, the contrivance seems adapted solely to ensure self-fertilisation; and no doubt it is useful for this end but the agency of insects is often required to cause the stamens to spring forward, as Kölreuter has shown to be the case with the barberry; and in this very genus, which seems to have a special contrivance for self-fertilisation, it is well known that, if closely-allied forms or varieties are planted near each other, it is hardly possible to raise pure seedlings, so largely do they naturally cross. In numerous other cases, far from self-fertilisation being favoured, there are special contrivances which effectually prevent the stigma receiving pollen from its own flower, as I could show from the works of Sprengel and others, as well as from my own observations for instance, in Lobelia fulgens, there is a really beautiful and elaborate contrivance by which all the infinitely numerous pollen granules are swept out of the conjoined anthers of each flower, before the stigma of that individual flower is ready to receive them; and as this flower is never visited, at least in my garden, by insects, it never sets a seed, though by placing pollen from one flower on the stigma of another, I raise plenty of seedlings. Another species of Lobelia, which is visited by bees, seeds freely in my garden. In very many other cases, though there is no special mechanical contrivance to prevent the stigma receiving pollen from the same flower, yet, as Sprengel, and more recently Hildebrand, and others, have shown, and as I can confirm, either the anthers burst before the stigma is ready for fertilisation, or the stigma is ready before the pollen of that flower is ready, so that these so-named dichogamous plants have in fact separated sexes, and must habitually be crossed. So it is with the reciprocally dimorphic and trimorphic plants previously alluded to. How strange are these facts! How strange that the pollen and stigmatic surface of the same flower, though placed so close together, as if for the very purpose of self-fertilisation, should be in so many cases mutually useless to each other? How simply are these facts explained on the view of an occasional cross with a distinct individual being advantageous or indispensable!

If several varieties of the cabbage, radish, onion, and of some other plants, be allowed to seed near each other, a large majority of the seedlings thus raised turn out, as I have found, mongrels: for instance, I raised 233 seedling cabbages from some plants of different varieties growing near each other, and of these only seventy-eight were true to their kind, and some even of these were not perfectly true. Yet the pistil of each cabbage-flower is surrounded not only by its own six stamens but by those of the many other flowers on the same plant; and the pollen of each flower readily gets on its own stigma without insect agency; for I have found that plants carefully protected from insects produce the full number of pods. How, then, comes it that such a vast number of the seedlings are mongrelized? It must arise from the pollen of a distinct *variety* having a prepotent effect over the flower's own pollen; and that this is part of the general law of good being derived from the intercrossing of distinct individuals of the same species. When distinct *species* are crossed the case is reversed, for a plant's own pollen is almost always prepotent over foreign pollen; but to this subject we shall return in a future chapter.

In the case of a large tree covered with innumerable flowers, it may be objected that pollen could seldom be carried from tree to tree, and at most only from flower to flower on the same tree; and flowers on the same tree can be considered as distinct individuals only in a limited sense. I believe this objection to be valid, but that nature has largely provided against it by giving to trees a strong tendency to bear flowers with separated sexes. When the sexes are separated, although the male and female flowers may be produced on the same tree, pollen must be regularly carried from flower to flower; and this · ill give a better chance of pollen being occasionally carried from tree to tree. That trees belonging to all orders have their sexes more often separated than other plants, I find to be the case in this country; and at my request Dr. Hooker tabulated the trees of New Zealand, and Dr. Asa Gray those of the United States, and the result was as I anticipated. On the other hand, Dr. Hooker informs me that the rule does not hold good in Australia: but if most of the Australian trees are dichogamous, the same result would follow as if they bore flowers with separated sexes. I have made these few remarks on trees simply to call attention to the subject.

Turning for a brief space to animals: various terrestrial species are hermaphrodites, such as the land Mollusca and earthworms; but these all pair. As yet I have not found a single terrestrial animal which can fertilise itself. This remarkable fact, which offers so strong a contrast with terrestrial plants, is intelligible on the view of an occasional cross being indispensable: for owing to the nature of the fertilising element there are no means. analogous to the action of insects and of the wind with plants. by which an occasional cross could be effected with terrestrial animals without the concurrence of two individuals. Of aquatic animals. there are many self-fertilising hermaphrodites but here the currents of water offer an obvious means for an occasional cross. As in the case of flowers, I have as yet failed, after consultation with one of the highest authorities, namely, Professor Huxley, to discover a single hermaphrodite animal with the organs of reproduction so perfectly enclosed that access from without, and the occasional influence of a distinct individual, can be shown to be physically impossible. Cirripedes long appeared to me to present, under this point of view, a case of great difficulty; but I have been enabled, by a fortunate chance, to prove that two individuals, though both are self fertilising hermaphrodites, do sometimes cross.

It must have struck most naturalists as a strange anomaly that, both with animals and plants, some species of the same family and even of the same genus, though agreeing closely with each other in their whole organisation, are hermaphrodites, and some unisexual But if, in fact, all hermaphrodites do occasionally intercross, the difference between them and unisexual species is, as far as function is concerned, very small.

From these several considerations and from the many special facts which I have collected, but which I am unable here to give, it appears that with animals and plants an occasional intercross between distinct individuals is a very general, if not universal, law of nature.

CIRCUMSTANCES FAVOURABLE FOR THE PRODUCTION OF NEW FORMS THROUGH NATURAL SELECTION

This is an extremely intricate subject. A great amount of variability, under which term individual differences are always included, will evidently be favourable. A large number of individuals, by giving a better chance within any given period for the appearance of profitable variations, will compensate for a lesser amount of variability in each individual and is, I believe, a highly important element of success. Though Nature grants long periods of time for the work of natural selection, she does not grant an indefinite period; for as all organic beings are striving to seize on each place in the economy of nature, if any one species does not become modified and improved in a corresponding degree with its competitors, it will be exterminated. Unless favourable variations be inherited by some at least of the offspring, nothing can be effected by natural selection. The tendency to reversion may often check or prevent the work; but as this tendency has not prevented man from forming by selection numerous domestic races, why should it prevail against natural selection?

In the case of methodical selection, a breeder selects for some definite object, and if the individuals be allowed freely to intercross, his work will completely fail. But when many men, without intending to alter the breed, have a nearly common standard of perfection, and all try to procure and breed from the best animals, improvement surely but slowly follows from this unconscious process of selection, notwithstanding that there is no separation of selected individuals. Thus it will be under nature; for within a confined area, with some place in the natural polity not perfectly occupied, all the individuals varying in the right direction, though in different degrees, will tend to be preserved. But if the area be large, its several districts will almost certainly present different conditions of life; and then, if the same species undergoes modification in different districts, the newly-formed varieties will intercross on the confines of each. But we shall see in the sixth chapter that intermediate varieties, inhabiting intermediate districts, will in the long run generally be supplanted by one of the adjoining varieties. Intercrossing will chiefly affect those animals which unite for each birth and wander much, and which do not breed at a very quick rate. Hence with animals of this nature, for instance, birds, varieties will generally be confined to separated countries; and this I find to be the case. With hermaphrodite organisms which cross only occasionally, and likewise with animals which unite for each birth, but which wander little and can increase at a rapid rate, a new and improved variety might be quickly formed on any one spot, and might there maintain itself in a body and afterwards spread, so that the individuals of the new variety would chiefly cross together. On this principle, nurserymen always prefer saving seed from a large body of plants, as the chance of intercrossing is thus lessened.

Even with animals which unite for each birth, and which do not propagate rapidly, we must not assume that free intercrossing would always eliminate the effects of natural selection; for I can bring forward a considerable body of facts showing that within the same area, two varieties of the same animal may long remain distinct, from haunting different stations, from breeding at slightly different seasons, or from the individuals of each variety preferring to pair together.

Intercrossing plays a very important part in nature by keeping the individuals of the same species, or of the same variety, true and uniform in character. It will obviously thus act far more efficiently with those animals which unite for each birth; but, as already stated, we have reason to believe that occasional intercrosses take place with all animals and plants. Even if these take place only at long intervals of time, the young thus produced will gain so much in vigour and fertility over the offspring from long-continued self-fertilisation, that they will have a better chance of surviving and propagating their kind; and thus in the long run the influence of crosses, even at rare intervals, will be great. With respect to organic beings extremely low in the scale, which do not propagate sexually, nor conjugate, and which cannot possibly intercross, uniformity of character can be retained by them under the same conditions of life, only through the principle of inheritance, and through natural selection which will destroy any individuals departing from the proper type. If the conditions of life change and the form undergoes modification, uniformity of character can be given to the modified offspring, solely by natural selection preserving similar favourable variations.

Isolation, also, is an important element in the modification of species through natural selection. In a confined or isolated area, if not very large, the organic and inorganic conditions of life will generally be almost uniform; so that natural selection will tend to modify all the varying individuals of the same species in the same manner. Intercrossing with the inhabitants of the surrounding districts will, also, be thus prevented. Moritz Wagner has lately published an interesting essay on this subject, and has shown that the service rendered by isolation in preventing crosses between newly-formed varieties is probably greater even than I supposed. But from reasons already assigned I can by no means agree with this naturalist, that migration and isolation are necessary elements for the formation of new species. The importance of isolation is likewise great in preventing, after any physical change in the conditions such as of climate, elevation of the land, etc., the immigration of better adapted organisms; and thus new places in the natural economy of the district will be left open to be filled up by the modification of the old inhabitants. Lastly, isolation will give time for a new variety to be improved at a slow rate; and this may sometimes be of much importance. If, however, an isolated area be very small, either from being surrounded by barriers, or from having very peculiar physical conditions, the total number of the inhabitants will be small; and this will retard the production of new species through natural selection, by decreasing the chances of favourable variations arising.

The mere lapse of time by itself does nothing, either for or against natural selection. I state this because it has been erroneously asserted that the element of time has been assumed by me to play an all-important part in modifying species, as if all the forms of life were necessarily undergoing change through some innate law. Lapse of time is only so far important, and its importance in this respect is great, that it gives a better chance of beneficial variations arising and of their being selected, accumulated, and fixed. It likewise tends to increase the direct action of the physical conditions of life, in relation to the constitution of each organism.

If we turn to nature to test the truth of these remarks, and look at any small isolated area, such as an oceanic island, although the number of species inhabiting it is small, as we shall see in our chapter on Geographical Distribution; yet of these species a very large proportion are endemic,—that is, have been produced there and nowhere else in the world. Hence an oceanic island at first sight seems to have been highly favourable for the production of new species. But we may thus deceive ourselves, for to ascertain whether a small isolated area, or a large open area like a continent, has been most favourable for the production of new organic forms, we ought to make the comparison within equal times; and this we are incapable of doing.

Although isolation is of great importance in the production of new species, on the whole I am inclined to believe that largeness of area is still more important, especially for the production of species which shall prove capable of enduring for a long period, and of spreading widely. Throughout a great and open area, not only will there be a better chance of favourable variations, arising from the large number of individuals of the same species there supported, but the conditions of life are much more complex from the large number of already existing species; and if some of these many species become modified and improved, others will have to be improved in a corresponding degree, or they will be exterminated. Each new form, also, as soon as it has been much improved, will be able to spread over the open and continuous area, and will thus come into competition with many other forms. Moreover, great areas, though now continuous, will often, owing to former oscillations of level, have existed in a broken condition; so that the good effects of isolation will generally, to a certain extent, have concurred. Finally, I conclude that, although small isolated areas have been in some respects highly favourable for the production of new species, yet that the course of modification will generally have been more rapid on large areas; and what is more important, that the new forms produced on large areas, which already have been victorious over many competitors, will be those that will spread most widely, and will give rise to the greatest number of new varieties and species. They will thus play a more important part in the changing history of the organic world.

In accordance with this view, we can, perhaps, understand some facts which will be again alluded to in our chapter on Geographical Distribution; for instance, the fact of the productions of the smaller continent of Australia now yielding before those of the larger Europæo-Asiatic area. Thus, also, it is that continental productions have everywhere become so largely naturalised on islands. On a small island, the race for life will have been less severe, and there will have been less modification and less extermination. Hence, we can understand how it is that the flora of Madeira, according to Oswald Heer, resembles to a certain extent the extinct tertiary flora of Europe. All fresh-water basins, taken together, make a small area compared with that of the sea or of the land. Consequently, the competition between fresh-water productions will have been less severe than elsewhere; new forms will have been then more slowly produced, and old forms more slowly exterminated. And it is in fresh-water basins that we find seven genera of Ganoid fishes, remnants of a once preponderant order: and in fresh water we find some of the most anomalous forms now known in the world as the Ornithorhynchus and Lepidosiren. which, like fossils, connect to a certain extent orders at present widely sundered in the natural scale. These anomalous forms may be called living fossils; they have endured to the present day, from having inhabited a confined area, and from having been exposed to less varied, and therefore less severe, competition.

To sum up, as far as the extreme intricacy of the subject permits, the circumstances favourable and unfavourable for the production of new species through natural selection. I conclude that for terrestrial productions a large continental area, which has undergone many oscillations of level, will have been the most favourable for the production of many new forms of life, fitted to endure for a long time and to spread widely. Whilst the area existed as a continent, the inhabitants will have been numerous in individuals and kinds, and will have been subjected to severe competition. When converted by subsidence into large separate islands, there will still have existed many individuals of the same species on each island; intercrossing on the confines of the range of each new species will have been checked; after physical changes of any kind, immigration will have been prevented, so that new places in the polity of each island will have had to be filled up by the modification of the old inhabitants; and time will have been allowed for the varieties in each to become well modified and perfected. When, by renewed elevation, the islands were reconverted into a continental area, there will again have been very severe competition: the most favoured or improved varieties will have been enabled to spread: there will have been much extinction of the less improved forms, and the relative proportional numbers of the various inhabitants of the reunited continent will again have been changed; and again there will have been a fair field for natural selection to improve still further the inhabitants, and thus to produce new species.

That natural selection generally acts with extreme slowness I fully admit. It can act only when there are places in the natural polity of a district which can be better occupied by the modification of some of its existing inhabitants. The occurrence of such places will often depend on physical changes, which generally take place very slowly, and on the immigration of better adapted forms being prevented. As some few of the old inhabitants become modified, the mutual relations of others will often be disturbed; and this will create new places, ready to be filled up by better adapted forms; but all this will take place very slowly. Although all the individuals of the same species differ in some slight degree from each other, it would often be long before differences of the right nature in various parts of the organisation might occur. The result would often be greatly retarded by free intercrossing. Many will exclaim that these several causes are amply sufficient to neutralise the power of natural selection. I do not believe so. But I do believe that natural selection will generally act very slowly, only at long intervals of time, and only on a few of the inhabitants of the same region. I further believe that these slow, intermittent results accord well with what geology tells us of the rate and manner at which the inhabitants of the world have changed.

Slow though the process of selection may be, if feeble man can do much by artificial selection, I can see no limit to the amount of change, to the beauty and complexity of the co-adaptations between all organic beings, one with another and with their physical conditions of life, which may have been affected in the long course of time through nature's power of selection, that is by the survival of the fittest.

EXTINCTION CAUSED BY NATURAL SELECTION

This subject will be more fully discussed in our chapter on Geology; but it must here be alluded to from being intimately connected with natural selection. Natural selection acts solely through the preservation of variations in some way advantageous, which consequently endure. Owing to the high geometrical rate of increase of all organic beings, each area is already fully stocked with inhabitants; and it follows from this, that as the favoured forms increase in number, so, generally, will the less favoured decrease and become rare. Rarity, as geology tells us, is the precursor to extinction. We can see that any form which is represented by few individuals will run a good chance of utter extinction, during great fluctuations in the nature of the seasons, or from a temporary increase in the number of its enemies. But we may go further than this; for, as new forms are produced, unless we admit that specific forms can go on indefinitely increasing in number, many old forms must become extinct. That the number of specific forms has not indefinitely increased, geology plainly tells us; and we shall presently attempt to show why it is that the number of species throughout the world has not become immeasurably great.

We have seen that the species which are most numerous in individuals have the best chance of producing favourable variations within any given period. We have evidence of this, in the facts stated in the second chapter, showing that it is the common and diffused or dominant species which offer the greatest number of recorded varieties. Hence, rare species will be less quickly modified or improved within any given period; they will consequently be beaten in the race for life by the modified and improved descendants of the commoner species.

From these several considerations I think it inevitably follows, that as new species in the course of time are formed through natural selection, others will become rarer and rarer, and finally extinct. The forms which stand in closest competition with those undergoing modification and improvement, will naturally suffer most. And we have seen in the chapter on the Struggle for Existence, that it is the most closely-allied forms,—varieties of the same species, and species of the same genus or of related genera,—which, from having nearly the same structure, constitution, and habits, generally come into the severest competition with each other; consequently, each new variety or species, during the progress of its formation, will generally press hardest on its nearest kindred, and tend to exterminate them. We see the same process of extermination amongst our domesticated productions, through the selection of improved forms by man. Many curious instances could be given showing how quickly new breeds of cattle, sheep, and other animals, and varieties of flowers, take the place of older and inferior kinds. In Yorkshire, it is historically known that the ancient black cattle were displaced by the longhorns, and that these "were swept away by the shorthorns" (I quote the words of an agricultural writer) "as if by some murderous pestilence."

DIVERGENCE OF CHARACTER

The principle, which I have designated by this term, is of high importance, and explains, as I believe, several important facts. In the first place, varieties, even strongly marked ones, though having somewhat of the character of species—as is shown by the hopeless doubts in many cases how to rank them—yet certainly differ far less from each other than do good and distinct species. Nevertheless, according to my view, varieties are species in the process of formation, or are, as I have called them, incipient species. How, then, does the lesser difference between varieties become augmented into the greater difference between species? That this does habitually happen, we must infer from most of the innumerable species throughout nature presenting well-marked differences; whereas varieties, the supposed prototypes and parents of future well-marked species, present slight and ill-defined differences. Mere chance, as we may call it, might cause one variety to differ in some character from its parents, and the offspring of this variety again to differ from its parent in the very same character and in a greater degree; but this alone would never account for so habitual and large a degree of difference as that between the species of the same genus.

As has always been my practice, I have sought light on this head from our domestic productions. We shall here find something analogous. It will be admitted that the production of races so different as shorthorn and Hereford cattle, race and cart horses, the several breeds of pigeons, etc., could never have been effected by the mere chance accumulation of similar variations during many successive generations. In practice, a fancier is, for instance, struck by a pigeon having a slightly shorter beak; another fancier is struck by a pigeon having a rather longer beak; and on the acknowledged principle that "fanciers do not and will not admire a medium standard, but like extremes," they both go on (as has actually occurred with the sub-breeds of the tumbler pigeon) choosing and breeding from birds with longer and longer beaks, or with shorter and shorter beaks. Again, we may suppose that at an early period of history, the men of one nation or district required swifter horses, whilst those of another required stronger and bulkier horses. The early differences would be very slight; but, in the course of time, from the continued selection of swifter horses in the one case, and of stronger ones in the other, the differences would become greater, and would be noted as forming two sub-breeds. Ultimately, after the lapse of centuries, these sub-breeds would become converted into two well-established and distinct breeds. As the differences became greater, the inferior animals with intermediate characters, being neither very swift nor very strong, would not have been used for breeding, and will thus have tended to disappear. Here, then, we see in man's productions the action of what may be called the principle of divergence, causing differences, at first barely appreciable, steadily to increase, and the breeds to diverge in character, both from each other and from their common parent.

But how, it may be asked, can any analogous principle apply in nature? I believe it can and does apply most efficiently (though it was a long time before I saw how), from the simple circumstance that the more diversified the descendants from any one species become in structure, constitution, and habits, by so much will they be better enabled to seize on many and widely diversified places in the polity of nature, and so be enabled to increase in numbers.

We can clearly discern this in the case of animals with simple habits. Take the case of a carnivorous quadruped, of which the number that can be supported in any country has long ago arrived at its full average. If its natural power of increase be allowed to act, it can succeed in increasing (the country not undergoing any change in conditions) only by its varying descendants seizing on places at present occupied by other animals; some of them, for instance, being enabled to feed on new kinds of prey, either dead or alive; some inhabiting new stations, climbing trees, frequenting water, and some perhaps becoming less carnivorous. The more diversified in habits and structure the descendants of our carnivorous animals become, the more places they will be enabled to occupy. What applies to one animal will apply throughout all time to all animals—that is, if they vary—for otherwise natural selection can effect nothing. So it will be with plants. It has been experimentally proved, that if a plot of ground be sown with one species of grass, and a similar plot be sown with several distinct genera of grasses, a greater number of plants and a greater weight of dry herbage can be raised in the latter than in the former case. The same has been found to hold good when one variety and several mixed varieties of wheat have been sown on equal spaces of ground. Hence, if any one species of grass were to go on varying, and the varieties were continually selected which differed from each other in the same manner, though in a very slight degree, as do the distinct species and genera of grasses, a greater number of individual plants of this species, including its modified descendants, would succeed in living on the same piece of ground. And we know that each species and each variety of grass is annually sowing almost countless seeds; and is thus striving, as it may be said, to the utmost to increase in number. Consequently, in the course of many thousand generations, the most distinct varieties of any one species of grass would have the best chance of succeeding and of increasing in numbers, and thus of supplanting the less distinct varieties; and varieties, when rendered very distinct from each other, take the rank of species.

The truth of the principle that the greatest amount of life can be supported by great diversification of structure, is seen under many natural circumstances. In an extremely small area, especially if freely open to immigration, and where the contest between individual and individual must be very severe, we always find great diversity in its inhabitants. For instance, I found that a piece of turf, three feet by four in size, which had been exposed for many years to exactly the same conditions, supported twenty species of plants, and these belonged to eighteen genera and to eight orders, which shows how much these plants differed from each other. So it is with the plants and insects on small and uniform islets: also in small ponds of fresh water. Farmers find that they can raise most food by a rotation of plants belonging to the most different orders; nature follows what may be called a simultaneous rotation. Most of the animals and plants which live close round any small piece of ground, could live on it (supposing its nature not to be in any way peculiar), and may be said to be striving to the utmost to live there; but, it is seen, that where they come into the closest competition, the advantages of diversification of structure, with the accompanying differences of habit and constitution, determine that the inhabitants, which thus jostle each other most closely, shall, as a general rule, belong to what we call different genera and orders.

The same principle is seen in the naturalisation of plants through man's agency in foreign lands. It might have been expected that the plants which would succeed in becoming naturalised in any land would generally have been closely allied to the indigenes; for these are commonly looked at as specially created and adapted for their own country. It might also, perhaps, have been expected that naturalised plants would have belonged to a few groups more especially adapted to certain stations in their new homes. But the case is very different; and Alph. de Candolle has well remarked, in his great and admirable work, that floras gain by naturalisation, proportionally with the number of the native genera and species, far more in new genera than in new species. To give a single instance: in the last edition of Dr. Asa Gray's 'Manual of the Flora of the Northern United States,' 260 naturalised plants are enumerated, and these belong to 162 genera. We thus see that these naturalised plants are of a highly diversified nature. They differ, moreover, to a large extent, from the indigenes, for out of the 162 naturalised genera, no less than 100 genera are not there indigenous, and thus a large proportional addition is made to the genera now living in the United States.

By considering the nature of the plants or animals which have in any country struggled successfully with the indigenes, and have there become naturalised, we may gain some crude idea in what manner some of the natives would have to be modified, in order to gain an advantage over their compatriots; and we may at least infer that diversification of structure, amounting to new generic differences, would be profitable to them.

The advantage of diversification of structure in the inhabitants of the same region is, in fact, the same as that of the physiological division of labor in the organs of the same individual body—a subject so well elucidated by Milne Edwards. No physiologist doubts that a stomach adapted to digest vegetable matter alone, or flesh alone, draws most nutriment from these substances. So in the general economy of any land, the more widely and perfectly the animals and plants are diversified for different habits of life, so will a greater number of individuals be capable of there supporting themselves. A set of animals, with their organisation but little diversified, could hardly compete with a set more perfectly diversified in structure. It may be doubted, for instance, whether the Australian marsupials, which are divided into groups differing but little from each other, and feebly representing, as Mr. Waterhouse and others have remarked, our carnivorous, ruminant, and rodent mammals, could successfully compete with these well-developed orders. In the Australian mammals, we see the process of diversification in an early and incomplete stage of development.

THE PROBABLE EFFECTS OF THE ACTION OF NATURAL SELECTION THROUGH DIVERGENCE OF CHARACTER AND EXTINCTION, ON THE DESCENDANTS OF A COMMON ANCESTOR

After the foregoing discussion, which has been much compressed, we may assume that the modified descendants of any one species will succeed so much the better as they become more diversified in structure, and are thus enabled to encroach on places occupied by other beings. Now let us see how this principle of benefit being derived from divergence of character, combined with the principles of natural selection and of extinction, tends to act.

The accompanying diagram will aid us in understanding this rather perplexing subject. Let A to L represent the species of a genus large in its own country; these species are supposed to resemble each other in unequal degrees, as is so generally the case in nature, and as is represented in the diagram by the letters standing at unequal distances. I have said a large genus, because, as we saw in the second chapter, on an average more species vary in large genera than in small genera; and the varying species of the large genera present a greater number of varieties. We have, also, seen that the species, which are the commonest and the most widely diffused, vary more than do the rare and restricted species. Let (A) be a common, widely-diffused, and varying species, belonging to a genus large in its own country. The branching and diverging dotted lines of unequal lengths proceeding from (A), may represent its varying offspring. The variations are supposed to be extremely slight, but of the most diversified nature; they are not supposed all to appear simultaneously, but often after long intervals of time; nor are they all supposed to endure for equal periods. Only those variations which are in some way profitable will be preserved or naturally selected. And here the importance of the principle of benefit derived from divergence of character comes in; for this will generally lead to the most different or divergent variations (represented by the outer dotted lines) being preserved and accumulated by natural selection. When a dotted line reaches one of the horizontal lines, and is there marked by a small numbered letter, a sufficient amount of variation is supposed to have been accumulated to form it into a fairly well-marked variety, such as would be thought worthy of record in systematic work.

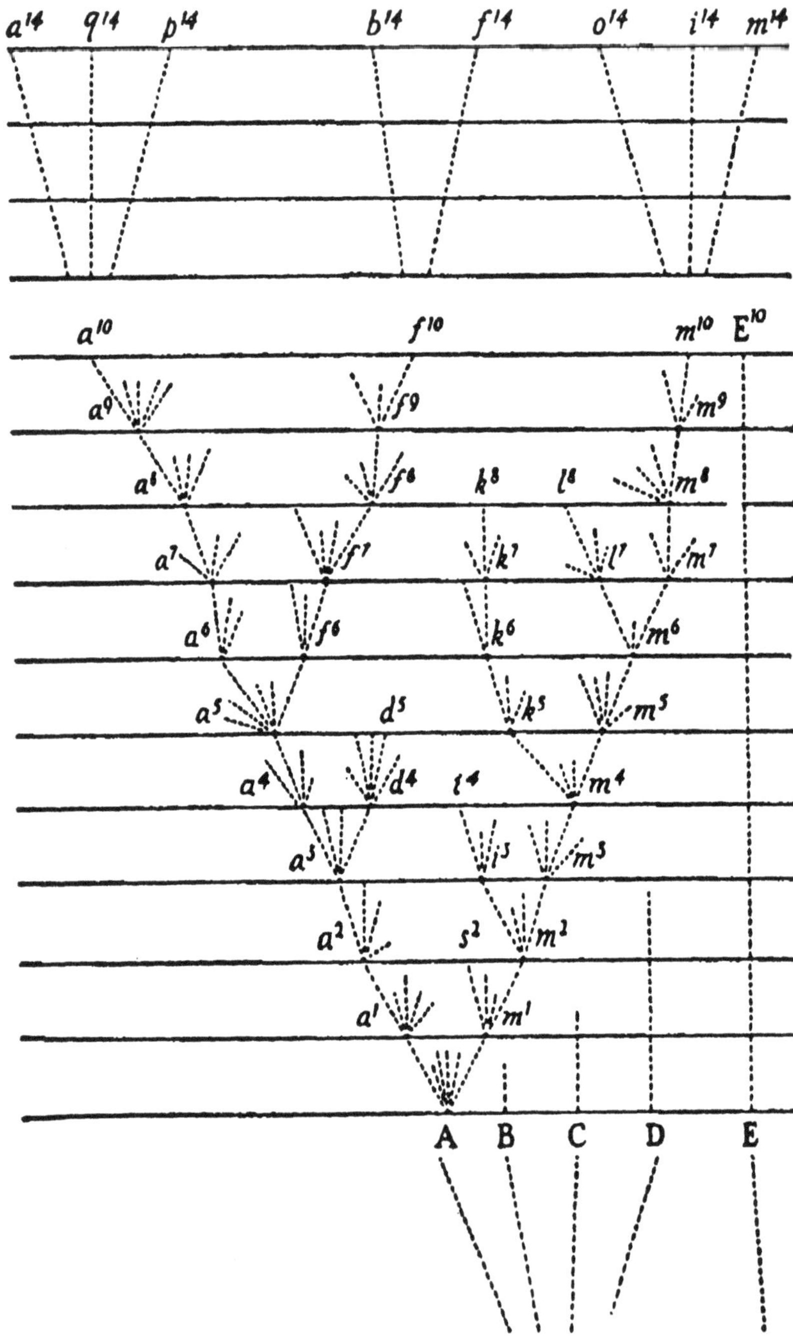

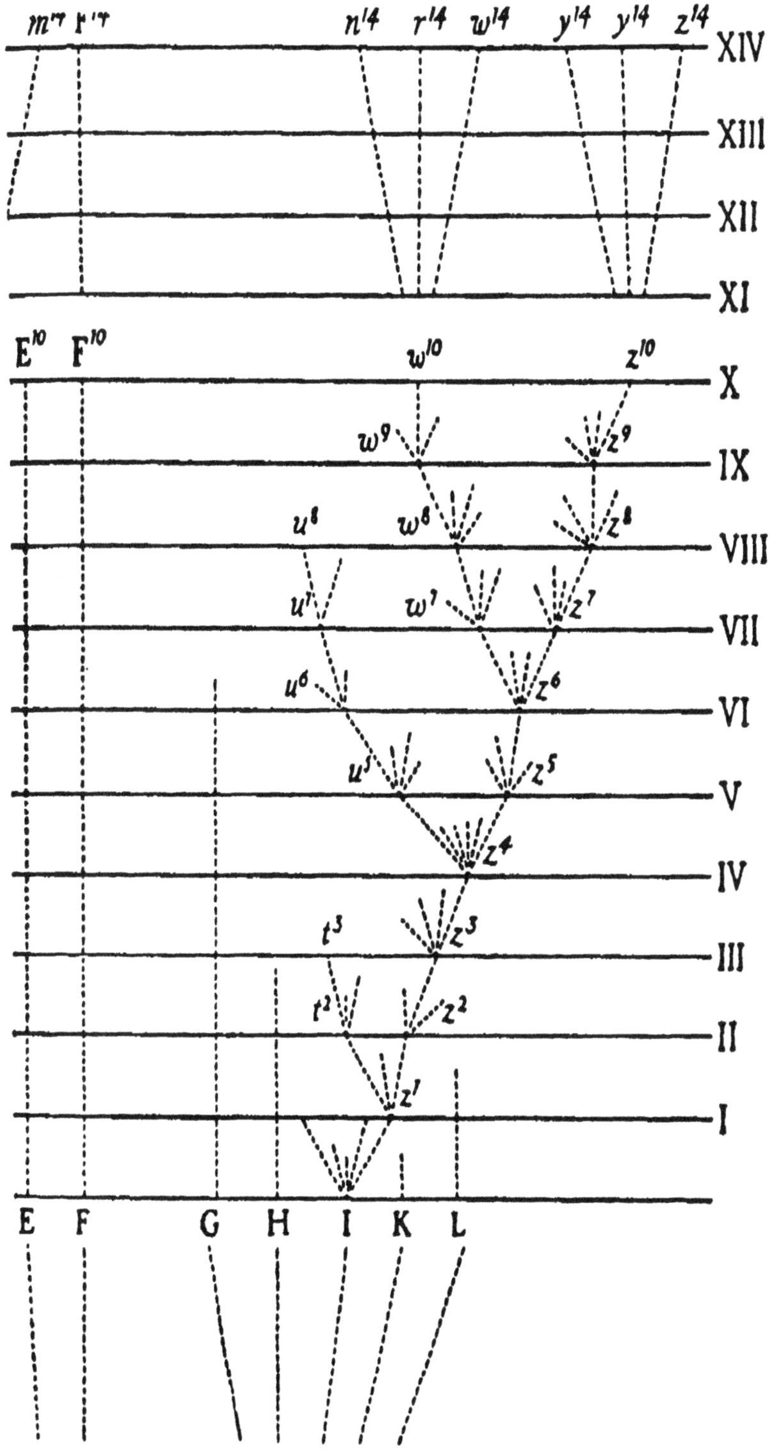
n14 r14 w14 y14 y14 z14
XIV
XIII
XII
XI
E10 F10 w10 z10
X
w9 z9
IX
u8 w8 z8
VIII
u7 w7 z7
VII
u6 z6
VI
u5 z5
V
z4
IV
t3 z3
III
t2 z2
II
z1
I
E F G H I K L

The intervals between the horizontal lines in the diagram, may represent each a thousand or more generations. After a thousand generations, species (A) is supposed to have produced two fairly well-marked varieties, namely a^1 and m^1. These two varieties will generally still be exposed to the same conditions which made their parents variable, and the tendency to variability is in itself hereditary; consequently they will likewise tend to vary, and commonly in nearly the same manner as did their parents. Moreover, these two varieties, being only slightly modified forms, will tend to inherit those advantages which made their parent (A) more numerous than most of the other inhabitants of the same country; they will also partake of those more general advantages which made the genus to which the parent-species belonged, a large genus in its own country. And all these circumstances are favourable to the production of new varieties.

If, then, these two varieties be variable, the most divergent of their variations will generally be preserved during the next thousand generations. And after this interval, variety a^1 is supposed in the diagram to have produced variety a^2, which will, owing to the principle of divergence, differ more from (A) than did variety a^1. Variety m^1 is supposed to have produced two varieties, namely m^2 and s^2, differing from each other, and more considerably from their common parent (A). We may continue the process by similar steps for any length of time; some of the varieties, after each thousand generations, producing only a single variety, but in a more and more modified condition, some producing two or three varieties, and some failing to produce any. Thus the varieties or modified descendants of the common parent (A), will generally go on increasing in number and diverging in character. In the diagram the process is represented up to the ten-thousandth generation, and under a condensed and simplified form up to the fourteen-thousandth generation.

But I must here remark that I do not suppose that the process ever goes on so regularly as is represented in the diagram, though in itself made somewhat irregular, nor that it goes on continuously; it is far more probable that each form remains for long periods unaltered, and then again undergoes modification. Nor do I suppose that the most divergent varieties are invariably preserved: a medium form may often long endure, and may or may not produce more than one modified descendant; for natural selection will always act according to the nature of the places which are either unoccupied or not perfectly occupied by other beings; and this will depend on infinitely complex relations. But as a general rule, the more diversified in structure the descendants from any one species can be rendered, the more places they will be enabled to seize on, and the more their modified progeny will increase. In our diagram the line of succession is broken at regular intervals by small numbered letters marking the successive forms which have become sufficiently distinct to be recorded as varieties. But these breaks are imaginary, and might have been inserted anywhere, after intervals long enough to allow the accumulation of a considerable amount of divergent variation.

As all the modified descendants from a common and widely-diffused species, belonging to a large genus, will tend to partake of the same advantages which made their parent successful in life, they will generally go on multiplying in number as well as diverging in character: this is represented in the diagram by the several divergent branches proceeding from (A). The modified offspring from the later and more highly improved branches in the lines of descent, will, it is probable, often take the place of and so destroy, the earlier and less improved branches: this is represented in the diagram by some of the lower branches not reaching to the upper horizontal lines. In some cases no doubt the process of modification will be confined to a single line of descent and the number of modified descendants will not be increased; although the amount of divergent modification may have been augmented. This case would be represented in the diagram, if all the lines proceeding from (A) were removed, excepting that from a^1 to a^{10}. In the same way the English racehorse and English pointer have apparently both gone on slowly diverging in character from their original stocks, without either having given off any fresh branches or races.

After ten thousand generations, species (A) is supposed to have produced three forms, a^{10}, f^{10}, and m^{10}, which, from having diverged in character during the successive generations, will have come to differ largely, but perhaps unequally, from each other and from their common parent. If we suppose the amount of change between each horizontal line in our diagram to be excessively small, these three forms may still be only well-marked varieties; but we have only to suppose the steps in the process of modification to be more numerous or greater in amount, to convert these three forms into doubtful or at least into well-defined species. Thus the diagram illustrates the steps by which the small differences distinguishing varieties are increased into the larger differences distinguishing species. By continuing the same process for a greater number of generations (as shown in the diagram in a condensed and simplified manner), we get eight species, marked by the letters between a^{14} and m^{14}, all descended from (A). Thus, as I believe, species are multiplied, and genera are formed.

In a large genus it is probable that more than one species would vary. In the diagram I have assumed that a second species (I) has produced, by analogous steps, after ten thousand generations, either two well-marked varieties (w^{10} and z^{10}) or two species, according to the amount of change supposed to be represented between the horizontal lines. After fourteen thousand generations, six new species, marked by the letters n^{14} to z^{14}, are supposed to have been produced. In any genus, the species which are already very different in character from each other, will generally tend to produce the greatest number of modified descendants; for these will have the best chance of seizing on new and widely different places in the polity of nature: hence in the diagram I have chosen the extreme species (A), and the nearly extreme species (I), as those which have largely varied, and have given rise to new varieties and species. The other nine species (marked by capital letters) of our original genus, may for long but unequal periods continue to transmit unaltered descendants; and this is shown in the diagram by the dotted lines unequally prolonged upwards.

But during the process of modification, represented in the diagram, another of our principles, namely that of extinction, will have played an important part. As in each fully stocked country natural selection necessarily acts by the selected form having some advantage in the struggle for life over other forms, there will be a constant tendency in the improved descendants of any one species to supplant and exterminate in each stage of descent their predecessors and their original progenitor. For it should be remembered that the competition will generally be most severe between those forms which are most nearly related to each other in habits, constitution, and structure. Hence all the intermediate forms between the earlier and later states, that is between the less and more improved states of the same species, as well as the original parent-species itself, will generally tend to become extinct. So it probably will be with many whole collateral lines of descent which will be conquered by later and improved lines.

If, however, the modified offspring of a species get into some distinct country, or become quickly adapted to some quite new station, in which offspring and progenitor do not come into competition, both may continue to exist.

If, then, our diagram be assumed to represent a considerable amount of modification, species (A) and all the earlier varieties will have become extinct, being replaced by eight new species (a^{14} to m^{14}); and species (I) will be replaced by six (n^{14} to z^{14}) new species.

But we may go further than this. The original species of our genus were supposed to resemble each other in unequal degrees, as is so generally the case in nature; species (A) being more nearly related to B, C, and D, than to the other species; and species (I) more to G, H, K, L, than to the others. These two species (A) and (I) w ere also supposed to be very common and widely diffused species, so that they must originally have had some advantage over most of the other species of the genus. Their modified descendants, fourteen in number at the fourteen-thousandth generation, will probably have inherited some of the same advantages; they have also been modified and improved in a diversified manner at each stage of descent, so as to have become adapted to many related places in the natural economy of their country. It seems, therefore, extremely probable that they will have taken the places of, and thus exterminated, not only their parents (A) and (I), but likewise some of the original species which were most nearly related to their parents. Hence very few of the original species will have transmitted offspring to the fourteen-thousandth generation. We may suppose that only one, (F), of the two species (E) and (F) which were least closely related to the other nine original species, has transmitted descendants to this late stage of descent.

The new species in our diagram, descended from the original eleven species, will now be fifteen in number. Owing to the divergent tendency of natural selection, the extreme amount of difference in character between species a^{14} and z^{14} will be much greater than that between the most distinct of the original eleven species. The new species, moreover, will be allied to each other in a widely different manner. Of the eight descendants from (A) the three marked a^{14}, q^{14}, p^{14}, will be nearly related from having recently branched off from a^{10}; b^{14}, and f^{14}, from having diverged at an earlier period from a^{5} , will be in some degree distinct from the three first-named species; and lastly, o^{14}, i^{14} and m^{14} will be nearly related one to the other, but, from having diverged at the first commencement of the process of modification, will be widely different from the other five species, and may constitute a sub-genus or a distinct genus.

The six descendants from (I) will form two sub-genera or genera. But as the original species (I) differed largely from (A), standing nearly at the extreme end of the original genus, the six descendants from (I) will, owing to inheritance alone, differ considerably from the eight descendants from (A); the two groups, moreover, are supposed to have gone on diverging in different directions. The intermediate species, also (and this is a very important consideration), which connected the original species (A) and (I), have all become, excepting (F), extinct, and have left no descendants. Hence the six new species descended from (I), and the eight descendants from (A), will have to be ranked as very distinct genera, or even as distinct sub-families.

Thus it is, as I believe, that two or more genera are produced by descent with modification, from two or more species of the same genus. And the two or more parent-species are supposed to be descended from some one species of an earlier genus. In our diagram, this is indicated by the broken lines, beneath the capital letters, converging in sub-branches downwards towards a single point; this point represents a species, the supposed progenitor of our several sub-genera and genera.

It is worth while to reflect for a moment on the character of the new species F^{14} which is supposed not to have diverged much in character, but to have retained the form of (F), either unaltered or altered only in a slight degree. In this case, its affinities to the other fourteen new species will be of a curious and circuitous nature. Being descended from a form which stood between the parent-species (A) and (I), now supposed to be extinct and unknown, it will be in some degree intermediate in character between the two groups descended from these two species. But as these two groups have gone on diverging in character from the type of their parents, the new species (F^{14}) will not be directly intermediate between them, but rather between types of the two groups; and every naturalist will be able to call such cases before his mind.

In the diagram, each horizontal line has hitherto been supposed to represent a thousand generations, but each may represent a million or more generations; it may also represent a section of the successive strata of the earth's crust including extinct remains. We shall, when we come to our chapter on Geology, have to refer again to this subject, and I think we shall then see that the diagram throws light on the affinities of extinct beings, which, though generally belonging to the same orders, families, or genera, with those now living, yet are often, in some degree, intermediate in character between existing groups; and we can understand this fact, for the extinct species lived at various remote epochs when the branching lines of descent had diverged less.

I see no reason to limit the process of modification, as now explained, to the formation of genera alone. If, in the diagram, we suppose the amount of change represented by each successive group of diverging dotted lines to be great, the forms marked a^{14} to p^{14}, those marked b^{14} and f^{14}, and those marked o^{14} to m^{14}, will form three very distinct genera. We shall also have two very distinct genera descended from (I), differing widely from the descendants of (A). Those two groups of genera will thus form two distinct families, or orders, according to the amount of divergent modification supposed to be represented in the diagram. And the two new families, or orders, are descended from two species of the original genus, and these are supposed to be descended from some still more ancient and unknown form.

We have seen that in each country it is the species belonging to the larger genera which oftenest present varieties or incipient species. This, indeed, might have been expected; for, as natural selection acts through one form having some advantage over other forms in the struggle for existence, it will chiefly act on those which already have some advantage; and the largeness of any group shows that its species have inherited from a common ancestor some advantage in common. Hence, the struggle for the production of new and modified descendants will mainly lie between the larger groups which are all trying to increase in number. One large group will slowly conquer another large group, reduce its numbers, and thus lessen its chance of further variation and improvement. Within the same large group, the later and more highly perfected sub-groups, from branching out and seizing on many new places in the polity of Nature, will constantly tend to supplant and destroy the earlier and less improved sub-groups. Small and broken groups and sub-groups will finally disappear. Looking to the future, we can predict that the groups of organic beings which are now large and triumphant, and which are least broken up, that is, which have as yet suffered least extinction, will, for a long period, continue to increase. But which groups will ultimately prevail, no man can predict; for we know that many groups, formerly most extensively developed, have now become extinct. Looking still more remotely to the future, we may predict that, owing to the continued and steady increase of the larger groups, a multitude of smaller groups will become utterly extinct, and leave no modified descendants; and consequently, that, of the species living at any one period extremely few will transmit descendants to a remote futurity. I shall have to return to this subject in the chapter on Classification, but I may add that as, according to this view, extremely few of the more ancient species have transmitted descendants to the present day, and, as all the descendants of the same species form a class, we can understand how it is that there exist so few classes in each main division of the animal and vegetable kingdoms. Although few of the most ancient species have left modified descendants, yet, at remote geological periods, the earth may have been almost as well peopled with species of many genera, families, orders, and Classes, as at the present time.

ON THE DEGREE TO WHICH ORGANISATION TENDS TO ADVANCE

Natural Selection acts exclusively by the preservation and accumulation of variations, which are beneficial under the organic and inorganic conditions to which each creature is exposed at all periods of life. The ultimate result is that each creature tends to become more and more improved in relation to its conditions. This improvement inevitably leads to the gradual advancement of the organisation of the greater number of living beings throughout the world. But here we enter on a very intricate subject, for naturalists have not defined to each other's satisfaction what is meant by an advance in organisation. Amongst the vertebrata the degree of intellect and an approach in structure to man clearly come into play. It might be thought that the amount of change which the various parts and organs pass through in their development from the embryo to maturity would suffice as a standard of comparison; but there are cases, as with certain parasitic crustaceans, in which several parts of the structure become less perfect, so that the mature animal cannot be called higher than its larva. Von Baer's standard seems the most widely applicable and the best, namely, the amount of differentiation of the parts of the same organic being, in the adult state, as I should be inclined to add, and their specialisation for different functions; or, as Milne Edwards would express it, the completeness of the division of physiological labour. But we shall see how obscure this subject is if we look, for instance, to fishes, amongst which some naturalists rank those as highest which, like the sharks, approach nearest to amphibians; whilst other naturalists range the common bony or teleostean fishes as the highest, inasmuch as they are most strictly fish-like, and differ most from the other vertebrate classes. We see still more plainly the obscurity of the subject by turning to plants, amongst which the standard of intellect is of course quite excluded; and here some botanists rank those plants as highest which have every organ, as sepals, petals, stamens, and pistils, fully developed in each flower; whereas other botanists, probably with more truth, look at the plants which have their several organs much modified and reduced in number, as the highest.

If we take as the standard of high organisation, the amount of differentiation and specialisation of the several organs in each being when adult (and this will include the advancement of the brain for intellectual purposes), natural selection clearly leads towards this standard: for all physiologists admit that the specialisation of organs, inasmuch as in this state they perform their functions better, is an advantage to each being; and hence the accumulation of variations tending towards specialisation is within the scope of natural selection. On the other hand, we can see, bearing in mind that all organic beings are striving to increase at a high ratio and to seize on every unoccupied or less well occupied place in the economy of nature, that it is quite possible for natural selection gradually to fit a being to a situation in which several organs would be superfluous or useless: in such cases there would be retrogression in the scale of organisation. Whether organisation on the whole has actually advanced from the remotest geological periods to the present day will be more conveniently discussed in our chapter on Geological Succession.

But it may be objected that if all organic beings thus tend to rise in the scale, how is it that throughout the world a multitude of the lowest forms still exist; and how is it that in each great class some forms are far more highly developed than others? Why have not the more highly developed forms everywhere supplanted and exterminated the lower? Lamarck, who believed in an innate and inevitable tendency towards perfection in all organic beings, seems to have felt this difficulty so strongly, that he was led to suppose that new and simple forms are continually being produced by spontaneous generation. Science has not as yet proved the truth of this belief, whatever the future may reveal. On our theory the continued existence of lowly organisms offers no difficulty; for natural selection, or the survival of the fittest, does not necessarily include progressive development—it only takes advantage of such variations as arise and are beneficial to each creature under its complex relations of life. And it may be asked what advantage, as far as we can see, would it be to an infusorian animalcule—to an intestinal worm—or even to an earthworm, to be highly organised. If it were no advantage, these forms would be left, by natural selection, unimproved or but little improved, and might remain for indefinite ages in their present lowly condition. And geology tells us that some of the lowest forms, as the infusoria and rhizopods, have remained for an enormous period in nearly their present state. But to suppose that most of the many now existing low forms have not in the least advanced since the first dawn of life would be extremely rash; for every naturalist who has dissected some of the beings now ranked as very low in the scale, must have been struck with their really wondrous and beautiful organisation.

Nearly the same remarks are applicable if we look to the different grades of organisation within the same great group; for instance, in the Vertebrata, to the co-existence of mammals and fish—amongst Mammalia, to the co existence of man and the ornithorhynchus—amongst fishes, to the coexistence of the shark and the lancelet (Amphioxus), which latter fish in the extreme simplicity of its structure approaches the invertebrate classes. But mammals and fish hardly come into competition with each other; the advancement of the whole class of mammals, or of certain members in this class, to the highest grade, would not lead to their taking the place of fishes. Physiologists believe that the brain must be bathed by warm blood to be highly active, and this requires aerial respiration; so that warm-blooded mammals when inhabiting the water lie under a disadvantage in having to come continually to the surface to breathe. With fishes, members of the shark family would not tend to supplant the lancelet; for the lancelet, as I hear from Fritz Müller, has as sole companion and competitor on the barren sandy shore of South Brazil, an anomalous annelid. The three lowest orders of mammals, namely, marsupials, edentata, and rodents, co-exist in South America in the same region with numerous monkeys, and probably interfere little with each other. Although organisation, on the whole, may have advanced and be still advancing throughout the world, yet the scale will always present many degrees of perfection; for the high advancement of certain whole classes, or of certain members of each class, does not at all necessarily lead to the extinction of those groups with which they do not enter into close competition. In some cases, as we shall hereafter see, lowly organised forms appear to have been preserved to the present day, from inhabiting confined or peculiar stations, where they have been subjected to less severe competition, and where their scanty numbers have retarded the chance of favorable variations arising.

Finally, I believe that many lowly organised forms now exist throughout the world, from various causes. In some cases variations or individual differences of a favorable nature may never have arisen for natural selection to act on and accumulate. In no case, probably, has time sufficed for the utmost possible amount of development. In some few cases there has been what we must call retrogression of organisation. But the main cause lies in the fact that under very simple conditions of life a high organisation would be of no service, —possibly would be of actual disservice, as being of a more delicate nature, and more liable to be put out of order and injured.

Looking to the first dawn of life, when all organic beings, as we may believe, presented the simplest structure, how, it has been asked, could the first steps in the advancement or differentiation of parts have arisen? Mr. Herbert Spencer would probably answer, that, as soon as simple unicellular organism came by growth or division to be compounded of several cells, or became attached to any supporting surface, his law "that homologous units of any order became differentiated in proportion as their relations to incident forces became different" would come into action. But as we have no facts to guide us, speculation on the subject is almost useless. It is, however, an error to suppose that there would be no struggle for existence, and, consequently, no natural selection, until many forms had been produced: variations in a single species inhabiting an isolated station might be beneficial, and thus the whole mass of individuals might be modified, or two distinct forms might arise. But, as I remarked towards the close of the Introduction, no one ought to feel surprise at much remaining as yet unexplained on the origin of species, if we make due allowance for our profound ignorance on the mutual relations of the inhabitants of the world at the present time, and still more so during past ages.

CONVERGENCE OF CHARACTER

Mr. H. C. Watson thinks that I have overrated the importance of divergence of character (in which, however, he apparently believes), and that convergence, as it may be called, has likewise played a part. If two species, belonging to two distinct though allied genera, had both produced a large number of new and divergent forms, it is conceivable that these might approach each other so closely that they would have all to be classed under the same genus; and thus the descendants of two distinct genera would converge into one. But it would in most cases be extremely rash to attribute to convergence a close and general similarity of structure in the modified descendants of widely distinct forms. The shape of a crystal is determined solely by the molecular forces, and it is not surprising that dissimilar substances should sometimes assume the same form; but with organic beings we should bear in mind that the form of each depends on an infinitude of complex relations, namely on the variations which have arisen, these being due to causes far too intricate to be followed out,—on the nature of the variations which have been preserved or selected, and this depends on the surrounding physical conditions, and in a still higher degree on the surrounding organisms with which each being has come into competition,—and lastly, on inheritance (in itself a fluctuating element) from innumerable progenitors, all of which have had their forms determined through equally complex relations. It is incredible that the descendants of two organisms, which had originally differed in a marked manner, should ever afterwards converge so closely as to lead to a near approach to identity throughout their whole organisation. If this had occurred, we should meet with the same form, independently of genetic connection, recurring in widely separated geological formations; and the balance of evidence is opposed to any such an admission.

Mr. Watson has also objected that the continued action of natural selection, together with divergence of character, would tend to make an indefinite number of specific forms. As far as mere inorganic conditions are concerned, it seems probable that a sufficient number of species would soon become adapted to all considerable diversities of heat, moisture, etc.; but I fully admit that the mutual relations of organic beings are more important; and as the number of species in any country goes on increasing, the organic conditions of life must become more and more complex. Consequently there seems at first sight no limit to the amount of profitable diversification of structure, and therefore no limit to the number of species which might be produced. We do not know that even the most prolific area is fully stocked with specific forms: at the Cape of Good Hope and in Australia, which support such an astonishing number of species, many European plants have become naturalised. But geology shows us, that from an early part of the tertiary period the number of species of shells, and that from the middle part of this same period the number of mammals, has not greatly or at all increased. What then checks an indefinite increase in the number of species? The amount of life (I do not mean the number of specific forms) supported on an area must have a limit, depending so largely as it does on physical conditions; therefore, if an area be inhabited by very many species, each or nearly each species will be represented by few individuals; and such species will be liable to extermination from accidental fluctuations in the nature of the seasons or in the number of their enemies. The process of extermination in such cases would be rapid, whereas the production of new species must always be slow. Imagine the extreme case of as many species as individuals in England, and the first severe winter or very dry summer would exterminate thousands on thousands of species. Rare species, and each species will become rare if the number of species in any country becomes indefinitely increased, will, on the principle often explained, present within a given period few favorable variations; consequently, the process of giving birth to new specific forms would thus be retarded. When any species becomes very rare, close interbreeding will help to exterminate it; authors have thought that this comes into play in accounting for the deterioration of the aurochs in Lithuania, of red deer in Scotland, and of bears in Norway, etc. Lastly, and this I am inclined to think is the most important element, a dominant species, which has already beaten many competitors in its own home, will tend to spread and supplant many others. Alph. de Candolle has shown that those species which spread widely, tend generally to spread *very* widely; consequently, they will tend to supplant and exterminate several species in several areas, and thus check the inordinate increase of specific forms throughout the world. Dr. Hooker has recently shown that in the southeast corner of Australia, where, apparently, there are many invaders from different quarters of the globe, the endemic Australian species have been greatly reduced in number. How much weight to attribute to these several considerations I will not pretend to say; but conjointly they must limit in each country the tendency to an indefinite augmentation of specific forms.

SUMMARY OF CHAPTER

If under changing conditions of life organic beings present individual differences in almost every part of their structure, and this cannot be disputed; if there be, owing to their geometrical rate of increase, a severe struggle for life at some age, season, or year, and this certainly cannot be disputed; then, considering the infinite complexity of the relations of all organic beings to each other and to their conditions of life, causing an infinite diversity in structure, constitution, and habits, to be advantageous to them, it would be a most extraordinary fact if no variations had ever occurred useful to each being's own welfare, in the same manner as so many variations have occurred useful to man. But if variations useful to any organic being ever do occur, assuredly individuals thus characterised will have the best chance of being preserved in the struggle for life; and from the strong principle of inheritance, these will tend to produce offspring similarly characterised. This principle of preservation, or the survival of the fittest, I have called Natural Selection. It leads to the improvement of each creature in relation to its organic and inorganic conditions of life; and consequently, in most cases, to what must be regarded as an advance in organisation. Nevertheless, low and simple forms will long endure if well fitted for their simple conditions of life.

Natural selection, on the principle of qualities being inherited at corresponding ages, can modify the egg, seed, or young, as easily as the adult. Amongst many animals, sexual selection will have given its aid to ordinary selection, by assuring to the most vigorous and best adapted males the greatest number of offspring. Sexual selection will also give characters useful to the males alone, in their struggles or rivalry with other males; and these characters will be transmitted to one sex or to both sexes, according to the form of inheritance which prevails.

Whether natural selection has really thus acted in adapting the various forms of life to their several conditions and stations, must be judged by the general tenor and balance of evidence given in the following chapters. But we have already seen how it entails extinction; and how largely extinction has acted in the world's history, geology plainly declares. Natural selection, also, leads to divergence of character; for the more organic beings diverge in structure, habits, and constitution, by so much the more can a large number be supported on the area,—of which we see proof by looking to the inhabitants of any small spot, and to the productions naturalised in foreign lands. Therefore, during the modification of the descendants of any one species, and during the incessant struggle of all species to increase in numbers, the more diversified the descendants become, the better will be their chance of success in the battle for life. Thus the small differences distinguishing varieties of the same species, steadily tend to increase, till they equal the greater differences between species of the same genus, or even of distinct genera.

We have seen that it is the common, the widely diffused and widely ranging species, belonging to the larger genera within each class, which vary most; and these tend to transmit to their modified offspring that superiority which now makes them dominant in their own countries. Natural selection, as has just been remarked, leads to divergence of character and to much extinction of the less improved and intermediate forms of life. On these principles, the nature of the affinities, and the generally well-defined distinctions between the innumerable organic beings in each class throughout the world, may be explained. It is a truly wonderful fact—the wonder of which we are apt to overlook from familiarity—that all animals and all plants throughout all time and space should be related to each other in groups, subordinate to groups, in the manner which we everywhere behold—namely, varieties of the same specie most closely related, species of the same genus less closely and unequally related, forming sections and sub-genera, species of distinct genera much less closely related, and genera related in different degrees, forming subfamilies, families, orders, sub-classes, and classes. The several subordinate groups in any class cannot be ranked in a single file, but seem clustered round points, and these round other points, and so on in almost endless cycles. If species had been independently created, no explanation would have been possible of this kind of classification; but it is explained through inheritance and the complex action of natural selection, entailing extinction and divergence of character, as we have seen illustrated in the diagram.

The affinities of all the beings of the same class have sometimes been represented by a great tree. I believe this simile largely speaks the truth. The green and budding twigs may represent existing species; and those produced during former years may represent the long succession of extinct species. At each period of growth all the growing twigs have tried to branch out on all sides, and to overtop and kill the surrounding twigs and branches, in the same manner as species and groups of species have at all times overmastered other species in the great battle for life. The limbs divided into great branches, and these into lesser and lesser branches, were themselves once, when the tree was young, budding twigs; and this connection of the former and present buds, by ramifying branches, may well represent the classification of all extinct and living species in groups subordinate to groups. Of the many twigs which flourished when the tree was a mere bush, only two or three, now grown into great branches, yet survive and bear the other branches; so with the species which lived during long-past geological periods, very few have left living and modified descendants. From the first growth of the tree, many a limb and branch has decayed and dropped off; and these fallen branches of various sizes may represent those whole orders, families, and genera which have now no living representatives, and which are known to us only in a fossil state. As we here and there see a thin straggling branch springing from a fork low down in a tree, and which by some chance has been favoured and is still alive on its summit, so we occasionally see an animal like the Ornithorhynchus or Lepidosiren, which in some small degree connects by its affinities two large branches of life, and which has apparently been saved from fatal competition by having inhabited a protected station. As buds give rise by growth to fresh buds, and these, if vigorous, branch out and overtop on all sides many a feebler branch, so by generation I believe it has been with the great Tree of Life, which fills with its dead and broken branches the crust of the earth, and covers the surface with its ever-branching and beautiful ramifications.

METHODS OF STUDY IN NATURAL HISTORY

LOUIS AGASSIZ

Louis Agassiz (1807-1873) became the greatest science teacher and public lecturer in the United States in the 19th century while continuing his fundamental research. The method of teaching that he developed at Harvard revolutionized the teaching of science in the whole country, and many students spread his influence from the Anderson School of Natural History that he founded and operated during the summer on an island in Buzzards Bay, Massachusetts.

Agassiz's greatest contributions in research were in the classification, first of fish, and then in zoological classification in general, and in the study of glaciers. It was he who postulated that much of Europe and North America had been subject to glaciation during prehistoric ice ages. Ironically, the work of Agassiz facilitated Darwin's development of the theory of natural selection, though Agassiz himself never accepted Darwin's theory of evolution.

Agassiz was born in a village on the shore of Lake Morat in the canton of Fribourg, Switzerland. His father was a Protestant minister. After attending the gymnasium in Bienne and the academy at Lausanne, he went on for university studies at Zurich and Heidelberg, received a doctor of philosophy degree from Erlangen, and doctor of medicine from Munich.

He went to Paris where he studied with George Cuvier, French paleontologist, and with Alexander Humbolt, a German scholar who was one of the founders of scientific geography. In 1832 Agassiz became professor of natural history at the University of Neuchâtel in Switzerland.

On the recommendation of British geologist Charles Lyell, Agassiz was invited to give a series of lectures in Boston in 1846. He then decided to remain in the United States permanently and in 1848 became professor of natural history at Harvard. His first wife had died in Europe, and in 1850 he married Elizabeth Cabot Cary of Boston, a well-known writer and promoter of women's education.

In 1858, Agassiz founded the Museum of Comparative Geology at Harvard and served as its curator during the last ten years of his life. Upon his death, his son succeeded him as curator.

1. Why do you think Louis Agassiz is regarded as one of the greatest teachers of science?

2. What is Agassiz's criticism of Aristotle's writings on natural history?

3. What is the importance of the classification of animals?

4. What were the contributions of Lamarck, Cuvier, and Baer to zoological studies? What was their significance?

METHODS OF STUDY IN NATURAL HISTORY

LOUIS AGASSIZ

Chapter I.
GENERAL SKETCH OF THE EARLY PROGRESS IN NATURAL HISTORY.

It is my intention, in this series of papers, to give the history of the progress in Natural History from the beginning,—to show how men first approached Nature,—how the facts of Natural History have been accumulated, and how these facts have been converted into science. In so doing, I shall present the methods followed in Natural History on a wider scale and with broader generalizations than if I limited myself to the study as it exists to-day. The history of humanity, in its efforts to understand the Creation, resembles the development of any individual mind engaged in the same direction. It has its infancy, with the first recognition of surrounding objects; and, indeed, the early observers seem to us like children in their first attempts to understand the world in which they live. But these efforts, that appear childish to us now, were the first steps in that field of knowledge which is so extensive that all our progress seems only to show us how much is left to do.

Aristotle is the representative of the learning of antiquity in Natural Science. The great mind of Greece in his day, and a leader in all the intellectual culture of his time, he was especially a naturalist, and his work on Natural History is a record not only of his own investigations, but of all preceding study in this department. It is evident that even then much had been done, and, in allusion to certain peculiarities of the human frame, which he does not describe in full, he refers his readers to familiar works, saying, that illustrations in point may be found in anatomical text-books.

Strange that in Aristotle's day, two thousand years ago, such books should have been in general use, and that in our time we are still in want of elementary text-books of Natural History, having special reference to the animals of our own country, and adapted to the use of schools. One fact in Aristotle's "History of Animals" is very striking, and makes it difficult for us to understand much of its contents. It never occurs to him that a time may come when the Greek language—the language of all culture and science in his time—would not be the language of all cultivated men. He took, therefore, little pains to characterize the animals he alludes to, otherwise than by their current names; and of his descriptions of their habits and peculiarities, much is lost upon us from their local character and expression. There is also a total absence of systematic form, of any classification or framework to express the divisions of the animal kingdom into larger or lesser groups. His only divisions are genera and species: classes, orders, and families, as we understand them now, are quite foreign to the Greek conception of the animal kingdom. Fishes and birds, for instance, they considered as genera, and their different representatives as species. They grouped together quadrupeds also, in contradistinction to animals with legs and wings, and they distinguished those that bring forth living young from those that lay eggs. But though a system of Nature was not familiar even to their great philosopher, and Aristotle had not arrived at the idea of a classification on general principles, he yet stimulated a search into the closer affinities among animals by the differences he pointed out. He divided the animal kingdom into two groups, which he called *Enaima* and *Anaima*, or animals with blood and animals without blood. We must remember, however, that by the word *blood* he designated only the red fluid circulating in the higher animals; whereas a fluid akin to blood exists in all animals, variously colored in some, but colorless in a large number of others.

After Aristotle, a long period elapsed without any addition to the information he left us. Rome and the Middle Ages gave us nothing, and even Pliny added hardly a fact to those that Aristotle recorded. And though the great naturalists of the sixteenth century gave a new impulse to this study, their investigations were chiefly directed towards a minute acquaintance with the animals they had an opportunity of observing, mingled with commentaries upon the ancients. Systematic Zoölogy was but little advanced by their efforts.

We must come down to the last century, to Linnæus, before we find the history taken up where Aristotle had left it, and some of his suggestions carried out with new freshness and vigor. Aristotle had already distinguished between genera and species; Linnæus took hold of this idea, and gave special names to other groups of different weight and value. Besides species and genera, he gives us orders and classes,—considering classes the most comprehensive, then orders, then genera, then species He did not, however, represent these groups as distinguished by their nature, but only by their range; they were still to him, as genera and species had been to Aristotle, only larger or smaller groups, not founded upon and limited by different categories of structure. He divided the animal kingdom into six classes, which I give here, as we shall have occasion to compare them with other classifications:—*Mammalia, Birds, Reptiles, Fishes, Insects,* and *Worms.*

That this classification should have expressed all that was known, in the last century, of the most general relations among animals, only shows how difficult it is to generalize on such a subject; nor should we expect to find it an easy task, when we remember the vast number of species (about a quarter of a million) already noticed by naturalists. Linnæus succeeded, however, in finding a common character on which to unite most of his classes; but his definition of the class of Mammalia, that group to which we ourselves belong, remained very imperfect. Indeed, in the earlier editions of his classification, he does not apply the name of Mammalia to this class, but calls the higher animals *Quadrupedia*, characterizing them as the animals with four legs and covered with fur or hair, that bring forth living young and nurse them with milk. In thus admitting external features as class characters, he excluded many animals which by their mode of reproduction, as well as by their respiration and circulation, belong to this class as much as the Quadrupeds,—as, for instance, all the Cetaceans (Whales, Porpoises, and the like), which, though they have not legs, nor are their bodies covered with hair or fur, yet bring forth living young, nurse them with milk, are warm-blooded and air-breathing. As more was learned of these animals, there arose serious discussion and criticism among contemporary naturalists respecting the classification of Linnæus, all of which led to a clearer insight into the true relations among animals. Linnæus himself, in his last edition of the "Systema Naturæ," shows us what important progress he had made since he first announced his views; for he there substitutes for the name of *Quadrupedia* that of *Mammalia*, including among them the Whales, which he characterizes as air-breathing, warm-blooded, and bringing forth living young which they nurse with milk. Thus the very deficiencies of his classification stimulated naturalists to new criticism and investigation into the true limits of classes, and led to the recognition of one most important principle,—that such groups are founded, not on external appearance, but on internal structure, and that internal structure, therefore, is the thing to be studied. The group of Quadrupeds was not the only defective one in this classification of Linnæus; his class of Worms, also, was most heterogeneous, for he included among them Shell-Fishes, Slugs, Star-Fishes, Sea-Urchins, Corals, and other animals that bear no relation whatever to the class of Worms as now defined.

But whatever its defects, the classification of Linnæus was the first attempt at grouping animals together according to certain common structural characters. His followers and pupils engaged at once in a scrutiny of the differences and similarities among animals, which soon led to a great increase in the number of classes; instead of six, there were presently nine, twelve, and more. But till Cuvier's time there was no great principle of classification. Facts were accumulated and more or less systematized, but they were not yet arranged according to law; the principle was still wanting by which to generalize them and give meaning and vitality to the whole. It was Cuvier who found the key. He himself tells us how he first began, in his investigations upon the internal organization of animals, to use his dissections with reference to finding the true relations between animals, and how ever after his knowledge of anatomy assisted him in his classifications, while his classifications threw new light again on his anatomical investigations,—each science thus helping to fertilize the other. He was not one of those superficial observers who are in haste to announce every new fact that they chance to find, and his first paper specially devoted to classification gave to the world the ripe fruit of years of study. This was followed by his great work, "Le Règne Animal." He said that animals were united in their most comprehensive groups, not on special characters, but on different *plans of structure*,—moulds, he called them, in which all animals had been cast. He tells us this in such admirable language, that I must, to do justice to his thought, give it in his own words:—

"Si l'on considère le règne animal d'après les principes que nous venons de poser en se débarrassant des préjugés établis sur les divisions anciennement admises, en n'ayant égard qu'à l'organisation et à la nature des animaux, et non pas à leur grandeur, à leur utilité, au plus ou moins de connaissance que nous en avons, ni à toutes les autres circonstances accessoires, on trouvera qu'il existe quatre formes principales, quatre plans généraux, si l'on peut s'exprimer ainsi, d'après lesquels tous les animaux semblent avoir été modelés, et dont les divisions ultérieures, de quelque titre que les naturalistes les aient décorées, ne sont que des modifications assez légères, fondées sur le développement ou l'addition de quelques parties, qui ne changent rien à l'essence du plan." [If we consider the animal kingdom according to the principles advanced above,—freeing ourselves at the same time from prejudices founded on previously established divisions, and looking at animals only with reference to their nature and organization, excluding their size, their utility, our greater or less familiarity with them, and all other accessory circumstances,—we shall find that there exist four principal forms, four general plans, if we may so express it, in accordance with which all animals seem to have been modelled, and the ulterior divisions of which, by whatever title naturalists may have dignified them, are only comparatively light modifications, founded on the development or the addition of some parts not affecting the essential elements of the plan.]

The value of this principle was soon tested by its application to facts already known, and it was found that animals whose affinities had been questionable before were now at once referred to their true relations with other animals by ascertaining whether they were build on one or another of these plans. Of such plans or structural conceptions Cuvier found in the whole animal kingdom only four, which he called *Vertebrates, Mollusks, Articulates*, and *Radiates.*

With this new principle as the basis of investigation, it was no longer enough for the naturalist to know a certain amount of feature characteristic of a certain number of animals,—he must penetrate deep enough into their organization to find the secret of their internal structure. Till he can do this, he is like the traveller in a strange city, who looks on the exterior of edifices entirely new to him, but knows nothing of the plan of their internal architecture. To be able to read in the finished structure the plan on which the whole is built is now essential to every naturalist.

Each of these plans may be stated in the most general terms. In the *Vertebrates* there is a vertebral column terminating in a prominent head; this column has an arch above and an arch below, forming a double internal cavity. The parts are symmetrically arranged on either side of the longitudinal axis of the body. In the *Mollusks*, also, the parts are arranged according to a bilateral symmetry on either side of the body, but the body has but one cavity, and is a soft, concentrated mass, without a distinct individualization of parts. In the *Articulates* there is but one cavity, and the parts are here again arranged on either side of the longitudinal axis, but in these animals the whole body is divided from end to end into transverse rings or joints movable upon each other. In the *Radiates* we lose sight of the bilateral symmetry so prevalent in the other three, except as a very subordinate element of structure; the plan of this lowest type is an organic sphere, in which all parts bear definite relations to a vertical axis.

It is not upon any special features, then, that these largest divisions of the animal kingdom are based, but simply upon the general structural idea. Striking as this statement was, it was coldly received at first by contemporary naturalists: they could hardly grasp Cuvier's wide generalizations, and perhaps there was also some jealousy of the grandeur of his views. Whatever the cause, his principle of classification was not fully appreciated; but it opened a new road for study, and gave us the key-note to the natural affinities among animals. Lamarck, his contemporary, not recognizing the truth of this principle, distributed the animal kingdom into two great divisions, which he calls *Vertebrates* and *Invertebrates*. Ehrenberg also, at a later period, announced another division under two head,—those with a continuous solid nervous centre, and those with merely scattered nervous swellings. But there was no real progress in either of these latter classifications, so far as the primary divisions are concerned; for they correspond to the old division of Aristotle, under the head of animals with or without blood, the *Enaima* and *Anaima*.

This coincidence between systems based on different foundations may teach us that every structural combination includes certain inherent necessities which will bring animals together on whatever set of features we try to classify them; so that the division of Aristotle, founded on the circulating fluids, or that of Lamarck, founded on the absence or presence of a backbone, or that of Ehrenberg, founded on the differences of the nervous system, covers the same ground. Lamarck attempted also to make the faculties of animals a basis for division among them. But our knowledge of the psychology of animals is still too imperfect to justify any such use of it. His divisions into Apathetic, Sensitive, and Intelligent animals are entirely theoretical. He places, for instance, Fishes and Reptiles among the Intelligent animals, as distinguished from Crustacea and Insects, which he refers to the second division. But one would be puzzled to say how the former manifest more intelligence than the latter, or why the latter should be placed among the Sensitive animals. Again, some of the animals that he calls Apathetic have been proved by later investigators to show an affection and care for their young, seemingly quite inconsistent with the epithet he has applied to them. In fact, we know so little of the faculties of animals that any classification based upon our present information about them must be very imperfect.

Many modifications of Cuvier's great divisions have been attempted; but though some improvements have been made in the details of his classification, all departures from its great fundamental principle are errors, and do but lead us away from the recognition of the true affinities among animals. Some naturalists, for instance, have divided off a part of the Radiate and Articulates, insisting upon some special feature of structure, and mistaking these for the more important and general characteristics of their respective plans. Subsequent investigations have shown these would-be improvements to be retrograde movements, only proving more clearly that Cuvier detected in his four plans all the great structural ideas on which the vast variety of animals is founded. This result is of greater importance than may at first appear. Upon it depends the question, whether all such classifications represent merely individual impressions and opinions of men, or whether there is really something in Nature that presses upon us certain divisions among animals, certain affinities, certain limitations, founded upon essential principles of organization. Are our systems the inventions of naturalists, or only their reading of the Book of Nature? and can that book have more than one reading? If these classifications are not mere inventions, if they are not an attempt to classify for our own convenience the objects we study, then they are thoughts which, whether we detect them or not, are expressed in Nature,—then Nature is the work of thought, the production of intelligence, carried out according to plan, therefore premeditated,—and in our study of natural objects we are approaching the thoughts of the Creator, reading his conceptions, interpreting a system that is his and not ours.

All the divergence from the simplicity and grandeur of the division of the animal kingdom first recognized by Cuvier arises from an inability to distinguish between the essential features of a plan and its various modes of execution. We allow the details to shut out the plan itself, which exists quite independent of special forms. I hope we shall find a meaning in all these plans that will prove them to be the parts of one great conception and the work of one Mind.

CHAPTER II.
NOMENCLATURE AND CLASSIFICATION.

PROCEEDING upon the view that there is a close analogy between the way in which every individual student penetrates into Nature and the progress of science as a whole in the history of humanity, I continue my sketch of the successive steps that have led to our present state of knowledge. I began with Aristotle, and showed that this great philosopher, though he prepared a digest of all the knowledge belonging to his time, yet did not feel the necessity of any system or of any scientific language differing from the common mode of expression of his day. He presents his information as a man with his eyes open narrates in a familiar style what he sees. As civilization spread and science had its representatives in other countries besides Greece, it became indispensable to have a common scientific language, a technical nomenclature, combining many objects under common names, and enabling every naturalist to express the results of his observations readily and simply in a manner intelligible to all other students of Natural History.

Linnæus devised such a system, and to him we owe a most simple and comprehensive scientific mode of designating animals and plants. It may at first seem no advantage to give up the common names of the vernacular and adopt the unfamiliar ones, but a word of explanation will make the object clear. Perceiving, for instance, the close relations between certain members of the larger groups, Linnæus gave to them names that should be common to all, and which are called generic names,—as we speak of Ducks, when we would designate in one word the Mallard, the Widgeon, the Canvas-Back, etc.; but to these generic names he added qualifying epithets, called specific names, to indicate the different kinds in each group. For example, the Lion, the Tiger, the Panther, the Domestic Cat constitute such a natural group, which Linnæus called *Felis*, Cat, indicating the whole genus; but the species he designates as *Felis catus*, the Domestic Cat,—*Felis leo*, the Lion,—*Felis tigris*, the Tiger,—*Felis panthera*, the Panther. So he called all the Dogs *Canis*; but for the different kinds we have *Canis familiaris*, the Domestic Dog,—*Canis* lupus, the Wolf—*Canis vulpes*, the Fox, etc.

In some families of the vegetable kingdom we can appreciate better application of this nomenclature, because we have something corresponding to it in the vernacular. We have, for instance, one name for all the Oaks, but we call the different kinds Swamp Oak, Red Oak, White Oak, Chestnut Oak, etc. So Linnæus in his botanical nomenclature, called all the Oaks by the generic name *Quercus*, (characterizing them by their fruit, the acorn, common to all,) and qualified them as *Quercus bicolor, Quercus rubra, Quercus alba, Quercus castanea*, etc., etc. His nomenclature, being so easy of application, became at once exceedingly popular, and made him the great scientific legislator of his century. He insisted on Latin names, because, if every naturalist should use his own language, it must lead to great confusion, and this Latin nomenclature of double significance was adopted by all. Another advantage of this binominal Latin nomenclature consists in preventing the confusion frequently arising from the use of the same name to designate different animals in different parts of the world,—as, for instance, the name of Robin, used in America to designate a bird of the Thrush family, which is entirely different form the Robin of the Old

World, one of the warblers,—or of different names for the same animal, as Perch or Chogset or Burgall for our Cunner. Nothing is more to be deprecated than an over-appreciation of technicalities, valuing the name more highly than the thing; but some knowledge of this scientific nomenclature is necessary to every student of Nature.

While Linnæus pointed out classes, orders, genera, and species, other naturalists had detected other divisions among animals, called families. Lamarck, who had been a distinguished botanist before he began his study of the animal kingdom, brought to his zoölogical researches his previous methods of investigation. Families in the vegetable kingdom had long been distinguished by French botanists; and one cannot examine the groups they call by this name, without perceiving, that, though they bring them together and describe them according to other characters, they have been unconsciously led to unite them from the general similarity of their port and bearing. Take, for instance, the families of Pines, Oaks, Beeches, Maples, etc., and you feel at once, that, besides the common characters given in the technical descriptions of these different groups of trees, there is also a general resemblance among them together, even if we knew nothing of the special features of their structure. By an instinctive recognition of this family likeness between plants, botanists have been led to seek for structural characters on which to unite them, and the groups so founded generally correspond with the combinations suggested by their appearance.

By a like process Lamarck combined animals into families. His method was adopted by French naturalists generally, and found favor especially with Cuvier, who was particularly successful in limiting families among animals, and in naming them happily, generally selecting names expressive of the features on which the groups were founded, or borrowing them from familiar animals. Much, indeed, depends upon the pleasant sound and the significance of a name; for an idea reaches the mind more easily when well expressed, and Cuvier's names were both simple and significant. His descriptions are also remarkable for their graphic precision,—giving all that is essential, omitting all that is merely accessory. He has given us the keynote to his progress in his own expressive language:—

"Je dus donc, et cette obligation me prit un temps considérable, je dus faire marcher de front l'anatomie et la zoologie, les dissections et le classement; chercher dans mes premières remarques sur l'organisation des distributions meilleures; m'en servir pour arriver à des remarques nouvelles; employer encore ces remarques à perfectionner les distributions; faire sortir enfin de cette fécondation mutuelle des deux sciences, l'une par l'autre, un système zoologique propre à servir d'introducteur et de guide dans le champ de l'anatomie, et un corps de doctrine anatomiue propre à servir de développement et d'explication au système zoologique."* [*"I therefore felt myself obliged, and this obligation cost me no little time, to make my studies in anatomy and zoölogy, dissection and classification, keep pace with each other; to seek in my earlier investigations upon organization a better distribution of groups; to employ these again as a means of perfecting my classification; to arrive, in short, by this mutual fecundation of the two sciences at a zoölogical system which might serve as a pioneer and guide in the field of anatomy, and an anatomical method which would aid in the development and explanation of the zoölogical system."]

It is deeply to be lamented that so many naturalists have entirely overlooked this significant advice of Cuvier's, with respect to combining zoölogical and anatomical studies in order to arrive at a clearer perception of the true affinities among animals. To sum it up in one word, he tells us that the secret of his method is "comparison,"—ever comparing and comparing throughout the enormous range of his knowledge of the organization of animals, and found upon the differences as well as the similarities those broad generalizations under which he has included all animal structure. And this method, so prolific in his hands, has also a lesson for us all. In this country there is a growing interest in the study of Nature; but while there exist hundreds of elementary works illustrating the native animals of Europe, there are few such books here to satisfy the demand for information respecting the animals of our land and water. We are thus forced to turn more and more to out own investigations and less to authority; and the true method of obtaining independent knowledge is this very method of Cuvier's,—comparison.

Let us make the most common application of it to natural objects. Suppose we see together a Dog, a Cat, A Bear, a Horse, a Cow, and a Deer. The first feature that strikes us as common to any two of them is the horn in the Cow and Deer. But how shall we associate either of the others with these? We examine the teeth, and find those of the Dog, the Cat, and the Bear sharp and cutting, while those of the Cow, the Deer, and the Horse have flat surfaces, adapted to grinding and chewing, rather than cutting and tearing. We compare these features of their structure with the habits of these animals, and find that the first are carnivorous, that they seize and tear their prey, while the others are herbivorous or grazing animals, living only on vegetable substances, which they chew and grind. We compare further the Horse and Cow, and find that the Horse has front teeth both in the upper and lower jaw, while the Cow has them only in the lower; and going still further, and comparing the internal with the external features, we find this arrangement of the teeth in direct relation to the different structure of the stomach in the two animals,—the Cow having a stomach with four pouches, adapted to a mode of digestion by which the food is prepared for the second mastication, while the Horse has a simple stomach. Comparing the Cow and the Deer, we find that the digestive apparatus is the same in both; but though they both have horns, in the Cow the horn is hollow, and remains through life firmly attached to the bone, while in the Deer it is solid and is shed every year. With these facts before us, we cannot hesitate to place the Dog, the Cat, and the Bear in one division, as carnivorous animals, and the other three in another division as herbivorous animals,—and looking a little further, we perceive, that, in common with the Cow and the Deer, the Goat and the Sheep have cloven feet, and that they are all ruminants, while the Horse has a single hoof, does not ruminate, and must therefore be separated from them, even though, like them, he is herbivorous.

This is but the simplest illustration taken from the most familiar objects, of this comparative method; but the same process is equally applicable to the most intricate problems in animal structures, and will give us the clew to all true affinities between animals. The education of a naturalist now consists chiefly in learning how to compare. If he have any power of generalization, when he has collected his facts, this habit of mental comparison will lead him up to principles, and to the great laws of combination. It must not discourage us, that the process is a slow and laborious one, and the results of one lifetime after all very small. It might seem invidious, were I to show here how small is the sum total of the work accomplished even by the great exceptional men, whose names are known throughout the civilized world. But I may at least be permitted to speak disparagingly of my own efforts, and to sum up in the fewest words the result of my life's work. I have devoted my whole life to the study of Nature, and yet a single sentence may express all that I have done. I have shown that there is a correspondence between the succession of Fishes in geological times and the different stages of their growth in the egg,—this is all. It chanced to be a result that was found to apply to other groups and has led to other conclusions of a like nature. But, such as it is, it has been reached by this system of comparison, which, though I speak of it now in its application to the study of Natural History, is equally important in every other branch of knowledge. By the same process the most mature results of scientific research in Philology, in Ethnology, and in Physical Science are reached. And let me say that the community should foster the purely intellectual efforts of scientific men as carefully as they do their elementary schools and their practical institutions, generally considered so much more useful and important to the public. For from what other source shall we derive the higher results that are gradually woven into the practical resources of our life, except from the researches of those very men who study science, not for its uses, but for its truth? It is this that gives it its noblest interest: it must be for truth's sake, and not even for the sake of its usefulness to humanity, that the scientific man studies Nature. The application of science to the useful arts requires other abilities, other qualities, other tools than his; and therefore I say that the man of science who follows his studies into their practical application is false to his calling. The practical man stands ever ready to take up the work where the scientific man leaves it, and to adapt it to the material wants and uses of daily life.

The publication of Cuvier's proposition, that the animal kingdom is built on four plans, however imperfectly understood and appreciated at first, created, nevertheless, an extraordinary excitement throughout the scientific world. All naturalists proceeded to test it, and some among them soon recognized in it a great scientific truth,—while others, who thought more of making themselves prominent than of advancing science, proposed poor amendments, that were sure to be rejected on further investigations. Some of these criticisms and additions, however, were truly improvements, and touched upon points overlooked by Cuvier. Blainville, especially, took up the element of form among animals,—whether divided on two sides, whether radiated, whether irregular, etc. He, however, made the mistake of giving very elaborate names to animals already known under simpler ones. Why, for instance, call all animals with parts radiating in every direction *Actinomorpha* or *Actinozoaria*, when they had received the significant name of *Radiates*? It seemed to be a new system, when in fact it was only a new name. Ehrenberg, likewise, made an important distinction, when he united the animals according to the difference in their nervous systems; but he also encumbered the nomenclature unnecessarily, when he added to the names *Anaima* and *Enaima* of Aristotle those of *Myeloneura* and *Ganglioneura*.

But it is not my object to give all the classifications of different authors here, and I will therefore pass over many noted ones, as those of Burmeister, Milne-

Edwards, Siebold and Stannius, Owen, Leuckart, Vogt, Van Beneden, and others, and proceed to give some account of one investigator who did as much for the progress of Zoölogy as Cuvier, though he is comparatively little known among us.

Karl Ernst von Baer proposed a classification based, like Cuvier's, upon plan; but he recognized what Cuvier failed to perceive,—namely, the importance of distinguishing between type (by which he means exactly what Cuvier means by plan) and complication of structure,—in other words, between plan and the execution of the plan. He recognized four types, which correspond exactly to Cuvier's four plans, though he calls them by different names. Let us compare them.

Cuvier.	*Baer.*
Radiates,	Peripheric,
Mollusks,	Massive,
Articulates,	Longitudinal,
Vertebrates.	Doubly Symmetrical.

Though perhaps less felicitous, the names of Baer express the same ideas as those of Cuvier. By the *Peripheric* type he signified those animals in which all parts converge from the periphery or circumference of the animal to its centre. Cuvier only reverses this definition in this name of *Radiates*, signifying the animals in which all parts radiate from the centre to the circumference. By *Massive*, Baer indicated those animals in which the body is undivided, soft and concentrated, without a very distinct individualization of parts,—exactly the animals included by Cuvier under his name of *Mollusks*, or soft-bodied animals. In his selection of the epithet *Longitudinal*, Baer was less fortunate; for all animals have a longitudinal diameter, and this word was not, therefore, sufficiently special. Yet his *Longitudinal* type answers exactly to Cuvier's *Articulates*,—animals in which all parts are arranged in a succession of articulated joints along a longitudinal axis. Cuvier has expressed this jointed structure in the name *Articulates*; whereas Baer, in his name of *Longitudinal*, referred only to the arrangement of joints in longitudinal succession, in a continuous string, as it were, one after another, indicating thus the prevalence of length as the predominant diameter of the body. For the *Doubly Symmetrical* type his name is the better of the two; since Cuvier's name of *Vertebrates* alludes only to the backbone,—while Baer, who is an embryologist, signified in his their mode of growth also. He knew what Cuvier did not know, when he first proposed his classification, that in its first formation the germ of the Vertebrate divides in two folds; one turning up above the backbone, to form and enclose all the sensitive organs,—the spinal marrow, the organs of sense, all those organs by which life is expressed; the other turning down below the backbone, and enclosing all those organs by which life is maintains,—the organs of digestion, of respiration, or circulation, of reproduction, etc. So there is in this type not only an equal division of parts on either side, but also a division above and below, making thus a double symmetry in the plan, expressed by Baer in the name he gave it. Baer was perfectly original in his conception of these four types, for his paper was published in the very same year with that of Cuvier. But even in Germany, his native land, his ideas were not fully appreciated: strange that it should be so,—for, had his countrymen recognized his genius, they might have earlier claimed him as the compeer of the great French naturalist.

Baer also founded the science of Embryology, under the guidance of his teacher, Döllinger. His researches in this direction showed him that animals were not only built on four plans, but that they grew according to four modes of development. The Vertebrate arises from the egg differently from the Articulate,—the Articulate differently from the Mollusk,—the Mollusk differently from the Radiate. Cuvier only showed us the four plans as they exist in the adult; Baer went a step further, and showed us the four plans in the process of formation.

But his greatest scientific achievement is perhaps the discovery that all animals originate form eggs, and that all these eggs are at first identical in substance and structure. The wonderful and untiring research condensed into this simple statement, that all animals arise from eggs, and that all those eggs are identical in the beginning, may well excite our admiration. This egg consists of an outer envelope, the vitelline membrane, containing a fluid more or less dense, and variously colored, the yolk; within this is a second envelope, the so-called germinative vesicle, containing a somewhat different and more transparent fluid, and in the fluid of this second envelope float one or more so-called germinative specks. At this stage of their growth all eggs are microscopically small, yet each one has such tenacity of its individual principle of life that no egg was ever known to swerve from the pattern of the parent animals that gave it birth.

* * * * *

THE SECOND LAW OF THERMODYNAMICS

CLAUSIUS

Rudolph Julius Emmanuel Clausius was born on January 2, 1822, at Koslin in Pomerania. He studied for four years in the University of Berlin and was graduated at Halle in 1848. In 1850 he became professor of physics at the Royal Artillery and Engineering School in Berlin. He then passed successively as professor of physics to Zurich, Warzburg, and Bonn. He was professor of physics at Bonn from 1869 until his death, which occurred on August 24, 1888.

Clausius was one of the founders of the kinetic theory of gases and of the science of thermodynamics. He and Lord Kelvin at about the same time and independently announced the Second Law of thermodynamics. Clausius particularly developed the theory of thermodynamics by applying it to the study of gases and vapors.

The first extract which follows is taken from a paper entitled "Ueber die bewegende Kraft der Wärme," published in the *Annalen der Pbysik und Cbemie* Vol. 79, pp. 368 and 500, 1850. The translation by W. F. Magie appears in Harper's *Scientific Memoirs.*

The function which was introduced into thermodynamics by Clausius under the name of entropy had already been used and some of its properties studied by Rankine and Lord Kelvin, but owing to his giving it its name and to his development of its properties it is commonly ascribed to him. It is not necessary to present the thermodynamic and mathematic arguments by which Clausius arrived at the equation with which the extract opens. The extract as given deals with the application of entropy to the universe. The paper from which it is taken appeared in the *Annalen der Physik und Cbemie*, Vol. 125, p. 353, 1865, under the title "Ueber verschiedene für die Anwendung bequeme Formen der Hauptgleichungen der mechanischen Wärmetheorie."

1. What is the second law of thermodynamics? How did Clausius arrive at it?

2. What was the major contribution of Carnot to the thinking of Clausius?

3. What does Clausius mean by "entropy"?

4. What is Clausius' model of the universe?

THE SECOND LAW OF THERMODYNAMICS

CLAUSIUS

Since heat was first used as a motive power in the steam-engine, thereby suggesting from practice that a certain quantity of work may be treated as equivalent to the heat needed to produce it, it was natural to assume also in theory a definite relation between a quantity of heat and the work which in any possible way can be produced by it, and to use this relation in drawing conclusions about the nature and the laws of heat itself. In fact, several fruitful investigations of this sort have already been made; yet I think that the subject is not yet exhausted, but on the other hand deserves the earnest attention of physicists, partly because serious objections can be raised to the conclusions that have already been reached, partly because other conclusions, which may readily be drawn and which will essentially contribute to the establishment and completion of the theory of heat, still remain entirely unnoticed or have not yet been stated with sufficient definiteness.

The most important of the researches here referred to was that of S. Carnot, and the ideas of this author were afterwards given analytical form in a very skilful way by Clapeyron. Carnot showed that whenever work is done by heat and no permanent change occurs in the condition of the working body, a certain quantity of heat passes from a hotter to a colder body. In the steam-engine, for example, by means of the steam which is developed in the boiler and precipitated in the condenser, heat is transferred from the grate to the condenser. This transfer he considered as the heat change, corresponding to the work done. He says expressly that no heat is lost in the process, but that the quantity of heat remains unchanged, and adds: "This fact is not doubted; it was assumed at first without investigation, and then established in many cases by calorimetric measurements. To deny it would overthrow the whole theory of heat, of which it is the foundation." I am not aware, however, that it has been sufficiently proved by experiment that no loss of heat occurs when work is done; it may, perhaps, on the contrary, be asserted with more correctness that even if such a loss has not been proved directly, it has yet been shown by other facts to be not only admissible, but even highly probable. If it be assumed that heat, like a substance, can not diminish in quantity, it must also be assumed that it can not increase. It is, however, almost impossible to explain the heat produced by friction except as an increase in the quantity of heat. The careful investigations of Joule, in which heat is produced in several different ways by the application of mechanical work, have almost certainly proved not only the possibility of increasing the quantity of heat in any circumstances but also the law that the quantity of heat developed is proportional to the work expended in the operation. To this it must be added that other facts have lately become known which support the view, that heat is not a substance, but consists in a motion of the least parts of bodies. If this view is correct, it is admissible to apply to heat the general mechanical principle that a motion may be transformed into work, and in such a manner that the loss of *vis viva is* proportional to the work accomplished.

These facts, with which Carnot also was well acquainted and the importance of which he has expressly recognized, almost compel us to accept the equivalence between heat and work, on the modified hypothesis that the accomplishment of work; requires not merely a change in the distribution of heat, but also an actual consumption of heat, and that, conversely, heat can be developed again by the expenditure of work.

* * * * *

If any body changes its volume, mechanical work will in general be either produced or expended. It is, however, in most cases impossible to determine this exactly, since besides the external work there is generally an unknown amount of internal work done. To avoid this difficulty, Carnot employed the ingenious method already referred to of allowing the body to undergo its various changes in succession, which are so arranged that it returns at last exactly to its original condition. In this case, If internal work is done in some of the changes, it is exactly compensated for in the others, and we may be sure that the external work, which remains over after the changes are completed, is all the work that has been done. Clapeyron has represented this process graphically in a very clear way, and we shall follow his presentation now for the permanent gases, with a slight alteration rendered necessary by our principle.

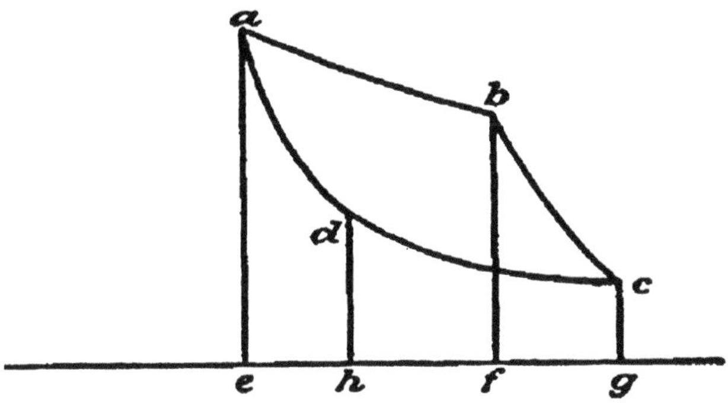

Figure 43

In the figure (Fig. 43), let the abscissa *oe* represent the volume and the ordinate *ea* the pressure on a unit weight of gas, in a condition in which its temperature = *t*. We assume that the gas is contained in an extensible envelope, which, however, cannot exchange heat with it. If, now, it is allowed to expand in this envelope, its temperature would fall if no heat were imparted to it. To avoid this, let it be put in contact, during its expansion, with a body *A*, which is kept at the constant temperature *t*, and which imparts just so much heat to the gas that its temperature also remains equal to *t*. During this expansion at constant temperature, its pressure diminishes according to Mariotte's law, and may be represented by the ordinate of a curve, *ab*, which is a portion of an equilateral hyperbola. When the volume of the gas has increased in this way from *oe* to *of*, the body *A* is removed, and the expansion is allowed to continue without the introduction of more heat. The temperature will then fall, and the pressure diminish more rapidly than before. The law which is followed in this part of the process may be represented by the curve *be*. When the volume of the gas has increased in this way from *of* to *og*, and its temperature has fallen from t to τ, we begin to compress it, in order to restore it again to its original volume *oe*. If it were left to itself its temperature would again rise. This, however, we do not permit, but bring it in contact with a body *B*, at the constant temperature τ, to which it at once gives up the heat that is produced, so that it keeps the temperature τ; and while it is in contact with this body we compress it so far (by the amount *gb)* that the remaining compression *be* is exactly sufficient to raise its temperature from τ to t, if during this last compression it gives up no heat. During the former compression the pressure increases according to Mariotte's law, and is represented by the portion *cd* of an equilateral hyperbola. During the latter, on the other hand, the increase is more rapid and is represented by the curve *da*. This curve must end exactly at *a*, for since at the end of the operation the volume and temperature have again their original values, the same must be true of the pressure also, which is a function of them both. The gas is therefore in the same condition again as it was at the beginning.

Now, to determine the work produced by these changes, for the reasons already given, we need to direct our attention only to the external work. During the expansion the gas does work, which is determined by the integral of the product of the differential of the volume into the corresponding pressure, and is therefore represented geometrically by the quadrilaterals *eabf* and *fbcg*. During the compression, on the other hand, work is expended, which is represented similarly by the quadrilaterals *gcdb* and *bdae*. The excess of the former quantity of work over the latter is to be looked on as the whole work produced during the changes, and this is represented by the quadrilateral *abcd*.

If the process above described is carried out in the reverse order, the same magnitude, *abcd, is* obtained as the excess of the work expended over the work done.

* * * * *

Carnot assumed, as has already been mentioned, that *the equivalent of the work done by beat is found in the mere transfer of beat from a hotter to a colder body, while the quantity of heat remains undiminished.*

The latter part of this assumption—namely, that the quantity of heat remains undiminished—contradicts our former principle, and must therefore be rejected if we are to retain that principle. On the other hand, the first part may still obtain in all its essentials. For though we do not need a special equivalent for the work done, since we have assumed as such an actual consumption of heat, it still may well be possible that such a transfer of heat occurs at the same time as the consumption of heat, and also stands in a definite relation to the work done. It becomes important, therefore, to consider whether this assumption, besides the mere possibility, has also a sufficient probability in its favor.

A transfer of heat from a hotter to a colder body always occurs in those cases in which work is done by heat, and in which also the condition is fulfilled that the working substance is in the same state at the end as at the beginning of the operation. . . . Yet, in order to establish a relation between the heat transferred and the work done, a certain restriction is necessary. For since a transfer of heat can take place without mechanical effect if a hotter and a colder body are immediately in contact and heat passes from one to the other by conduction, the way in which the transfer of a certain quantity of heat between two bodies at the temperatures t and τ can be made to do the maximum of work is so to carry out the process, as was done in the above cases, that two bodies of different temperatures never come in contact.

It is this maximum of work which must be compared with the heat transferred. When this is done it appears that there is in fact ground for asserting, with Carnot, that it depends only on the quantity of the heat transferred and on the temperatures t and τ of the two bodies A and B, but not on the nature of the substance by means of which the work is done This maximum has, namely, the property that by expending it as great a quantity of heat can be transferred from the cold body B to the hot body A as passes from A to B when it is produced. This may easily be seen, if we think of the whole process formerly described as carried out in the reverse order, so that, for example, in the first case the gas first expands by itself, until its temperature falls from t to τ, is then expanded in connection with B, is then compressed by itself until its temperature is again t, and finally is compressed in connection with A. In this case more work will be employed during the compression than is produced during the expansion, so that on the whole there is a loss of work, which is exactly as great as the gain of work in the former process. Further, there will be just as much heat taken from the body B as was before given to it, and just as much given to the body A as was before taken from it, whence it follows not only that the same amount of heat is produced as was formerly consumed, but also that the heat which in the former process was transferred from A to B now passes from B to A.

If we now suppose that there are two substances of which the one can produce more work than the other by the transfer of a given amount of heat, or, what comes to the same thing, needs to transfer less heat from A to B to produce a given quantity of work, we may use these two substances alternately by producing work with one of them in the above process, and by expending work upon the other in the reverse process. At the end of the operations both bodies are in their original condition; further, the work produced will have exactly counterbalanced the work done, and therefore, by our former principle, the quantity of heat can have neither increased nor diminished. The only change will occur in the distribution of the heat, since more heat will be transferred from B to A than from A to B, and so on the whole heat will be transferred from B to A. By repeating these two processes alternately it would be possible, without any expenditure of force or any other change, to transfer as much heat as we please from a cold to a hot body, and this is not in accord with the other relations of heat, since it always shows a tendency to equalize temperature differences and therefore to pass from hotter to colder bodies.

It seems, therefore, to be theoretically admissible to retain the first and the really essential part of Carnot' s assumptions, and to apply it as a second principle in conjunction with the first; and the correctness of this method is, as we shall soon see, established already in many cases by its consequences.

ENTROPY

[The symbols in the first equation represent quality of heat (Q), absolute temperature (T), and entropy (S).]

We obtain the equation

$$\int \frac{dQ}{T} = S - S_0$$

which, while somewhat differently arranged, is the same as that which was formerly used to determine S.

If we wish to designate S by a proper name we can say of it that it is the *transformation content* of the body, in the same way that we say of the quantity U that it is the *heat and work content* of the body. However, since I think it is better to take the names of such quantities as these, which are important for science, from the ancient languages, so that they can be introduced without change into all the modern languages, I propose to name the magnitude the *entropy* of the body, from the Greek word ἡ τροπή, a transformation. I have intentionally formed the word *entropy so* as to be as similar as possible to the word *energy*, since both these quantities which are to be known by these names, are so nearly related to each other in their physical significance that a certain similarity in their names seemed to me advantageous.

* * * * *

Finally I may allow myself to touch on a matter whose complete treatment would not be in place here, because the statements necessary for that purpose would take up too much room, but of which I believe that even the following short indication will nol be without interest, in that it will contribute to the recognition of the importance of the quantities which I have introduced into the formulation of the second law of the mechanical theory of heat

The second law, in the form which I have given it, states the fact that all transformations which occur in nature occur in a certain sense which I have taken as positive, of themselves, that is, without compensation, but that they can only occur in the opposite or negative sense in such a way that they are compensated by positive transformations which occur at the same time. The application of this law to the universe leads to a conclusion to which W. Thomson first called attention and about which I have already spoken in a recently published paper. This conclusion is that if among all the changes of state which occur in the universe the transformations in one sense exceed in magnitude those in the opposite sense, then the general condition of the universe will change more and more in the former sense, and the universe will thus persistently approach a final state.

The question now arises whether this final state can be characterized in a simple and also a definite way. This can be done by treating the transformations, as I have done, as mathematical quantities, whose equivalent values can be calculated and united in a sum by algebraic addition.

In my papers so far published I have carried out such calculations with respect to the heat present in bodies and to the arrangement of the constituents of the bodies. For each body there are found two quantities, the transformation value of its heat content and its disgregration, the sum of which is its entropy. This however does not complete the business. The discussion must also be extended to the radiant heat, or otherwise expressed, to the heat transmitted through the universe in the form of advancing vibrations of the ether, and also to such motions as cannot be comprehended under the name heat.

The treatment of these latter motions, at least as far as they are the motions of ponderable masses, can be briefly settled, since we come by a simple argument to the following conclusion: If a mass, which is so great that in comparison with it an atom may be considered as vanishingly small, moves as a whole, the transformation value of this motion is to be looked on as vanishingly small in the same way in comparison with its kinetic energy; from which it follows that if such a motion is transformed into heat by a passive resistance, then the equivalent value of the uncompensated transformation which then occurs is simply represented by the transformation value of the heat produced. The radiant heat, however, cannot be treated so briefly, since there is need still of a certain special treatment in order to find out how its transformation value is to be determined. Although, in the paper which was recently published and to which I have previously referred, I have already discussed radiant heat in its connection with the mechanical theory of heat, yet I have not as yet treated the question which has here come up, since it was then only my purpose to prove that there was no contradiction between the laws of radiant heat and a fundamental law which I assumed in the mechanical theory of heat. I reserve for future consideration the more particular application of the mechanical theory of heat and especially of the law of equivalents of transformation to radiant heat.

For the present I will confine myself to announcing as a result of my argument that if we think of that quantity which with reference to a single body I have called its entropy, as formed in a consistent way, with consideration of all the circumstances, for the whole universe, and if we use in connection with it the other simpler concept of energy, we can express the fundamental laws of the universe which correspond to the two fundamental laws of the mechanical theory of heat in the following simple form.

1. The energy of the universe is constant.
2. The entropy of the universe tends toward a maximum.

Look for knowledge not in books but in things themselves.

William Gilbert, *De Magnete*

Science is the creation of scientists, and every scientific advance bears somehow the mark of the man who made it. The artist exposes himself in his work; the scientist seems rather to hide in his, but he is there.

Anne Roe, The Psychology of the Scientist, 1961

The brightest flashes in the world of thought are incomplete until they have been proved to have their counterparts in the world of fact.

John Tyndall, *Fragments of Science*

Entia non sunt multiplicanda praeter necessitatem.
It is vain to do more with what can be done with less.

William of Occam

Errors using inadequate data are much less than those using no data at all.

Charles Babbage

No amount of experimentation can ever prove me right; a single experiment can prove me wrong.

Albert Einstein

Dans le champs de l'observation, l'hasard ne favorise que las esprits préparés.
In the field of observation, luck favors only those minds which have been prepared.

Louis Pasteur, *Encyclopaedia Britannica*, 1911

THE GERM THEORY AND ITS APPLICATIONS TO MEDICINE AND SURGERY

LOUIS PASTEUR

Louis Pasteur was born at Dôle, Jura, France, December 27, 1822, and died near Saint-Cloud, September 28, 1895. His interest in science, and especially in chemistry, developed early, and by the time he was twenty-six he was professor of the physical sciences at Dijon. The most important academic positions held by him later were those as professor of chemistry at Strasburg, 1849; dean of the Faculty of Sciences at Lille, 1854; science director of the École Normale Supérieure, Paris, 1857; professor of geology, physics, and chemistry at the École des Beaux Arts; professor of chemistry at the Sorbonne, 1867. After 1875 he carried on his researches at the Pasteur Institute. He was a member of the Institute, and received many honors from learned societies at home and abroad.

In respect of the number and importance, practical as well as scientific, of his discoveries, Pasteur has hardly a rival in the history of science. He may be regarded as the founder of modern stereo-chemistry; and his discovery that living organisms are the cause of fermentation is the basis of the whole modern germ-theory of disease and of the antiseptic method of treatment. His investigations of the diseases of beer and wine; of pébrine, a disease affecting silk-worms; of anthrax, and of fowl cholera, were of immense commercial importance and led to conclusions which have revolutionized physiology, pathology, and therapeutics. By his studies in the culture of bacteria of attenuated virulence he extended widely the practice of inoculation with a milder form of various diseases, with a view to producing immunity.

The following paper presents some of the most important of his contributions, and exemplify his extraordinary powers of lucid exposition and argument.

1. How did Pasteur determine that bacteria—or germs or microbes—were the cause of various diseases?

2. What was the significance of Pasteur's discoveries for surgery? For the treatment of illness?

3. What is "pasteurization"?

THE GERM THEORY AND ITS APPLICATIONS TO MEDICINE AND SURGERY

LOUIS PASTEUR

THE Sciences gain by mutual support. When, as the result of my first communications on the fermentations in 1857-1858, it appeared that the ferments, properly so-called, are living beings, that the germs of microscopic organisms abound in the surface of all objects, in the air and in water; that the theory of spontaneous generation is chimerical; that wines, beer, vinegar, the blood, urine and all the fluids of the body undergo none of their usual changes in pure air, both Medicine and Surgery received fresh stimulation. A French physician, Dr. Davaine, was fortunate in making the first application of these principles to Medicine, in 1863.

Our researches of last year, left the etiology of the putrid disease, or septicemia, in a much less advanced condition than that of anthrax. We had demonstrated the probability that septicemia depends upon the presence and growth of a microscopic body, but the absolute proof of this important conclusion was not reached. To demonstrate experimentally that a microscopic organism actually is the cause of a disease and the agent of contagion, I know no other way, in the present state of Science, than to subject the *microbe* (the new and happy term introduced by M. Sedillot) to the method of cultivation out of the body. It may be noted that in twelve successive cultures, each one of only ten cubic centimeters volume, the original drop will be diluted as if placed in a volume of fluid equal to the total volume of the earth. It is just this form of test to which M. Joubert and I subjected the anthrax bacteridium. Having cultivated it a great number of times in a sterile fluid, each culture being started with a minute drop from the preceding, we then demonstrated that the product of the last culture was capable of further development and of acting in the animal tissues by producing anthrax with all its symptoms. Such is—as we believe—the indisputable proof that *anthrax is a bacterial disease.*

Our researches concerning the septic vibrio had not so far been convincing, and it was to fill up this gap that we resumed our experiments. To this end, we attempted the cultivation of the septic vibrio from an animal dead of septicemia. It is worth noting that all of our first experiments failed, despite the variety of culture media we employed—urine, beer yeast water, meat water, etc. Our culture media were not sterile, but we found—most commonly—a microscopic organism showing no relationship to the septic vibrio, and presenting the form, common enough elsewhere, of chains of extremely minute spherical granules possessed of no virulence whatever. This was an impurity, introduced, unknown to us, at the same time as the septic vibrio; and the germ undoubtedly passed from the intestines—always inflamed and distended in septicemic animals—into the abdominal fluids from which we took our original cultures of the septic vibrio. If this explanation of the contamination of our cultures was correct, we ought to find a pure culture of the septic vibrio in the heart's blood of an animal recently dead of septicemia. This was what happened, but a new difficulty presented itself; all our cultures remained sterile. Furthermore this sterility was accompanied by loss in the culture media of (the original) virulence.

It occurred to us that the septic vibrio might be an obligatory anaerobe and that the sterility of our inoculated culture fluids might be due to the destruction of the septic vibrio by the atmospheric oxygen dissolved in the fluids. The Academy may remember that I have previously demonstrated facts of this nature in regard to the vibrio of butyric fermentation, which not only lives without air but is killed by the air.

It was necessary therefore to attempt to cultivate the septic vibrio either in a vacuum or in the presence of inert gases—such as carbonic acid.

Results justified our attempt; the septic vibrio grew easily in a complete vacuum, and no less easily in the presence of pure carbonic acid.

These results have a necessary corollary. If a fluid containing septic vibrios be exposed to pure air, the vibrios should be killed and all virulence should disappear. This is actually the case. If some drops of septic serum be spread horizontally in a tube and in a very thin layer, the fluid will become absolutely harmless in less than half a day, even if at first it was so virulent as to produce death upon the inoculation of the smallest portion of a drop.

Furthermore all the vibrios, which crowded the liquid as motile threads, are destroyed and disappear. After the action of the air, only fine amorphous granules can be found, unfit for culture as well as for the transmission of any disease whatever. It might be said that the air burned the vibrios.

If it is a terrifying thought that life is at the mercy of the multiplication of these minute bodies, it is a consoling hope that Science will not always remain powerless before such enemies, since for example at the very beginning of the study we find that simple exposure to air is sufficient at times to destroy them.

But, if oxygen destroys the vibrios, how can septicemia exist, since atmospheric air is present everywhere? How can such facts be brought in accord with the germ theory? How can blood, exposed to air, become septic through the dust the air contains?

All things are hidden, obscure and debatable if the cause of the phenomena be unknown, but everything is clear if this cause be known. What we have just said is true only of a septic fluid containing adult vibrios, in active development by fission: conditions are different when the vibrios are transformed into their germs, that is into the glistening corpuscles first described and figured in my studies on silk-worm disease, in dealing with worms dead of the disease called "flachérie." Only the adult vibrios disappear, burn up, and lose their virulence in contact with air: the germ corpuscles, under these conditions, remain always ready for new cultures, and for new inoculations.

All this however does not do away with the difficulty of understanding how septic germs can exist on the surface of objects, floating in the air and in water.

Where can these corpuscles originate? Nothing is easier than the production of these germs, in spite of the presence of air in contact with septic fluids.

If abdominal serous exudate containing septic vibrios actively growing by fission be exposed to the air, as we suggested above, but with the precaution of giving a substantial thickness to the layer, even if only one centimeter be used, this curious phenomenon will appear in a few hours. The oxygen is absorbed in the upper layers of the fluid—as is indicated by the change of color. Here the vibrios are dead and disappear. In the deeper layers, on the other hand, towards the bottom of this centimeter of septic fluid we suppose to be under observation, the vibrios continue to multiply by fission— protected from the action of oxygen by those that have perished above them: little by little they pass over to the condition of germ corpuscles with the gradual disappearance of the thread forms. So that instead of moving threads of varying length, sometimes greater than the field of the microscope, there is to be seen only a number of glittering points, lying free or surrounded by a scarcely perceptible amorphous mass. Thus is formed, containing the latent germ life, no longer in danger from the destructive action of oxygen, thus, I repeat, is formed the septic dust, and we are able to understand what has before seemed so obscure; we can see how putrescible fluids can be inoculated by the dust of the air, and how it is that putrid diseases are permanent in the world.

The Academy will permit me, before leaving these interesting results, to refer to one of their main theoretical consequences. At the very beginning of these researches, for they reveal an entirely new field, what must be insistently demanded? The absolute proof that there actually exist transmissible, contagious, infectious diseases of which the cause lies essentially and solely in the presence of microscopic organisms. The proof that for at least some diseases, the conception of spontaneous virulence must be forever abandoned—as well as the idea of contagion and an infectious element suddenly originating in the bodies of men or animals and able to originate diseases which propagate themselves under identical forms: and all of those opinions fatal to medical progress, which have given rise to the gratuitous hypotheses of spontaneous generation, of albuminoid ferments, of hemiorganisms, of archebiosis, and many other conceptions without the least basis in observation. What is to be sought for in this instance is the proof that along with our vibrio there does not exist an independent virulence belonging to the surrounding fluids or solids, in short that the vibrio is not merely an epiphenomenon of the disease of which it is the obligatory accompaniment. What then do we see, in the results that I have just brought out? A septic fluid, taken at the moment that the vibrios are not yet changed into germs, loses its virulence completely upon simple exposure to the air, but preserves this virulence, although exposed to air on the simple condition of being in a thick layer for some hours. In the first case, the virulence once lost by exposure to air, the liquid is incapable of taking it on again upon cultivation: but, in the second case, it preserves its virulence and can propagate, even after exposure to air. It is impossible, then, to assert that there is a separate virulent substance, either fluid or solid, existing, apart from the adult vibrio or its germ. Nor can it be supposed that there is a virus which loses its virulence at the moment that the adult vibrio dies; for such a substance should also lose its virulence when the vibrios, changed to germs, are exposed to the air. Since the virulence persists under these conditions it can only be due to the germ corpuscles—the only thing present. There is only one possible hypothesis as to the existence of a virus in solution, and that is that such a substance, which was present in our experiment in non-fatal amounts, should be continuously furnished by the vibrio itself, during its growth in the body of the living animal. But it is of little importance since the hypothesis supposes the forming and necessary existence of the vibrio.

I hasten to touch upon another series of observations which are even more deserving the attention of the surgeon than the preceding: I desire to speak of the effects of our microbe of pus when associated with the septic vibrio. There is nothing more easy to superpose—as it were—two distinct diseases and to produce what might be called a *septicemic purulcnt infection,* or a *purulent septicemia.* Whilst the microbe-producing pus, when acting alone, gives rise to a thick pus, white, or sometimes with a yellow or bluish tint, not putrid, diffused or enclosed by the so-called *pyogenic membrane,* not dangerous, especially if localized in cellular tissue, ready, if the expression may be used for rapid resorption; on the other hand the smallest abscess produced by this organism when associated with the septic vibrio takes on a thick gangrenous appearance, putrid, greenish and infiltrating the softened tissues. In this case the microbe of pus carried so to speak by the septic vibrio, accompanies it throughout the body: the highly-inflamed muscular tissues, full of serous fluid, showing also globules of pus here and there, are like a kneading of the two organisms.

By a similar procedure the effects of the anthrax bacteridium and the microbe of pus may be combined and the two diseases may be superposed, so as to obtain a purulent anthrax or an anthracoid purulent infection. Care must be taken not to exaggerate the predominance of the new microbe over the bacteridium. If the microbe be associated with the latter in sufficient amount it may crowd it out completely—prevent it from growing in the body at all. Anthrax does not appear, and the infection, entirely local, becomes merely an abscess whose cure is easy. The microbe-producing pus and the septic vibrio (not) being both anaerobes, as we have demonstrated, it is evident that the latter will not much disturb its neighbor. Nutrient substances, fluid or solid, can scarcely be deficient in the tissues from such minute organisms. But the anthrax bacteridium is exclusively aerobic, and the proportion of oxygen is far from being equally distributed throughout the tissues: innumerable conditions can diminish or exhaust the supply here and there, and since the microbe-producing pus is also aerobic, it can be understood how, by using a quantity slightly greater than that of the bacteridium it might easily deprive the latter of the oxygen necessary for it. But the explanation of the fact is of little importance: it is certain that under some conditions the microbe we are speaking of entirely prevents the development of the bacteridium.

Summarizing—it appears from the preceding facts that it is possible to produce at will, purulent infections with no elements of putrescence, putrescent purulent infections, anthracoid purulent infections, and finally combinations of these types of lesions varying according to the proportions of the mixtures of the specific organisms made to act on the living tissues.

These are the principal facts I have to communicate to the Academy in my name and in the names of my collaborators, Messrs. Joubert and Charnberland. Some weeks ago (Session of the 11th of March last) a member of the Section of Medicine and Surgery, M. Sedillot, after long meditation on the lessons of a brilliant career, did not hesitate to assert that the successes as well as the failures of Surgery find a rational explanation in the principles upon which the germ theory is based, and that this theory would found a new Surgery—already begun by a celebrated English surgeon, Dr. Lister, who was among the first to understand its fertility. With no professional authority, but with the conviction of a trained experimenter, I venture here to repeat the words of an eminent *confrère.*

In scientific thought we adopt the simplest theory which will explain all of the facts under consideration and enable us to predict new facts of the same kind. The catch in this criterion lies in the word "Simplest." It is really an aesthetic canon such as we find implicit in our criticisms of poetry or painting. The layman finds such a law as

$$\frac{dx}{dt} = K\frac{d^2x}{dy^2}$$

much less simple than "it oozes," of which it is the mathematical statement. The physicist reverses this judgement, and his statement is certainly the more fruitful of the two, so far as prediction is concerned. It is, however, a statement about something very unfamiliar to the plain man, namely, the rate of change of a rate of change.

J. B. S. Haldane, *Possible Worlds*, 1927

Physicists always publish their results completely. If our discovery has a commercial future that is an accident from which we must not profit. And if radium is to be used in the treatment of disease, it seems to me impossible for us to take advantage of that.

Marie Curie

THE CHEMICAL HISTORY OF A CANDLE

MICHAEL FARADAY

The Faraday effect in physics, Faraday's law of induction, and Faraday's laws of electrolysis all represent discoveries by the English physicist and chemist, Michael Faraday (1791-1867). He discovered a number of new organic compounds and wrote a manual of practical chemistry. More important, he was the first to develop an electric current from a magnetic field, and he invented the dynamo and electric motor. He introduced the terms *electrode, cathode, ion, anion, cation, ionization, electrolyte,* and *electrolysis.*

Faraday was born to a poor family in Surrey. They belonged to a small Presbyterian sect called the Sandemanians, which had a great influence on Faraday throughout his life. He learned to read, write, and cipher in a church Sunday school. At the age of 14, he became an apprentice to a book dealer and bookbinder and he began to read the books that he found about him.

A turning point in his life was in attendance at a series of lectures on chemistry by one of the noted chemists of the day, Sir Humphry Davy at the Royal Institute of Great Britain. Faraday was completely fascinated by the subject. A little later, at age 21, he was able to obtain appointment as a laboratory assistant with Davy.

In 1821 he married Sarah Barnard who always gave him strong support. At that time, he became superintendent of the Royal Institution and, in 1825, succeeded Davy as director of the laboratory and became Fullerian professor of chemistry there in 1833. He was elected a member of the Royal Society, but declined many other honors, including a knighthood. In 1858 he retired to a house at Hampton Court presented to him by Queen Victoria.

The Chemical History of a Candle was a series of six lectures published in 1861.

1. Compare and contrast the style of Faraday's writing to that of Galileo and Newton. Whom would you have preferred to have heard lecture?

2. Faraday argued that "There is no more open door by which you can enter the study of natural philosophy than by considering the physical phenomena of a candle." Why do you suppose he chose a candle as the vehicle for a discussion of physical science?

3. How many distinct physical phenomena (e.g. convection of heated air, melting of wax) can you identify as you read this selection?

THE CHEMICAL HISTORY OF A CANDLE

MICHAEL FARADAY

A COURSE OF LECTURES DELIVERED BEFORE A JUVENILE AUDIENCE AT THE ROYAL INSTITUTION

LECTURE I
A CANDLE: THE FLAME–ITS SOURCE–STRUCTURE–MOBILITY–BRIGHTNESS

I PURPOSE, in return for the honor you do us by coming to see what are our proceedings here, to bring before you, in the course of these lectures, the Chemical History of a Candle. I have taken this subject on a former occasion, and, were it left to my own will, I should prefer to repeat it almost every year, so abundant is the interest that attaches itself to the subject, so wonderful are the varieties of outlet which it offers into the various departments of philosophy. There is not a law under which any part of this universe is governed which does not come into play and is touched upon in these phenomena. There is no better, there is no more open door by which you can enter into the study of natural philosophy than by considering the physical phenomena of a candle. I trust; therefore, I shall not disappoint you in choosing this for my subject rather than any newer topic, which could not be better, were it even so good.

And before proceeding, let me say this also: that, though out subject be so great and our intention that of treating it honestly, seriously, and philosophically, yet I mean to pass away from all those who are seniors among us. I claim the privilege of speaking to juveniles as a juvenile myself. I have done so on former occasions, and, if you please, I shall do so again. And, though I stand here with the knowledge of having the words I utter given to the world, yet that shall not deter me from speaking in the same familiar way to those whom I esteem nearest to me on this occasion.

And now, my boys and girls, I must first tell you of what candles are made. Some are great curiosities. I have here some bits of timber, branches of trees, particularly famous for their burning. And here you see apiece of that very curious substance, taken out of some of the bogs in Ireland, called *candle-wood*; a hard, strong, excellent wood, evidently fitted for good work as a register of force, and yet, withal, burning so well that where it is found they make splinters of it, and torches, since it burns like a candle, and gives a very good light indeed. And in this wood we have one of the most beautiful illustrations of the general nature of a candle that I can possibly give. The fuel provided, the means of bringing that fuel to the place of chemical action, the regular and gradual supply of air to that place of action—heat and light—all produced by a little piece of wood of this kind, forming, in fact, a natural candle.

But we must speak of candles as they are in commerce. Here are a couple of candles commonly called dips. They are made of lengths of cotton cut off, hung up by a loop, dipped into melted tallow, taken out again and cooled, then redipped, until there is an accumulation of tallow round the cotton. In order that you may have an idea of the various characters of these candles, you see these which I hold in my hand—they are very small and very curious. They are, or were, the candles used by the miners in coal mines. In olden times the miner had to find his own candles, and it was supposed that a small candle would not so soon set fire to the fire-damp in the coal mines as a large one; and for that reason, as well as for economy's sake, he had candles made of this sort—20, 30, 40, or 60 to the pound. They have been replaced since then by the steel-mill, and then by the Davy lamp, and other safety lamps of various kinds. I have here a candle that was taken out of the *Royal George*, it is said, by Colonel Pasley. It has been sunk in the sea for many years, subject to the action of salt water. It shows you how well candles may be preserved; for, though it is cracked about and broken a great deal, yet when lighted it goes on burning regularly, and the tallow resumes its natural condition as soon as it is fused.

Mr. Field, of Lambeth, has supplied me abundantly with beautiful illustrations of the candle and its materials; I shall therefore now refer to them. And, first, there is the suet—that fat of the ox—Russian tallow, I believe, employed in the manufacture of these dips, which Gay-Lussac, or some one who intrusted him with his knowledge, converted into that beautiful substance, stearin, which you see lying beside it. A candle, you know, is not now a greasy thing like an ordinary tallow candle, but a clean thing, and you may almost scrape off and pulverize the drops which fall from it without soiling any thing. This is the process he adopted: The fat or tallow is first boiled with quick-lime, and made into a soap, and then the soap is decomposed by sulphuric acid, while a quantity of glycerin is produced at the same time. Glycerin—absolutely a sugar, or a substance similar to sugar—comes out of the tallow in this chemical change. The oil is then pressed out of it; and you see here this series of pressed cakes, showing how beautifully the impurities are carried out by the oily part as the pressure goes on increasing, and at last you have left that substance, which is melted, and cast into candles as here represented. The candle I have in my hand is a stearin candle, made of stearin form tallow in the way I have told you. Then here is a sperm candle, which comes from the purified oil of the spermaceti whale. Here, also, are yellow beeswax and refined beeswax, from which candles are made. Here, too, is that curious substance called paraffine, and some paraffine candles, made of paraffine obtained from the bogs of Ireland. I have here also a substance brought from Japan since we have forced an entrance into that out-of-the-way place—a sort of wax which a kind friend has sent me, and which forms a new material for the manufacture of candles.

And how are these candles made? I have told you about dips, and I will show you how moulds are made. Let us imagine any of these candles to be made of materials which can be cast. "Cast!" you say. "Why, a candle is a thing that melts, and surely if you can melt it you can cast it." Not so. It is wonderful, in the progress of manufacture, and in the consideration of the means best fitted to produce the required result, how things turn up which one would not expect beforehand. Candles can not always be cast. A wax candle can never be cast. It is made by a particular process which I can illustrate in a minute or two, but I must not spend much time on it. Wax is a thing which, burning so well, and melting so easily in a candle, can not be cast. However, let us take a material that can be cast. Here is a frame, with a number of moulds fastened in it. The first thing to be done is to put a wick through them. Here is one—a plaited wick, which does not require snuffing—supported by a little wire. It goes to the bottom, where it is pegged in—the little peg holding the cotton tight, and stopping the aperture so that nothing but fluid shall run out. At the upper part there is a little bar placed across, which stretches the cotton and holds it in the mould. The tallow is then melted, and the moulds are filled. After a certain time, when the moulds are cool, the excess of tallow is poured off at one corner, and then cleaned off altogether, and the ends of the wick cut away. The candles alone then remain in the mould, and you have only to upset them, as I am doing, when out they tumble, for the candles are made in the form of cones, being narrower at the top than at the bottom; so that, what with their form and their own shrinking, they only need a little shaking, and out they fall. In the same way are made these candles of stearin and of paraffine. It is a curious thing to see how wax candles are made. A lot of cottons are hung upon frames, as you see here, and covered with metal tags at the ends to keep the wax from covering the cotton in those places. These are carried to a heater, where the wax is melted. As you see, the frames can turn round; and, as they turn, a man takes a vessel of wax and pours it first down one, and then the next, and the next, and so on. When he has gone once round, if it is sufficiently cool, he gives the first a second coat, and so on until they are all of the required thickness. When they have been thus clothed, or fed, or made up to that thickness, they are taken off and placed elsewhere. I have here, by the kindness of Mr. Field, several specimens of these candles. Here is one only half finished. They are then taken down and well rolled upon a fine stone slab, and the conical top is moulded by properly shaped tubes, and the bottoms cut off and trimmed. This is done so beautifully that they can make candles in this way weighing exactly four or six to the pound, or any number they please.

We must not, however, take up more time about the mere manufacture, but go a little farther into the matter. I have not yet referred you to luxuries in candles (for there is such a thing as luxury in candles). See how beautifully these are colored; you see here mauve, magenta, and all the chemical colors recently introduced, applied to candles. You observe, also, different forms employed. Here is a fluted pillar most beautifully shaped; and I have also here some candles sent me by Mr. Pearsall, which are ornamented with designs upon them, so that, as they burn, you have, as it were, a glowing sun above, and bouquet of flowers beneath. All, however, that is fine and beautiful is not useful. These fluted candles, pretty as they are, are bad candles; they are bad because of their external shape. Nevertheless, I show you these specimens, sent to me from kind friends on all sides, that you may see what is done and what may be done in this or that direction; although, as I have said, when we come to these refinements, we are obliged to sacrifice a little in utility.

Now as to the light of the candle. We will light one or two, and set them at work in the performance of their proper functions. You observe a candles is a very different thing from a lamp. With a lamp you take a little oil, fill your vessel, put in a little moss or some cotton prepared by artificial means, and then light the top of the wick. When the flame runs down the cotton to the oil, it gets extinguished, but it goes on burning in the part above. Now I have no doubt you will ask how it is that the oil which will not burn of itself gets up to the top of the cotton, where it will burn. We shall presently examine that; but there is a much more wonderful thing about the burning of a candle than this. You have here a solid substance with no vessel to contain it; and how it is that this solid substance can get up to the place where the flame is? How is it that this solid get there, it not being a fluid? or, when it is made a fluid, then how is it that it keeps together? This is a wonderful thing about a candle.

We have here a good deal of wind, which will help us in some of our illustrations, but tease us in others; for the sake, therefore, of a little regularity, and to simplify the matter, I shall make a quiet flame, for who can study a subject when there are difficulties in the way not belonging to it? Here is a clever invention of some costermonger or street-stander in the market-place for the shading of their candles on Saturday nights, when they are selling their greens, or potatoes, or fish. I have very often admired it. They put a lamp-glass round the candle, supported on a kind of gallery, which clasps it, and it can be slipped up and down as required. By the use of this lamp-glass, employed in the same way, you have a steady flame, which you can look at, and carefully examine, as I hope you will do, at home.

You see, then, in the first instance, that a beautiful cup is formed. As the air comes to the candle, it moves upward by the force of the current which the heat of the candle produces, and it so cools all the sides of the wax, tallow, or fuel as to keep the edge much cooler than the part within; the part within melts by the flame that runs down the wick as far as it can go before it is extinguished, but the part on the outside does not melt. If I made a current in one direction, my cup would be lop-sided, and the fluid would consequently run over; for the same force of gravity which holds worlds together holds this fluid in a horizontal position, and if the cup be not horizontal, of course the fluid will run away in guttering. You see, therefore, that the cup is formed by this beautifully regular ascending current of air playing upon all sides, which keeps the exterior of the candle cool. No fuel would serve for a candle which has not the property of giving this cup, except such fuel as the Irish bogwood, where the material itself is like a sponge and holds its own fuel. You see now why you would have had such a bad result if you were to burn these beautiful candles that I have shown you, which are irregular, intermittent in their shape, and can not, therefore, have that nicely-formed edge to the cup which is the great beauty in a candle. I hope you will now see that the perfection of a process—that is, its utility—is the better point of beauty about it. It is not the best looking thing, but the best acting thing, which is the most advantageous to us. This good-looking candle is a bad-burning one. There will be a guttering round about it because of the irregularity of the stream of air and the badness of the cup which is formed thereby. You may see some pretty examples (and I trust you will notice these instances) of the action of the ascending current when you have a little gutter run down the side of a candle, making it thicker there than it is elsewhere. As the candle goes on burning, that keeps its place and forms a little pillar sticking up by the side, because, as it rises higher above the rest of the wax or fuel, the air gets better round it, and it is more cooled and better able to resist the action of the heat at a little distance. Now the greatest mistakes and faults with regard to candles, as in many other things, often bring with them instruction which we should not receive if they had not occurred. We come here to be philosophers, and I hope you will always remember that whenever a result happens, especially if it be new, you should say, "What is the cause? Why does it occur?" and you will, in the course of time, find out the reason.

Then there is another point about these candles which will answer a question—that is, as to the way in which this fluid gets out of the cup, up the wick, and into the place of combustion. You know that the flames on these burning wicks in candles made of beeswax, stearin, or spermaceti, do not run down to the wax or other matter, and melt it all away, but keep to their own right place. They are fenced off from the fluid below, and do not encroach on the cup at the sides. I can not imagine a more beautiful example than the condition of adjustment under which a candle makes one part subserve to the other to the very end of its action. A combustible thing like that, burning away gradually, never being intruded upon by the flame, is a very beautiful sight, especially when you come to learn what a vigorous thing flame is—what power it has of destroying the wax itself when it gets hold of it, and of disturbing its proper form if it come only too near.

But how does the flame get hold of the fuel? There is a beautiful point about that—*capillary attraction*. [Capillary attraction or repulsion is the cause which determines the ascent or descent of a fluid in a capillary tube. If a piece of thermometer tubing, open at each end, be plunged into water, the latter will instantly rise in the tube considerably above its external level. If, on the other hand, the tube be plunged into mercury, a repulsion instead of attraction will be exhibited, and the level of the mercury will be lower in the tube than it is outside.] "Capillary attraction!" you say—"the attraction of hairs." Well, never mind the name; it was given in old times, before we had a good understanding of what the real power was. It is by what is called capillary attraction that the fuel is conveyed to the part where combustion goes on, and is deposited there, not in a careless way, but very beautifully in the very midst of the centre of action, which takes place around it. Now I am going to give you one or two instances of capillary attraction. It is that kind of action or attraction which makes two things that do not dissolve in each other still hold together. When you wash your hands, you wet them thoroughly; you take a little soap to make the adhesion better, and you find your hands remain wet. This is by that kind of attraction of which I am about to speak. And, what is more, if your hands are not soiled (as they almost always are by the usages of life), if you put your finger into a little warm water, the water will creep a little way up the finger, though you may not stop to examine it. I have here a substance which is rather porous—a column of salt—and I will pour into the plate at the bottom, not water, as it appears, but a saturated solution of salt which can not absorb more, so that the action which you see will not be due to its dissolving any thing. We may consider the plate to be the candle, and the salt the wick, and this solution the melted tallow. (I have colored the fluid, that you may see the action better.) You observe that, now I pour in the fluid, it rises and gradually creeps up the salt higher and higher (Fig. 55); and provided the column does not tumble over, it will go to the top. If this blue solution were combustible, and we were to place a wick at the top of the salt, it would burn as it entered into the wick. It is a most curious thing to see this kind of action taking place, and to observe how singular some of the circumstances are about it. When you

Fig. 55

wash your hands, you take a towel to wipe off the water; and it is by that kind of wetting, or that kind of attraction which makes the towel become wet with water, that the wick is made wet with tallow. I have known some careless boys and girls (indeed, I have known it happen to careful people as well) who, having washed their hands and wiped them with a towel, have thrown the towel over the side of the basin, and before long it has drawn all the water out of the basin and conveyed it to the floor, because it happened to be thrown over the side in such a way as to serve the purpose of a siphon. That you may the better see the way in which the substances act one upon another, I have here a vessel made of wire gauze filled with water, and you may compare it in its action to the cotton in one respect, or to a piece of calico in the other. In fact, wicks are sometimes made of a kind of wire gauze. You will observe that this vessel is a porous thing; for if I pour a little water on to the top, it will run out at the bottom. You would be puzzled for a good while if I asked you what the state of this vessel is, what is inside it, and why is it there? The vessel is full of water, and yet you see the water goes in and runs out as if it were empty. In order to prove this to you, I have only to empty it. The reason is this: the wire, being once wetted, remains wet; the meshes are so small that the fluid is attracted so strongly from the one side to the other, as to remain in the vessel, although it is porous. In like manner, the particles of melted tallow ascend the cotton and get to the top; other particles then follow by their mutual attraction for each other, and as they reach the flame they are gradually burned.

Here is another application of the same principle. You see this bit of cane. I have seen boys about the streets, who are very anxious to appear like men, take a piece of cane, and light it, and smoke it, as an imitation of a cigar. They are enabled to do so by the permeability of the cane in one direction, and by its capillarity. If I place this piece of cane on a plate containing some camphene (which is very much like paraffine in its general character), exactly in the same manner as the blue fluid rose through the salt will this fluid rise through the piece of cane. There being no pores at the side, the fluid can not go in that direction, but must pass through its length. Already the fluid is at the top of the cane; now I can light it and make it serve as a candle. The fluid has risen by the capillary attraction of the piece of cane, just as it does through the cotton in the candle.

Now the only reason why the candle does not burn all down the side of the wick is that the melted tallow extinguishes the flame. You know that a candle, if turned upside down, so as to allow the fuel to run upon the wick, will be put out. The reason is, that the flame has not had time to make the fuel hot enough to burn, as it does above, where it is carried in small quantities into the wick, and has all the effect of the heat exercised upon it.

There is another condition which you must learn as regards the candle, without which you would not be able fully to understand the philosophy of it, and that is the vaporous condition of the fuel. In order that you may understand that, let me show you a very pretty but very commonplace experiment. If you blow a candle out cleverly, you will see the vapor rise from it. You have, I know, often smelt the vapor of a blown-out candle, and a very bad smell it is; but if you blow it out cleverly you will be able to see pretty well the vapor into which this solid matter is transformed. I will blow out one of these candles in such a way as not to disturb the air around it by the continuing action of my breath; and now, if I hold a lighted taper two or three inches from the wick, you will observe a train of fire going through the air till it reaches the candle (Fig. 56). I am obliged to be quick and ready, because if I allow the vapor time to cool, it becomes condensed into a liquid or solid, or the stream of combustible matter gets disturbed.

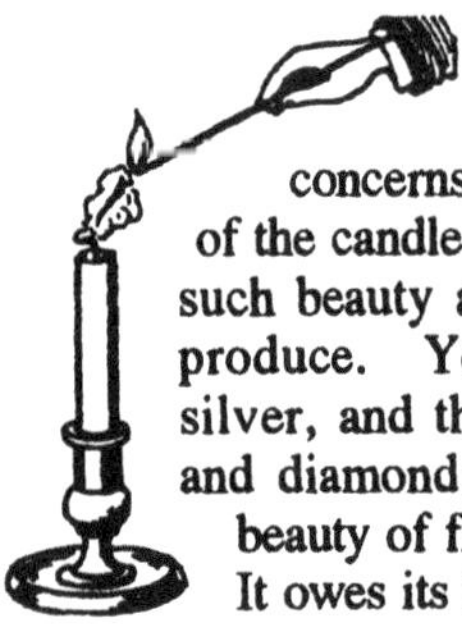

Fig. 56

Now as to the shape or form of the flame. It concerns us much to know about the condition which the matter of the candle finally assumes at the top of the wick, where you have such beauty and brightness as nothing but combustion or flame can produce. You have the glittering beauty of gold and silver, and the still higher lustre of jewels like the ruby and diamond; but none of these rival the brilliancy and beauty of flame. What diamond can shine like flame? It owes its lustre at nighttime to the very flame shining upon it. The flame shines in darkness, but the light which the diamond has is as nothing until the flame shines upon it, when it is brilliant again. The candle alone shines by itself and for itself, or for those who have arranged the materials. Now let us look a little at the form of the flame as you see it under the glass shade. It is steady and equal, and its general form is that which is represented in the diagram (Fig. 57), varying with atmospheric disturbances, and also varying according to the size of the candle. It is a bright oblong, brighter at the top than toward the bottom, with the wick in the middle, and, besides the wick in the middle, certain darker parts towards the bottom, where the ignition is not so perfect as in the part above. I have a drawing here, sketched many years ago by Hooker, when he made his investigations. It is the drawing of the flame of a lamp, but it will apply to the flame of a candle. The cup of the candle is the vessel or lamp; the melted spermaceti is the oil; and the wick is common to both. Upon that he sets this little flame, and then he represents what is true, a certain quantity of matter rising about it which you do not see, and which, if you have not been here before, or are not familiar with the subject, you will not know of. He has here represented the parts of the surrounding atmosphere that are very essential to the flame, and that are always present with it. There is a current formed, which draws the flame out; for the flame which you see is really drawn out by the current, and drawn upward to a great height, just as Hooker has here shown you by that prolongation of the current in the diagram. You may see this by taking a lighted candle, and putting it in the sun so as to get its shadow thrown on a piece of paper. How remarkable it is that that thing which is light enough to produce shadows of other objects can be made to throw its own shadow on a piece of white paper or card, so that you can actually see streaming round the flame something which is not part of the flame, but is ascending and drawing the flame upward. Now I am going to imitate the sunlight by applying the voltaic battery to the electric lamp. You now see our sun and its great luminosity; and by placing

Fig. 57

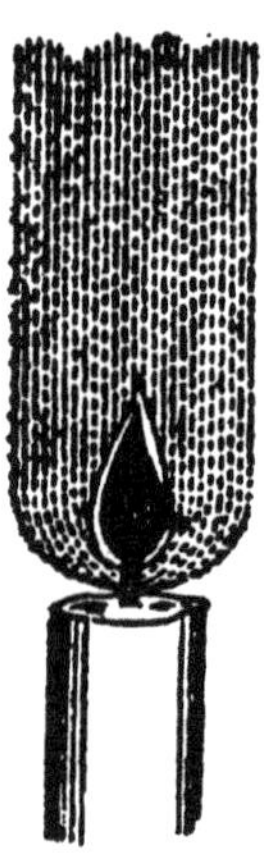

Fig. 58

a candle between it and the screen, we get the shadow of the wick; then there is a darkish part, as represented in the diagram (Fig. 58), and then a part which is more distinct. Curiously enough, however, what we see in the shadow as the darkest part of the flame is, in reality, the brightest part; and here you see streaming upward the ascending current of hot air, as shown by Hooker, which draws out the flame, supplies it with air, and cools the sides of the cup of melted fuel.

I can give you here a little farther illustration, for the purpose of showing you how flame goes up or down according to the current. I have here a flame—it is not a candle flame—but you can, no doubt, by this time generalize enough to be able to compare one thing with another. What I am about to do is to change the ascending current that takes the flame upward into a descending current. This I can easily do by the little apparatus you see before me. The flame, as I have said, is not a candle flame, but it is produced by alcohol, so that it shall not smoke too much. I will also color the flame with another substance, so that you may trace its course; for, with the spirit alone, you could hardly see well enough to have the opportunity of tracing its direction. [The alcohol had chloride of copper dissolved in it: this produces a beautiful green flame.] By lighting this spirit of wine we have then a flame produced, and you observe that when held in the air it naturally goes upward. You understand now, easily enough, why flames go up under ordinary circumstances: it is because of the draught of air by which the combustion is formed. But now, by blowing the flame down, you see I am enabled to make it go downward into this little chimney, the direction of the current being changed (Fig. 59). Before we have concluded this course of lectures we shall show you a lamp in which the flame goes up and the smoke goes down, or the flame goes down and the smoke goes up. You see, then, that we have the power in this way of varying the flame in different directions.

Fig. 59

There are now some other points that I must bring before you. Many of the flames you see here vary very much in their shape by the currents of air blowing around them in different directions; but we can, if we like, make flames so that they will look like fixtures, and we can photograph them—indeed, we have to photograph them—so that they become fixed to us, if we wish to find out every thing concerning them. That, however, is not the only thing I wish to mention. If I take a flame sufficiently large, it does not keep that homogeneous, that uniform condition of shape, but it breaks out with a power of life which is quite wonderful. I am about to use another kind of fuel, but one which is truly and fairly a representative of the wax or tallow of a candle. I have here a large ball of cotton, which will serve as a wick. And, now that I have immersed it in spirit and applied a light to it, in what way does it differ from an ordinary candle? Why, it differs very much in one respect, that we have a vivacity and power about it, a beauty and a life entirely different from the light presented by a candle. You see those fine tongues of flame rising up. You have the same general disposition of the mass of the flame from below upward; but, in addition to that, you have this remarkable breaking out into tongues which you do not perceive in the case of a candle. Now, why is this? I must explain it to you, because, when you understand that perfectly, you will be able to follow me better in what I have to say hereafter. I suppose some here will have made for themselves the experiment I am going to show you. Am I right in supposing that any body here has played at snapdragon? I do not know a more beautiful illustration of the philosophy of flame, as to a certain part of its history, than the game of snapdragon. First, here is the dish; and let me say, that when you play snapdragon properly you ought to have the dish well warmed; you ought also to have warm plums, and warm brandy, which, however, I have not got. When you have put the spirit into the dish, you have the cup and the fuel; and are not the raisins acting like the wicks? I now throw the plums into the dish, and light the spirit, and you see those beautiful tongues of flame that I refer to. You have the air creeping in over the edge of the dish forming these tongues. Why? Because, through the force of the current and the irregularity of the action of the flame, it can not flow in one uniform stream. The air flows in so irregularly that you have what would otherwise be a single image broken up into a variety of forms, and each of these little tongues has an independent existence of its own. Indeed, I might say, you have here a multitude of independent candles. You must not imagine, because you see these tongues all at once, that the flame is of this particular shape. A flame of that shape is never so at any one time.

Fig. 60

Never is a body of flame, like that which you just saw rising from the ball, of the shape it appears to you. It consists of a multitude of different shapes, succeeding each other so fast that the eye is only able to take cognizance of them all at once. In former times I purposely analyzed a flame of that general character, and the diagram shows you the different parts of which it is composed (Fig. 60). They do not occur all at once; it is only because we see these shapes in such rapid succession that they seem to us to exist all at one time.

It is too bad that we have not got farther than my game of snap dragon; but we must not, under any circumstances, keep you beyond your time. It will be a lesson to me in future to hold you more strictly to the philosophy of the thing than to take up your time so much with these illustrations.

LECTURE II

A CANDLE: BRIGHTNESS OF THE FLAME—AIR NECESSARY FOR COMBUSTION—PRODUCTION OF WATER

WE were occupied the last time we met in considering the general character and arrangement as regards the fluid portion of a candle, and the way in which that fluid got into the place of combustion. You see, when we have a candle burning fairly in a regular, steady atmosphere, it will have a shape something like the one shown in the diagram, and will look pretty uniform, although very curious in its character. And now I have to ask your attention to the means by which we are enabled to ascertain what happens in any particular part of the flame; why it happens; what it does in happening; and where, after all, the whole candle goes to; because, as you know very well, a candle being brought before us and burned, disappears, if burned properly, without the least trace of dirt in the candle stick; and this is a very curious circumstance. In order, then, to examine this candle carefully, I have arranged certain apparatus, the use of

Fig. 61

which you will see as I go on. Here is a candle; I am about to put the end of this glass tube into the middle of the flame—into that part which old Hooker has represented in the diagram as being rather dark, and which you can see at any time if you will look at a candle carefully, without blowing it about. We will examine this dark part first.

Now I take this bent glass tube, and introduce one end into that part of the flame, and you see at once that something is coming from the flame, out at the other end of the tube; and if I put a flask there, and leave it for a little while, you will see that something from the middle part of the flame is gradually drawn out, and goes through the tube, and into that flask, and there behaves very differently from what it does in the open air. It not only escapes from the end of the tube, but falls down to the bottom of the flask like a heavy substance, as indeed it is (Fig. 61). We find that this is the wax of the candle made into a vaporous fluid—not a gas. (You must learn the difference between a gas and a vapor: a gas remains permanent; a vapor is something that will condense.) If you blow out a candle, you perceive a very nasty smell, resulting from the condensation of this vapor. That is very different from what you have outside the flame; and, in order to make that more clear to you, I am about to produce and set fire to a larger portion of this vapor; for what we have in the small way in a candle, to understand thoroughly, we must, as philosophers, produce in a larger way, if needful, that we may examine the different parts. And now Mr. Anderson will give me a source of heat, and I am about to show you what that vapor is. Here is some wax in a glass flask, and I am going to make it hot, as the inside of that candle flame is hot, and the matter about the wick is hot. [The lecturer placed some wax in a glass flask, and heated it over a lamp.] Now I dare say that is hot enough for me. You see that the wax I put in it has become fluid, and there is a little smoke coming from it. We shall very soon have the vapor rising up. I will make it still hotter, and now we get more of it, so that I can actually pour the vapor out of the flask into that basin, and set it on fire there. This, then, is exactly the same kind of vapor as we have in the middle of the candle; and that you may be sure this is the case, let us try whether we have not got here, in this flask, a real combustible vapor out of the middle of the candle. [Taking the flask into which the tube from the candle proceeded, and introducing a lighted taper (Fig. 62).] See how it burns. Now this is the vapor from the middle of the candle, produced by its own heat; and that is one of the first things you have to consider with respect to the progress of the wax in the course of its combustion, and as regards the changes it under goes. I will arrange another tube carefully in the flame, and I should not wonder if we were able, by a little care, to get that vapor to pass through the tube to the other extremity, where we will light it, and obtain absolutely the flame of the candle at a place distant from it. Now, look at that. Is not I that a very pretty experiment? Talk about laying on gas—why, we can actually lay on a candle! And you see from this that there are clearly two different kinds of action—one the *production* of the vapor, and the other the *combustion* of it—both of which take place in particular parts of the candle.

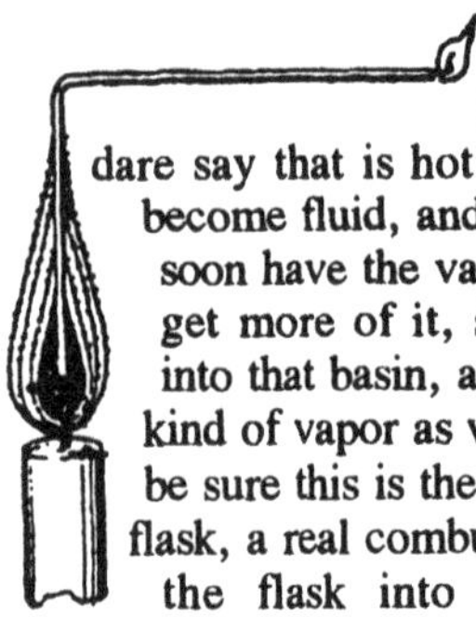

Fig. 62

I shall get no vapor from that part which is already burnt. If I raise the tube (Fig. 61) to the upper part of the flame, so soon as the vapor has been swept out what comes away will be no longer combustible; it is already burned. How burned? Why, burned thus: In the middle of the flame, where the wick is, there is this combustible vapor; on the outside of the flame is the air which we shall find necessary for the burning of the candle; between the two, intense chemical action takes place, whereby the air and the fuel act upon each other, and at the very same time that we obtain light the vapor inside is destroyed. If you examine where the heat of a candle is, you will find it very curiously arranged. Suppose I take this candle, and hold a piece of paper close upon the flame, where is the heat of that flame? Do you not see that it is not in the inside? It is in a ring, exactly in the place where I told you the chemical action was; and even in my irregular mode of making the experiment, if there is not too much disturbance, there will always be a ring. This is a good experiment for you to make at home. Take a strip of paper, have the air in the room quiet, and put the piece of paper right across the middle of the flame—(I must not talk while I make the experiment)—and you will find that it is burnt in two places, and that it is not burnt, or very little so, in the middle; and when you have tried the experiment once or twice, so as to make it nicely, you will be very interested to see where the heat is, and to find that it is where the air and the fuel come together.

This is most important for us as we proceed with our subject. Air is absolutely necessary for combustion; and, what is more, I must have you understand that *fresh* air is necessary, or else we should be imperfect in our reasoning and our experiments. Here is a jar of air; I place it over a candle, and it burns very nicely in it at first, showing that what I have said about it is true; but there will soon be a change. See how the flame is drawing upward, presently fading, and at last going out. And going out, why? Not because it wants air merely, for the jar is as full now as it was before; but it wants pure, fresh air. The jar is full of air, partly changed, partly not changed; but it does not contain sufficient of the fresh air which is necessary for the combustion of a candle. These are all points which we, as young chemists, have to gather up; and if we look a little more closely into this kind of action, we shall find certain steps of reasoning extremely interesting. For instance, here is the oil-lamp I showed you—an excellent lamp for our experiments—the old Argand lamp. I now make it like a candle [obstructing the passage of air into the centre of the flame]; there is the cotton; there is the oil rising up in it, and there is the conical flame. It burns poorly because there is a partial restraint of air. I have allowed no air to get to it save around the outside of the flame, and it does not burn well. I can not admit more air from the outside, because the wick is large; but if, as Argand did so cleverly, I open a passage to the middle of the flame, and so let air come in there, you will see how much more beautifully it burns. If I shut the air off, look how it smokes; and why? We have now some very interesting points to study: we have the case of the combustion of a candle; we have the case of a candle being put out by the want of air; and we have now the case of imperfect combustion, and this is to us so interesting that I want you to understand it as thoroughly as you do the case of a candle burning in its best possible manner. I will now make a great flame, because we need the largest possible illustrations. Here is a larger wick [burning turpentine on a ball of cotton]. All these things are the same as candles, after all. If we have larger wicks, we must have a larger *supply* of air, or we shall have less perfect combustion. Look, now, at this black substance going up into the atmosphere; there is a regular stream of it. I have provided means to carry off the imperfectly burned part, lest it should annoy you. Look at the soots that fly off from the flame; see what an imperfect combustion it is, because it can not get enough air. What, then, is happening? Why, certain things which are necessary to the combustion of a candle are absent, and very bad results are accordingly produced; but we see what happens to a candle when it is burnt in a pure and proper state of air. At the time when I showed you this charring by the ring of flame on the one side of the paper, I might have also shown you, by turning to the other side, that the burning of a candle produces the same kind of soot—charcoal, or carbon.

But, before I show that, let me explain to you, as it is quite necessary for our purpose, that, though I take a candle, and give you, as the general result, its combustion in the form of a flame, we must see whether combustion is always in this condition, or whether there are other conditions of flame; and we shall soon discover that there are, and that they are most important to us. I think, perhaps, the best illustration of such a point to us, as juveniles, is to show the result of strong contrast. Here is a little gunpowder. You know that gunpowder burns with flame; we may fairly call it flame. It contains carbon and other materials, which altogether cause it to burn with a flame. And here is some pulverized iron, or iron filings. Now I purpose burning these two things together. I have a little mortar in which I will mix them. (Before I go into these experiments, let me hope that none of you, by trying to repeat them for fun's sake, will do any harm. These things may all be very properly used if you take care, but without that much mischief will be done.) Well, then, here is a little gunpowder, which I put at the bottom of that little wooden vessel, and mix the iron filings up with it, my object being to make the gunpowder set fire to the filings and burn them in the air, and thereby show the difference between substances burning with flame and not with flame. Here is the mixture; and when I set fire to it you must watch the combustion, and you will see that it is of two kinds. You will see the gunpowder burning with a flame and the filings thrown up. You will see them burning, too, bu without the production of flame. They will each burn separately [The lecturer then ignited the mixture.] There is the gunpowder, which burns with a flame, and there are the fillings: they burn with a different kind of combustion. You see, then, these two great distinctions; and upon these differences depend all the utility and all the beauty of flame which we use for the purpose of giving out light When we use oil, or gas, or candle for the purpose of illumination, their fitness all depends upon these different kinds of combustion.

There are such curious conditions of flame that it requires some cleverness and nicety of discrimination to distinguish the kinds of combustion one from another. For instance, here is a powder which is very combustible, consisting, as you see, of separate little particles. It is called *lycopodium,* and each of these particles can produce a vapor, and produce its own flame; but, to see them burning, you would imagine it was all one flame. [Lycopodium is a yellowish powder found in the fruit of the club moss *(Lycopodium clavatum)*. It is used in fireworks.] I will now set fire to a quantity, and you will see the effect. We saw a cloud of flame, apparently in one body; but that rushing noise [referring to the sound produced by the burning] was a proof that the combustion was not a continuous or regular one. This is the lightning of the pantomimes, and a very good imitation. [The experiment was twice repeated by blowing lycopodium from a glass tube through a spirit flame.] This is not an example of combustion like that of the filings I have been speaking of, to which we must now return.

Suppose I take a candle and examine that part of it which appears brightest to our eyes. Why, there I get these black particles, which already you have seen many times evolved from the flame, and which I am now about to evolve in a different way. I will take this candle and clear away the gutterage, which occurs by reason of the currents of air; and if I now arrange the glass tube so as just to dip into this luminous part, as in our first experiment, only higher, you see the result. In place of having the same white vapor that you had before, you will now have a black vapor. There it goes, as black as ink;. It is certainly very different from the white vapor; and when we put a light to it we shall find that it does not burn, but that it puts the light out. Well, these particles, as I said before, are just the smoke of the candle; and this brings to mind that old employment which Dean Swift recommended to servants for their amusement, namely, writing on the ceiling of a room with a candle. But what is that black substance? Why, it is the same carbon which exists in the candle. How comes it out of the candle? It evidently existed in the candle, or else we should not have had it here. And now I want you to follow me in this explanation. You would hardly think that all those substances which fly about London, in the form of soots and blacks, are the very beauty and life of the flame, and which are burned in it as those iron filings were burned here. Here is a piece of wire gauze, which will not let the flame go through it; and I think you will see, almost immediately, that when I bring it low enough to touch that part of the flame which is otherwise so bright, it quells and quenches it at once, and allows a volume of smoke to rise up.

I want you now to follow me in this point—that whenever a substance burns, as the iron filings burnt in the flame of gunpowder, without assuming the vaporous state (whether it becomes liquid or remains solid), it becomes exceedingly luminous. I have here taken three or four examples apart from the candle on purpose to illustrate this point to you, because what I have to say is applicable to all substances, whether they burn or whether they do not burn—that they are exceedingly bright if they retain their solid state, and that it is to this presence of solid particles in the candle flame that it owes its brilliancy.

Here is a platinum wire, a body which does not change by heat. If I heat it in this flame, see how exceedingly luminous it becomes. I will make the flame dim for the purpose of giving a little light only, and yet you will see that the heat which it can give to that platinum wire, though far less than the heat it has itself, is able to raise the platinum wire to a far higher state of effulgence. This flame has carbon in it; but I will take one that has no carbon in it. There is a material, a kind of fuel—a vapor, or gas, whichever you like to call it—in that vessel, and it has no solid particles in it; sol take that because it is an example of flame itself burning without any solid matter whatever; and if I now put this solid substance in it, you see what an intense heat it has, and how brightly it causes the solid body to glow. This is the pipe through which we convey this particular gas, which we call hydrogen, and which you shall know all about the next time we meet. And here is a substance called oxygen, by means of which this hydrogen can burn; and although we produce, by their mixture, far greater heat than you can obtain from the candle, yet there is very little light. [Bunsen has calculated that the temperature of the oxyhydrogen blowpipe is 8061° Centigrade. Hydrogen burning in air has a temperature of 3259° C., and coal gas in air 2350° C.] If, however, I take a solid substance, and put that into it, we produce an intense light. If I take a piece of lime, a substance which will not burn, and which will not vaporize by the heat (and because it does not vaporize remains solid, and remains heated), you will soon observe what happens as to its glowing. I have here a most intense heat produced by the burning of hydrogen in contact with the oxygen; but there is as yet very little light—not for want of heat, but for want of particles which can retain their solid state; but when I hold this piece of lime in the flame of the hydrogen as it burns in the oxygen, see how it glows! This is the glorious lime light, which rivals the voltaic light, and which is almost equal to sunlight. I have here a piece of carbon or charcoal, which will burn and give us light exactly in the same manner as if it were burnt as part of a candle. The heat that is in the flame of a candle decomposes the vapor of the wax, and sets free the carbon particles; they rise up heated and glowing as this now glows, and then enter into the air. But the particles, when burnt, never pass off from a candle in the form of carbon. They go off into the air as a perfectly invisible substance, about which we shall know hereafter.

Is it not beautiful to think that such a process is going on, and that such a dirty thing as charcoal can become so incandescent? You see it comes to this—that all bright flames contain these solid particles; all things that burn and produce solid particles, either during the time they are burning, as in the candle or immediately after being burnt, as in the case of the gunpowder and iron filings—all these things give us this glorious and beautiful light.

Fig. 63

I will give you a few illustrations. Here is a piece of phosphorus, which burns with a bright flame. Very well; we may now conclude that phosphorus will produce, either at the moment that it is burning or afterwards, these solid particles. Here is the phosphorus lighted, and I cover it over with this glass for the purpose of keeping in what is produced. What is all that smoke? (Fig. 63) That smoke consists of those very particles which are produced by the combustion of the phosphorus. Here, again, are two substances. This is chlorate of potassa, and this other sulphuret of antimony. I shall mix these together a little, and then they may be burnt in many ways. I shall touch them with a drop of sulphuric acid, for the purpose of giving you an illustration of chemical action, and they will instantly burn. [The following is the action of the sulphuric in inflaming the mixture of sulphuret of antimony and chlorate of potassa. A portion of the latter is decomposed by the sulphuric acid into oxide of chlorine, bisulphate of potassa, and perchlorate of potassa. The oxide of chlorine inflames the sulphuret of antimony, which is a combustible body, and the whole mass instantly bursts into flame.] [The lecturer then ignited the mixture by means of sulphuric acid.] Now, from the appearance of things, you can judge for yourselves whether they produce solid matter in burning. I have given you the train of reasoning which will enable you to say whether they do or do not; for what is this bright flame but the solid particles passing off?

Mr. Anderson has in the furnace a very hot crucible. I am about to throw into it some zinc filings, and they will burn with a flame like gunpowder. I make this experiment because you can make it well at home. Now I want you to see what will be the result of the combustion of this zinc. Here it is burning—burning beautifully like a candle, I may say. But what is all that smoke, and what are those little clouds of wool which will come to you if you can not come to them, and make themselves sensible to you in the form of the old philosophic wool, as it was called? We shall have left in that crucible, also, a quantity of this woolly matter. But I will take a piece of this same zinc, and make an experiment a little more closely at home, as it were. You will have here the same thing happening. Here is the piece of zinc; there [pointing to a jet of hydrogen] is the furnace, and we will set to work and try and burn the metal. It glows, you see; there is the combustion; and there is the white substance into which it burns.

And so, if I take that flame of hydrogen as the representative of a candle, and show you a substance like zinc burning in the flame, you will see that it was merely during the action of combustion that this substance glowed—while it was kept hot; and if I take a flame of hydrogen and put this white substance from the zinc into it, look how beautifully it glows, and just because it is a solid substance.

I will now take such a flame as I had a moment since, and set free from it the particles of carbon. Here is some camphene, which will burn with a smoke; but if I send these particles of smoke through this pipe into the hydrogen flame you will see they will burn and become luminous, because we heat them a second time. There they are. Those are the particles of carbon reignited a second time. They are those particles which you can easily see by holding a piece of paper behind them, and which, while they are in the flame, are ignited by the heat produced, and, when so ignited, produce this brightness. When the particles are not separated you get no brightness. The flame of coal gas owes its brightness to the separation, during combustion, of these particles of carbon, which are equally in that as in a candle. I can very quickly alter that arrangement. Here, for instance, is a bright flame of gas. Supposing I add so much air to the flame as to cause it all to burn before those particles are set free, I shall not have this brightness; and I can do that in this way: If I place over the jet this wire-gauze cap, as you see, and then light the gas over it, it burns with a non-luminous flame, owing to its having plenty of air mixed with it before it burns; and if I raise the gauze, you see it does not burn below. [The "air-burner," which is of such value in the laboratory, owes its advantage to this principle. It consists of a cylindrical metal chimney, covered at the top with a piece of rather coarse iron wire gauze. This is supported over an Argand burner in such a manner that the gas may mix in the chimney with an amount of air sufficient to burn the carbon and hydrogen simultaneously, so that there may be no separation of carbon in the flame with consequent deposition of soot. The flame, being unable to pass through the wire gauze, burns in a steady, nearly invisible manner above.] There is plenty of carbon in the gas; but, because the atmosphere can get to it, and mix with it before it burns, you see how pale and blue the flame is. And if I blow upon a bright gas flame, so as to consume all this carbon before it gets heated to the glowing point, it will also burn blue. [The lecturer illustrated his remarks by blowing on the gas light.] The only reason why I have not the same bright light when I thus blow upon the flame is that the carbon meets with sufficient air to burn it before it gets separated in the flame in a free state. The difference is solely due to the solid particles not being separated before the gas is burnt.

You observe that there are certain products as the result of the combustion of a candle, and that of these products one portion may be considered as charcoal, or soot; that charcoal, when afterward burnt, produces some other product; and it concerns us very much now to ascertain what that other product is. We showed that something was going away; and I want you now to understand how much is going up into the air; and for that purpose we will have combustion on a little larger scale. From that candle ascends heated air, and two or three experiments will show you the ascending current; but, in order to give you a notion of the quantity of matter which ascends in this way, I will make an experiment by which I shall try to imprison some of the products of this combustion. For this purpose I have here what boys call a fire-balloon; I use this fire-balloon merely as a sort of measure of the result of the combustion we are considering; and I am about to make a flame in such an easy and simple manner as shall best serve my present purpose. This plate shall be the "cup," we will so say, of the candle; this spirit shall be our fuel; and I am about to place this chimney over it, because it is better for me to do so than to let things proceed at random. Mr. Anderson will now light the fuel, and here at the top we shall get the results of the combustion. What we get at the top of that tube is exactly the same, generally speaking, as you get from the combustion of a candle; but we do not get a luminous flame here, because we use a substance which is feeble in carbon. I am about to put this balloon—not into action, because that is not my object—but to show you the effect which results from the action of those products which arise from the candle, as they arise here from the furnace. [The balloon was held over the chimney (Fig. 64), when it immediately commenced to fill.] You see how it is disposed to ascend; but we must not let it up; because it might come in contact with those upper gaslights, and that would be very inconvenient. [The upper gaslights were turned out at the request of the lecturer, and the balloon was allowed to ascend.] Does not that show you what a large bulk of matter is being evolved? Now there is going through this tube [placing a large glass tube over a candle] all the products of that candle, and you will presently see that the tube will become quite opaque. Suppose I take another candle, and place it under a jar, and then put a light on the other side, just to show you what is going on. You see that the sides of the jar become cloudy, and the light begins to burn feebly. It is the products, you see, which make the light so dim, and this is the same thing which makes the sides of the jar so opaque. If you go home, and take a spoon that has been in the cold air, and hold it over a candle—not so as to soot it—you will find that it becomes dim just as that jar is dim. If you can get a silver dish, or something of that kind, you will make the experiment still better; and now, just to carry your thoughts forward to the time we shall next meet, let me tell you that it is *water* which causes the dimness, and when we next meet I will show you that we can make it, without difficulty, assume the form of a liquid.

Fig. 64

LECTURE III

PRODUCTS: WATER FROM THE COMBUSTION—NATURE OF WATER—A COMPOUND—HYDROGEN

I DARE say you well remember that when we parted we had just mentioned the word "products" from the candle; for when a candle burns we found we were able, by nice adjustment, to get various products from it. There was one substance which was not obtained when the candle was burning properly, which was charcoal or smoke, and there was some other substance that went upward from the flame which did not appear as smoke, but took some other form, and made part of that general current which, ascending from the candle upward, becomes invisible, and escapes. There were also other products to mention. You remember that in that rising current having its origin at the candle we found that one part was condensable against a cold spoon, or against a clean plate, or any other cold thing, and another part was incondensable.

We will first take the condensable part, and examine it, and, strange to say, we find that that part of the product is just water—nothing but water. On the last occasion I spoke of it incidentally, merely saying that water was produced among the condensable products of the candle; but to-day I wish to draw your attention to water, that we may examine it carefully, especially in relation to this subject, and also with respect to its general existence on the surface of the globe.

Now, having previously arranged an experiment for the purpose of condensing water from the products of the candle, my next point will be to show you this water; and perhaps one of the best means that I can adopt for showing its presence to so many at once is to exhibit a very visible action of water, and then to apply that test to what is collected as a drop at the bottom of that vessel. I have here a chemical substance, discovered by Sir Humphry Davy, which has a very energetic action upon water, which I shall use as a test of the presence of water. If I take a little piece of it—it is called potassium, as coming from potash—if I take a little piece of it, and throw it into that basin, you see how it shows the presence of water by lighting up and floating about, burning with a violet flame. I am now going to take away the candle which has been burning beneath the vessel containing ice and salt, and you see a drop of water—a condensed product of the candle—hanging from the under surface of the dish (Fig. 65). I will show you that potassium has the same action upon it as upon the water in that basin in the experiment we have just tried. See! it takes fire, and burns in just the same manner. I will take another drop upon this glass slab, and when I put the potassium on to it, you see at once, from its taking fire, that there is water present. Now that water was produced by the candle. In the same manner, if I put this spirit lamp under that jar, you will soon see the latter become damp from the dew which is deposited upon it—that dew being the result of combustion; and I have no doubt you will shortly see, by the drops of water which fall upon the paper below, that there is a good deal of water produced from the combustion of the lamp. I will let it remain, and you can afterward see how much water has been collected. So, if I take a gas lamp, and put any cooling arrangement over it, I shall get water—water being likewise produced from the combustion of gas. Here, in this bottle, is a quantity of water—perfectly pure, distilled water, produced from the combustion of a gas lamp—in no point different from the water that you distill from the river, or ocean, or spring, but exactly the same thing. Water is one individual

Fig. 65

thing; it never changes. We can add to it by careful adjustment for a little while, or we can take it apart and get other things from it; but water, as water, remains always the same, either in a solid, liquid, or fluid state. Here again [holding another bottle] is some water produced by the combustion of an oil lamp. A pint of oil, when burnt fairly and properly, produces rather more than a pint of water. Here, again, is some water, produced by a rather long experiment, from a wax candle. And so we can go on with almost all combustible substances, and find that if they burn with a flame, as a candle, they produce water. You may make these experiments yourselves: the head of a poker is a very good thing to try with, and if it remains cold long enough over the candle, you may get water condensed in drops on it; or a spoon, or ladle, or any thing else may be used, provided it be clean, and can carry off the heat, and so condense the water.

And now—to go into the history of this wonderful production of water from combustibles, and by combustion—I must first of all tell you that this water may exist in different conditions; and although you may now be acquainted with all its forms, they still require us to give a little attention to them for the present; so that we may perceive how the water, while it goes through its Protean changes, is entirely and absolutely the same thing, whether it is produced from a candle, by combustion, or from the rivers or ocean.

First of all, water, when at the coldest, is ice. Now we philosophers—I hope that I may class you and myself together in this case—speak of water as water, whether it be in its solid, or liquid, or gaseous state—we speak of it chemically as water. Water is a thing compounded of two substances, one of which we have derived from the candle, and the other we shall find elsewhere. Water may occur as ice; and you have had most excellent opportunities lately of seeing this. Ice changes back into water—for we had on our last Sabbath a strong instance of this change by the sad catastrophe which occurred in our own house, as well as in the houses of many of our friends—ice changes back into water when the temperature is raised; water also changes into steam when it is warmed enough. The water which we have here before us is in its densest state [water is in its densest state at a temperature of 39.1° Fahrenheit.]; and, although it changes in weight, in condition, in form, and in many other qualities, it still is water; and whether we alter it into ice by cooling, or whether we change it into steam by heat, it increases in volume—in the one case very strangely and powerfully, and in the other case very largely and wonderfully. For instance, I will now take this tin cylinder, and pour a little water into it, and, seeing how much water I pour in, you may easily estimate for yourselves how high it will rise in the vessel: it will cover the bottom about two inches. I am now about to convert the water into steam for the purpose of showing to you the different volumes which water occupies in its different states of water and steam.

Let us now take the case of water changing into ice: we can effect that by cooling it in a mixture of salt and pounded ice—and I shall do so to show you the expansion of water into a thing of larger bulk when it is so changed. [A mixture of salt and pounded ice reduces the temperature from 32 F. to zero the ice at the same time becoming fluid.] These bottles [holding one] are made of strong cast iron, very strong and very thick—I suppose they are the third of an inch in thickness; they are very carefully filled with water, so as to exclude all air, and then they are screwed down tight. We shall see that when we freeze the water in these iron vessels, they will not be able to hold the ice, and the expansion within them will break them in pieces as these [pointing to some fragments] are broken, which have been bottles of exactly the same kind. I am about to put these two bottles into that mixture of ice and salt for the purpose of showing that when water becomes ice it changes in volume in this extraordinary way.

In the mean time, look at the change which has taken place in the water to which we have applied heat; it is losing its fluid state. You may tell this by two or three circumstances. I have covered the mouth of this glass flask, in which water is boiling, with a watch glass. Do you see what happens? It rattles away like a valve chattering, because the steam rising from the boiling water sends the valve up and down, and forces itself out, and so makes it clatter. You can very easily perceive that the flask is quite full of steam, or else it would not force its way out. You see, also, that the flask contains a substance very much larger than the water, for it fills the whole of the flask over and over again, and there it is blowing away into the air; and yet you can not observe any great diminution in the bulk of the water, which shows you that its change of bulk is very great when it becomes steam.

I have put our iron bottles containing water into this freezing mixture, that you may see what happens. No communication will take place, you observe, between the water in the bottles and the ice in the outer vessel. But there will be a conveyance of heat from the one to the other, and if we are successful—we are making our experiment in very great haste—I expect you will by-and-by, so soon as the cold has taken possession of the bottles and their contents, hear a pop on the occasion of the bursting of the one bottle or the other, and, when we come to examine the bottles, we shall find their contents masses of ice, partly inclosed by the covering of iron which is too small for them, because the ice is larger n bulk than the water. You know very well that ice floats upon water; if a boy falls through a hole into the water, he tries to get on the ice again to float him up. Why does the ice float? Think of that, and philosophize. Because the ice is larger than the quantity of water which can produce it, and therefore the ice weighs the lighter and the water is the heavier.

To return now to the action of heat on water. See what a stream of vapor is issuing from this tin vessel! You observe, we must have made it quite full of steam to have it sent out in that great quantity. And now, as we can convert the water into steam by heat, we convert it back into liquid water by the application of cold. And if we take a glass, or any other cold thing, and hold it over this steam, see how soon it gets damp with water: it will condense it until the glass is warm—it condenses the water which is now running down the sides of it. I have here another experiment to show the condensation of water from a vaporous state back into a liquid state, in the same way as the vapor, one of the products of the candle, was condensed against the bottom of the dish and obtained in the form of water; and to show you how truly and thoroughly these changes take place, I will take this tin flask, which is now full of steam, and close the top. We shall see what takes place when we cause this water or steam to return back to the fluid state by pouring some cold water on the outside. [The

Fig. 66

lecturer poured the cold water over the vessel, when it immediately collapsed (Fig. 66).] You see what has happened. If I had closed the stopper, and still kept the heat applied to it, it would have burst the vessel; yet, when the steam returns to the state of water, the vessel collapses, there being a vacuum produced inside by the condensation of the steam. I show you these experiments for the purpose of pointing out that in all these occurrences there is nothing that changes the water into any other thing; it still remains water; and so the vessel is obliged to give way, and is crushed inward, as in the other case, by the farther application of heat, it would have been blown outward.

And what do you think the bulk of that water is when it assumes the vaporous condition? You see that cube [pointing to a cubic foot]. There, by its side, is a cubic inch (Fig. 67), exactly the same shape as the cubic foot, and that bulk of water [the cubic inch] is sufficient to expand into that bulk [the cubic foot] of steam; and, on the contrary, the application of cold will contract that large quantity of steam into this small quantity of water. [One of the iron bottles burst at that moment.] Ah! There is one of our bottles burst, and here, you see, is a crack down one side an eighth of an inch in width. [The other now exploded, sending the freezing mixture in all directions.] This other bottle is also broken; although the iron was nearly half an inch thick, the ice has burst it asunder. These changes always take place in water; they do not require to be always produced by artificial means; we only use them here because we want to produce a small winter round that little bottle instead of a long and severe one. But if you go to

Fig. 67

Canada, or to the North, you will find the temperature there out of doors will do the same thing as has been done here by the freezing mixture.

To return to our quiet philosophy. We shall not in future be deceived, therefore, by any changes that are produced in water. Water is the same every where, whether produced from the ocean or from the flame of the candle. Where, then, is this water which we get from a candle? I must anticipate a little, and tell you. It evidently comes, as to part of it, from the candle, but is it within the candle beforehand? No, it is not in the candle; and it is not in the air around about the candle which is necessary for its combustion. It is neither in one nor the other, but it comes from their conjoint action, a part from the candle, a part from the air; and this we have now to trace, so that we may understand thoroughly what is the chemical history of a candle when we have it burning on our table. How shall we get at this? I myself know plenty of ways, but I want you to get at it from the association in your own minds of what I have already told you.

I think you can see a little in this way. We had just now the case of a substance which acted upon the water in the way that Sir Humphry Davy showed us, and which I am now going to recall to your minds by making again an experiment upon that dish. [Potassium, the metallic basis of potash, was discovered by Sir Humphry Davy in 1807, who succeeded in separating it from potash by means of a powerful voltaic battery. Its great affinity for oxygen causes it to decompose water with evolution of hydrogen, which takes fire with the heat produced.] It is a thing which we have to handle very carefully; for you see, if I allow a little splash of water to come upon this mass, it sets fire to part of it; and if there were free access of air, it would quickly set fire to the whole. Now this is a metal—a beautiful and bright metal—which rapidly changes in the air, and, as you know, rapidly changes in water. I will put a piece on the water, and you see it burns beautifully, making a floating lamp, using the water in the place of air. Again, if we take a few iron filings or turnings and put them in water, we find that they likewise undergo an alteration. They do not change so much as this potassium does, but they change somewhat in the same way; they become rusty, and show an action upon the water, though in a different degree *of* intensity to what this beautiful metal does; but they act upon the water in the same manner generally as this potassium. I want you to put these different facts together in your minds. I have another metal here [zinc], and when we examined it with regard to the solid substance produced by its combustion, we had an opportunity of seeing that it burned; and I suppose, if I take a little strip of this zinc and put it over the candle, you will see something half way, as it were, between the combustion of potassium on the water and the action of iron—you see there is a sort of combustion. It has burned, leaving a white ash or residuum; and here also we find that the metal has a certain amount of action upon water.

By degrees we have learned how to modify the action of these different substances, and to make them tell us what we want to know. And now, first of all, I take iron. It is a common thing in all chemical reactions, where we get any result of this kind, to find that it is increased by the action of heat; and if we want to examine minutely and carefully the action of bodies one upon another, we often have to refer to the action of heat. You are aware, I believe, that iron filings burn beautifully in the air; but I am about to show you an experiment of this kind, because it will impress upon you what I am going to say about iron in its action on water. If I take a flame and make it hollow—you know why, because I want to get air to it and into it, and therefore I make it hollow—and then take a few iron filings and drop them into the flame, you see how well they burn. That combustion results from the chemical action which is going on when we ignite those particles. And so we proceed to consider these different effects, and ascertain what iron will do when it meets with water. It will tell us the story so beautifully, so gradually and regularly, that I think it will please you very much.

I have here a furnace with a pipe going through it like an iron gun barrel (Fig. 68), and I have stuffed that barrel full of bright iron turnings, and placed it across the fire to be made red-hot. We can either send air through the barrel to come in contact with the iron, or we can send steam from this little boiler at the

Fig. 68

end of the barrel. Here is a stop-cock which shuts off the steam from the barrel until we wish to admit it. There is some water in these glass jars, which I have colored blue, so that you may see what happens. Now you know very well that any steam I might send through that barrel, if it went through into the water, would be condensed; for you have seen that steam can not retain its gaseous form if it be cooled down; you saw it here [pointing to the tin flask] crushing itself into a small bulk, and causing the flask holding it to collapse; so that if I were to send steam through that barrel it would be condensed, supposing the barrel were cold; it is, therefore, heated to perform the experiment I am now about to show you. I am going to send the steam through the barrel in small quantities, and you shall judge for yourselves, when you see it issue from the other end, whether it still remains steam. Steam is condensable into water, and when you lower the temperature of steam you convert it back into fluid water; but I have lowered the temperature of the gas which I have collected in this jar by passing it through water after it has traversed the iron barrel, and still it does not change back into water. I will take another test and apply to this gas. (I hold the jar in an inverted position, or my substance would escape.) If I now apply a light to the mouth of the jar, it ignites with a slight noise. That tells you that it is not steam; steam puts out a fire: it does not burn; but you saw that what I had in that jar burnt. We may obtain this substance equally from water produced from the candle flame as from any other source. When it is obtained by the action of the iron upon the aqueous vapor, it leaves the iron in a state very similar to that in which these filings were after they were burnt. It makes the iron heavier than it was before. So long as the iron remains in the tube and is heated, and is cooled again without the access of air or water, it does not change in its weight; but after having had this current of steam passed over it, it then comes out heavier than it was before, having taken something out of the steam, and having allowed something else to pass forth, which we see here. And now, as we have another jar full, I will show you something most interesting. It is a combustible gas; and I might at once take this jar and set fire to the contents, and show you that it is combustible; but I intend to show you more, if I can. It is also a very light substance. Steam will condense; this body will rise in the air, and not condense. Suppose I take another glass jar, empty of all but air: if I examine it with a taper I shall find that it contains nothing but air. I will now take this jar full of the gas that I am speaking of, and deal with it as though it were a light body; I will hold both upside down, and turn the one up under the other (Fig. 69); and that which did contain the gas procured from the steam, what does it contain now? You will find it now only contains air. But look! Here is the combustible substance [taking the other jar] which I have poured out of the one jar into the other. It still preserves its quality, and condition, and independence, and therefore is the more worthy of our consideration, as belonging to the products of a candle.

Fig. 69

Now this substance which we have just prepared by the action of iron on the steam or water, we can also get by means of those other things which you have already seen act so well upon the water. If I take a piece of potassium, and make the necessary arrangements, it will produce this gas; and if, instead, a piece of zinc, I find, when I come to examine it very carefully, that the main reason why this zinc cannot act upon the water continuously as the other metal does, is because the result of the action of the water envelops the zinc in a kind of protecting coat. We have learned in consequence, that if we put into our vessel only the zinc and water, they, by themselves, do not give rise to much action, and we get no result. But suppose I proceed to dissolve off this varnish—this encumbering substance—which I can do by a little acid; the moment I do this 1 find the zinc acting upon the water exactly as the iron did, but at the common temperature. The acid in no way is altered, except in its combination with the oxide of zinc which is produced. I have now poured the acid into the glass, and the effect is as though I were applying heat to cause this boiling up. There is something coming off from the zinc very abundantly, which is not steam. There is a jar full of it; and you will find that I have exactly the same combustible substance remaining in the vessel, when I hold it upside down, that I produced during the experiment with the iron barrel. This is what we get from water, the same substance which is contained in the candle.

Let us now trace distinctly the connection between these two points. This is hydrogen—a body classed among those things which in chemistry we call elements, because we can get nothing else out of them. A candle is not an elementary body, because we can get carbon out of it; we can get this hydrogen out of it, or at least out of the water which it supplies. And this gas has been so named hydrogen, because it is that element which, in association with another, generates water. Mr. Anderson having now been able to get two or three jars of gas, we shall have a few experiments to make, and I want to show you the

Fig. 70

best way of making these experiments (Fig. 70). I am not afraid to show you, for I wish you to make experiments, if you will only make them with care and attention, and the assent of those around you. As we advance in chemistry we are obliged to deal with substances which are rather injurious if in their wrong places; the acids, and heat, and combustible things we use, might do harm if carelessly employed. If you want to make hydrogen, you can make it easily from bits of zinc, and sulphuric or muriatic acid. Here is what in former times was called the "philosopher's candle." It is a little phial with a cork and a tube or pipe passing through it. And I am now putting a few little pieces of zinc into it. This little instrument I am going to apply to a useful purpose in our demonstrations, for I want to show you that you can prepare hydrogen, and make some experiments with it as you please, at your own homes. Let me here tell you why I am so careful to fill this phial nearly, and yet not quite full. I do it because the evolved gas, which, as you have seen, is very combustible, is explosive to a considerable extent when mixed with air, and might lead to harm if you were to apply a light to the end of that pipe before all the air had been swept out of the space above the water. I am now about to pour in the sulphuric acid. I have used very little zinc and more sulphuric acid and water, because I want to keep it at work for some time. I therefore take care in this way to modify the proportions of the ingredients so that I may have a regular supply—not too quick and not too slow. Supposing I now take a glass and put it upside down over the end of the tube, because the hydrogen is light I expect that it will remain in that vessel a little while. We will now test the contents of our glass to see if there be hydrogen in it; I think I am safe in saying we have caught some [applying a light]. There it is, you see. I will now apply a light to the top of the tube. There is the hydrogen burning (Fig. 71). There is our philosophical candle. It is a foolish, feeble sort of a flame, you may say, but it is so hot that scarcely any common flame gives out so much heat. It goes on burning regularly, and I am now about to put that flame to burn under a certain arrangement, in order that we may examine its results and make use of the information which we may thereby acquire. Inasmuch as the candle produces water, and this gas comes out of the water, let us see what this gives us by the same process of combustion that the candle went through when it burnt in the

Fig. 71

atmosphere; and for that purpose I am going to put the lamp under this apparatus (Fig. 72), in order to condense whatever may arise from the combustion within it. In the course of a short time you will see moisture appearing in the cylinder, and you will get the water running down the side, and the water from this hydrogen flame will have absolutely the same effect upon all our tests, being obtained by the same general process as in the former case. This hydrogen is a very beautiful substance. It is so light that it carries things up; it is far lighter than the atmosphere; and I dare say I can show you this by an experiment which if you are very clever, some of you may even have skill enough to repeat. Here is our generator of hydrogen, and here are some soapsuds. I have an India-rubber tube connected with the hydrogen generator, and at the end of the tube is a tobacco pipe. I can thus put the pipe into the suds and blow bubbles by means of the hydrogen. You observe how the bubbles fall downward when I blow them with my warm breath; but notice the difference when I blow them with hydrogen. [The lecturer here blew bubbles with hydrogen, which rose to the roof of the theatre.] It shows you how light this gas must be in order to carry with it not merely the ordinary soap bubble, but the larger portion of a drop hanging to the bottom of it. I can show its lightness in a better way than this; larger bubbles than these may be so lifted up; indeed, in former times balloons used to be filled with this gas. Mr. Anderson will fasten this tube on to our generator, and we shall have a stream of hydrogen here with which we can charge this balloon made of collodion. I need not even be very careful to get all the air out, for I know the power of this gas to carry it up. [Two collodion balloons were inflated and sent up, one being held by a string.] Here is another larger one, made of thin membrane, which we will fill and allow to ascend; you will see they will all remain floating about until the gas escapes.

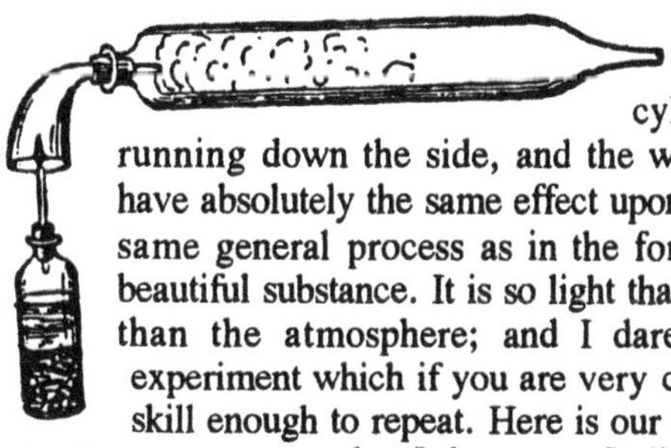

Fig. 72

What, then, are the comparative weights of these substances? I have a table here which will show you the proportion which their weights bear to each other. I have taken a pint and a cubic foot as the measures, and have placed opposite to them the respective figures. A pint measure of this hydrogen weighs three-quarters of our smallest weight, a grain, and a cubic foot weighs one-twelfth of an ounce; whereas a pint of water weighs 8,750 grains, and a cubic foot of water weighs almost 1,000 ounces. You see, therefore, what a vast difference there is between the weight of a cubic foot of water and a cubic foot of hydrogen.

Hydrogen gives rise to no substance that can become solid, either during combustion or afterward as a product of its combustion; but when it burns it produces water only; and if we take a cold glass and put it over the flame, it becomes damp, and you have water produced immediately in appreciable quantity; and nothing is produced by its combustion but the same water which you have seen the flame of the candle produce. It is important to remember that this hydrogen is the only thing in nature which furnishes water as the sole product of combustion.

And now we must endeavor to find some additional proof of the general character and composition of water, and for this purpose I will keep you a little longer, so that at our next meeting we may be better prepared for the subject. We have the power of arranging the zinc which you have seen acting upon the water by the assistance of an acid, in such a manner as to cause all the power to be evolved in the place where we require it. I have behind me a voltaic pile, and I am just about to show you, at the end of this lecture, its character and power, that you may see what we shall have to deal with when next we meet. I hold here the extremities of the wires which transport the power from behind me, and which I shall cause to act on the water.

We have previously seen what a power of combustion is possessed by the potassium, or the zinc, or the iron filings; but none of them show such energy as this. [The lecturer here made contact between the two terminal wires of the battery, when a brilliant flash of light was produced.] This light is, in fact, produced by a forty-zinc power of burning; it is a power that I can carry about in my hands through these wires at pleasure, although if I applied it wrongly to myself, it would destroy me in an instant, for it is a most intense thing, and the power you see here put forth while you count five [bringing the poles in contact and exhibiting the electric light] is equivalent to the power of several thunder-storms, so great is its force. [Professor Faraday has calculated that there is as much electricity required to decompose one grain of water as there is in a very powerful flash of lightning.] And that you may see what intense energy it has, I will take the ends of the wires which convey the power from the battery, and with it I dare say I can burn this iron file. Now this is a chemical power, and one which, when we next meet, I shall apply to water, and show you what results we are able to produce.

THE SPECIAL THEORY OF RELATIVITY

ALBERT EINSTEIN

Albert Einstein (1879-1955), regarded as one of the most creative intellects in all history, was the first to challenge seriously the Newtonian view of the universe. In the single year of 1905, at the age of twenty-six, he published five papers that profoundly affected the scientific world. A paper on molecular dimensions won for him a doctorate from the University of Zurich. Another was on the motion of small particles suspended in a stationary liquid that confirmed the existence of molecules. A third was on light, in which he developed the theory of quanta, or photons, and the photoelectric effect. Then came his special theory of relativity, and finally, a mathematical extension of that theory that established the equivalence of mass and energy, with the famous equation $E=mc^2$. He presented his general theory of relativity in 1916. In 1921 he was awarded the Nobel Prize in physics for his "photoelectric law" and "work in the field of theoretical physics." He devoted much of his later life to an unsuccessful attempt to establish a "unified field theory" that would bring together in a single system electromagnetic and gravitational phenomena.

Einstein was born in Ulm, in southern Germany, of Jewish parents. The family soon moved to Munich where the father and uncle ran a small electrochemical factory. Einstein disliked the elementary school and did poorly there. It was his uncle who introduced him to the world of mathematics. At age 14, he read with great interest a series of books on natural science.

After a second attempt at the entrance examination, he entered the Federal Polytechnic Institute in Zurich, Switzerland, in 1896 and graduated in 1900, though he spent much of his time studying on his own rather than attending lectures and doing assignments.

In 1902 he was appointed an examiner in the Swiss patent office in Bern, and shortly afterward married his college sweetheart, Mileva Maritsch. They had two sons. In 1909 he accepted an academic position at the University of Zurich. In succeeding years, he held research or teaching positions at Prague, back at Zurich, and in Berlin. After the break-up of his first marriage, he married his cousin, Elsa, in 1919.

Soon after Hitler's rise to power in 1933, Einstein renounced his German citizenship and left the country. He arrived in Princeton, New Jersey, in October of that year to become a full-time member of the newly formed Institute for Advanced Study.

It was a letter from Einstein, sent at the urging of Enrico Fermi and Leo Szilard, to President Franklin D. Roosevelt in 1939, that set in motion the inquiries that led to the Manhattan Project for making the atomic bomb.

1. Compare Einstein's view of space and time with that of Newton. Does the rigor used by Einstein in describing exactly what is meant by distance seem necessary to you?

2. What does Einstein mean by the principle of relativity?

3. Explain, in your own words, the conflict between the principle of relativity and the law of propagation of light. How does Einstein choose to resolve the conflict?

4. The implications of the special theory of relativity—that time and distance are not absolute quantities—seem to be in conflict with our preconceived notions of time and space. Why should such a deep-seated conflict exist?

5. "Everything is relative" is often used as a justification for almost any failure to adhere to a set of standards and Einstein is cited as the source for this quote. Do you believe that this is fair use of his work? Explain.

THE SPECIAL THEORY OF RELATIVITY

ALBERT EINSTEIN

I
PHYSICAL MEANING OF GEOMETRICAL PROPOSITIONS

IN your schooldays most of you who read this book made acquaintance with the noble building of Euclid's geometry, and you remember—perhaps with more respect than love—the magnificent structure, on the lofty staircase of which you were chased about for uncounted hours by conscientious teachers. By reason of your past experience, you would certainly regard everyone with disdain who should pronounce even the most out-of-the-way proposition of this science to be untrue. But perhaps this feeling of proud certainty would leave you immediately if some one were to ask you: "What, then, do you mean by the assertion that these propositions are true?" Let us proceed to give this question a little consideration.

Geometry sets out from certain conceptions such as "plane," "point," and "straight line," with which we are able to associate more or less definite ideas, and from certain simple propositions (axioms) which, in virtue of these ideas, we are inclined to accept as "true." Then, on the basis of a logical process, the justification of which we feel ourselves compelled to admit, all remaining propositions are shown to follow from those axioms, i.e. they are proven. A proposition is then correct ("true") when it has been derived in the recognized manner from the axioms. The question of the "truth" of the individual geometrical propositions is thus reduced to one of the "truth" of the axioms. Now it has long been known that the last question is not only unanswerable by the methods of geometry, but that it is in itself entirely without meaning. We cannot ask whether it is true that only one straight line goes through two points. We can only say that Euclidean geometry deals with things called "straight lines," to each of which is ascribed the property of being uniquely determined by two points situated on it. The concept "true" does not tally with the assertions of pure geometry, because by the word "true" we are eventually in the habit of designating always the correspondence with a "real" object; geometry, however, is not concerned with the relation of the ideas involved in it to objects of experience, but only with the logical connection of these ideas among themselves.

It is not difficult to understand why, in spite of this, we feel constrained to call the propositions of geometry "true." Geometrical ideas correspond to more or less exact objects in nature, and these last are undoubtedly the exclusive cause of the genesis of those ideas. Geometry ought to refrain from such a course, in order to give to its structure the largest possible logical unity. The practice, for example, of seeing in a "distance" two marked positions on a practically rigid body is something which is lodged deeply in our habit of thought. We are accustomed further to regard three points as being situated on a straight line, if their apparent positions can be made to coincide for observation with one eye, under suitable choice of our place of observation.

If, in pursuance of our habit of thought, we now supplement the propositions of Euclidean geometry by the single proposition that two points on a practically rigid body always correspond to the same distance (line-interval), independently of any changes in position to which we may subject the body, the propositions of Euclidean geometry then resolve themselves into propositions on the possible relative position of practically rigid bodies. Geometry which has been supplemented in this way is then to be treated as a branch of physics. We can now legitimately ask as to the "truth" of geometrical propositions interpreted in this way, since we are justified in asking whether these propositions are satisfied for those real things we have associated with the geometrical ideas. In less exact terms we can express this by saying that by the "truth" of a geometrical proposition in this sense we understand its validity for a construction with ruler and compasses.

It follows that a natural object is associated also with a straight line. Three points A, B and C on a rigid body thus lie in a straight line when, the points A and C being given, B is chosen such that the sum of the distances *AB* and *BE* is as short as possible. This incomplete suggestion will suffice for our present purpose.

Of course the conviction of the "truth" of geometrical propositions in this sense is founded exclusively on rather incomplete experience. For the present we shall assume the "truth" of the geometrical propositions, then at a later stage (in the general theory of relativity) we shall see that this "truth" is limited, and we shall consider the extent of its limitation.

II
THE SYSTEM OF CO-ORDINATES

ON the basis of the physical interpretation of distance which has been indicated, we are also in a position to establish the distance between two points on a rigid body by means of measurements. For this purpose we require a "distance" (rod S) which is to be used once and for all, and which we employ as a standard measure. If, now, A and B are two points on a rigid body, we can construct the line joining them according to the rules of geometry; then, starting from A, we can mark off the distance S time after time until we reach *B*. The number of these operations required is the numerical measure of the distance *AB*. This is the basis of all measurement of length.

Every description of the scene of an event or of the position of an object in space is based on the specification of the point on a rigid body (body of reference) with which that event or object coincides. This applies not only to scientific description, but also to everyday life. If I analyse the place specification "Trafalgar Square, London," I arrive at the following result. The earth is the rigid body to which the specification of place refers; "Trafalgar Square, London," is a well-defined point, to which a name has been assigned, and with which the event coincides in space.

This primitive method of place specification deals only with places on the surface of rigid bodies, and is dependent on the existence of points on this surface which are distinguishable from each other. But we can free ourselves from both of these limitations without altering the nature of our specification of position. If, for instance, a cloud is hovering over Trafalgar Square, then we can determine its position relative to the surface of the earth by erecting a pole perpendicularly on the Square, so that it reaches the cloud. The length of the pole measured with the standard measuring-rod, combined with the specification of the position of the foot of the pole, supplies us with a complete place specification. On the basis of this illustration, we are able to see the manner in which a refinement of the conception of position has been developed.

(a) We imagine the rigid body, to which the place specification is referred, supplemented in such a manner that the object whose position we require is reached by the completed rigid body.

(b) In locating the position of the object, we make use of a number (here the length of the pole measured with the measuring-rod) instead of designated points of reference.

(c) We speak of the height of the cloud even when the pole which reaches the cloud has not been erected. By means of optical observations of the cloud from different positions on the ground, and taking into account the properties of the propagation of light, we determine the length of the pole we should have required in order to reach the cloud.

From this consideration we see that it will be advantageous if, in the description of position, it should be possible by means of numerical measures to make ourselves independent of the existence of marked positions (possessing names) on the rigid body of reference. In the physics of measurement this is attained by the application of the Cartesian system of co-ordinates.

This consists of three plane surfaces perpendicular to each other and rigidly attached to a rigid body. Referred to a system of co-ordinates, the scene of any event will be determined (for the main part) by the specification of the lengths of the three perpendiculars or co-ordinates (x,y,z) which can be dropped from the scene of the event to those three plane surfaces. The lengths of these three perpendiculars can be determined by a series of manipulations with rigid measuring-rods performed according to the rules and methods laid down by Euclidean geometry.

In practice, the rigid surfaces which constitute the system of co-ordinates are generally not available; furthermore, the magnitudes of the co-ordinates are not actually determined by constructions with rigid rods, but by indirect means. If the results of physics and astronomy are to maintain their clearness, the physical meaning of specifications of position must always be sought in accordance with the above considerations.

We thus obtain the following result: Every description of events in space involves the use of a rigid body to which such events have to be referred. The resulting relationship takes for granted that the laws of Euclidean geometry hold for "distances," the "distance" being represented physically by means of the convention of two marks on a rigid body.

III
PLACE AND TIME IN CLASSICAL MECHANICS

THE purpose of mechanics is to describe how bodies change their position in space with "time." I should load my conscience with grave sins against the sacred spirit of lucidity were I to formulate the aims of mechanics in this way, without serious reflection and detailed explanations. Let us proceed to disclose these sins.

It is not clear what is to be understood here by "position" and "space." I stand at the window of a railway carriage which is travelling uniformly, and drop a stone on the embankment, without throwing it. Then, disregarding the influence of the air resistance, I see the stone descend in a straight line. A pedestrian who observes the misdeed from the footpath notices that the stone falls to earth in a parabolic curve. I now ask: Do the "positions" traversed by the stone lie "in reality" on a straight line or on a parabola? Moreover, what is meant here by motion "in space"? From the considerations of the previous section the answer is self-evident. In the first place we entirely shun the vague word "space," of which, we must honestly acknowledge, we cannot form the slightest conception, and we replace it by "motion relative to a practically rigid body of reference." The positions relative to the body of reference (railway carriage or embankment) have already been defined in detail in the preceding section. If instead of "body of reference" we insert "system of co-ordinates," which is a useful idea for mathematical description, we are in a position to say: The stone traverses a straight line relative to a system of co-ordinates rigidly attached to the carriage, but relative to a system of co-ordinates rigidly attached to the ground (embankment) it describes a parabola. With the aid of this example it is clearly seen that there is no such thing as an independently existing trajectory (lit. "path-curve"), but only a trajectory relative to a particular body of reference.

In order to have a *complete* description of the motion, we must specify how the body alters its position in time; i.e., for every point on the trajectory it must be stated at what time the body is situated there. These data must be supplemented by such a definition of time that, in virtue of this definition, these time-values can be regarded essentially as magnitudes (results of measurements) capable of observation. If we take our stand on the ground of classical mechanics, we can satisfy this requirement for our illustration in the following manner. We imagine two clocks of identical construction; the man at the railway-carriage window is holding one of them, and the man on the footpath the other. Each of the observers determines the position on his own reference-body occupied by the stone at each tick of the clock he is holding in his hand. In this connection we have not taken account of the inaccuracy involved by the finiteness of the velocity of propagation of light. With this and with a second difficulty prevailing here we shall have to deal in detail later.

IV
THE GALILEAN SYSTEM OF CO-ORDINATES

AS is well known, the fundamental law of the mechanics of Galilei-Newton, which is known as the law of *inertia*, can be stated thus: A body removed sufficiently far from other bodies continues in a state of rest or of uniform motion in a straight line. This law not only says something about the motion of the bodies, but it also indicates the reference-bodies or systems of coordinates, permissible in mechanics, which can be used in mechanical description. The visible fixed stars are bodies for which the law of inertia certainly holds to a high degree of approximation. Now if we use a system of co-ordinates which is rigidly attached to the earth, then, relative to this system, every fixed star describes a circle of immense radius in the course of an astronomical day, a result which is opposed to the statement of the law of inertia. So that if we adhere to this law we must refer these motions only to systems of coordinates relative to which the fixed stars do not move in a circle. A system of co-ordinates of which the state of motion is such that the law of inertia holds relative to it is called a "Galilean system of co-ordinates." The laws of the mechanics of Galilei-Newton can be regarded as valid only for a Galilean system of co-ordinates.

V
THE PRINCIPLE OF RELATIVITY (IN THE RESTRICTED SENSE)

IN order to attain the greatest possible clearness, let us return to our example of the railway carriage supposed to be travelling uniformly. We call its motion a uniform translation ("uniform" because it is of constant velocity and direction, "translation" because although the carriage changes its position relative to the embankment yet it does not rotate in so doing). Let us imagine a raven flying through the air in such a manner that its motion, as observed from the embankment, is uniform and in a straight line. If we were to observe the flying raven from the moving railway carriage, we should find that the motion of the raven would be one of different velocity and direction. but that it would still be uniform and in a straight line. Expressed in an abstract manner we may say: if a mass m is moving uniformly in a straight line with respect to a co-ordinate system K, then it will also be moving uniformly and in a straight line relative to a second co-ordinate system K', provided that the latter is executing a uniform translatory motion with respect to each. In accordance with the discussion contained in the preceding section, it follows that: if K is a Galilean co-ordinate system, then every other co-ordinate system R' is a Galilean one, when, in relation to *K*, it is in a condition of uniform motion of translation. Relative to *K'* the mechanical laws of Galilei-Newton hold good exactly as they do with respect to K.

We advance a step further in our generalisation when we express the tenet thus: If, relative to K, *K' is a* uniformly moving co-ordinate system devoid of rotation, then natural phenomena run their course with respect to *K'* according to exactly the same general laws as with respect to K. This statement is called the *principle of relativity* (in the restricted sense).

As long as one was convinced that all natural phenomena were capable of representation with the help of classical mechanics, there was no need to doubt the validity of this principle of relativity. But in view of the more recent development of electrodynamics and optics it became more and more evident that classical mechanics affords an insufficient foundation for the physical description of all natural phenomena. At this juncture the question of the validity of the principle of relativity became ripe for discussion, and it did not appear impossible that the answer to this question might be in the negative.

Nevertheless, there are two general facts which at the outset speak very much in favour of the validity of the principle of relativity. Even though classical mechanics does not supply us with a sufficiently broad basis for the theoretical presentation of all physical phenomena, still we must grant it a considerable measure of "truth," since it supplies us with the actual motions of the heavenly bodies with a delicacy of detail little short of wonderful. The principle of relativity must therefore apply with great accuracy in the domain of *mechanics.* But that a principle of such broad generality should hold with such exactness in one domain of phenomena, and yet should be invalid for another, is *a priori* not very probable.

We now proceed to the second argument, to which, moreover, we shall return later. If the principle of relativity (in the restricted sense) does not hold, then the Galilean co-ordinate systems *K, K', K',* etc., which are moving uniformly relative to each other, will not be *equivalent* for the description of natural phenomena. In this case we should be constrained to believe that natural laws are capable of being formulated in a particularly simple manner, and of course only on condition that, from amongst all possible Galilean co-ordinate systems, we should have chosen one (K_0) of a particular state of motion as our body of reference. We should then be justified (because of its merits for the description of natural phenomena) in calling this system "absolutely at rest," and all other Galilean systems K "in motion." If, for instance, our embankment were the system K_0 then our railway carriage would be a system K, relative to which less simple laws would hold than with respect to K,. This diminished simplicity would be due to the fact that the carriage K would be in motion (i.e, "really") with respect to K_0. In the general laws of nature which have been formulated with reference to K, the magnitude and direction of the velocity of the carriage would necessarily play a part. We should expect, for instance, that the note emitted by an organ pipe paced with its axis parallel to the direction of travel would be different from that emitted if the axis of the pipe were paced perpendicular to this direction. Now in virtue of its motion in an orbit round the sun, our earth is comparable with a railway carriage travelling with a velocity of about 30 kilometres per second. If the principle of relativity were not valid we should therefore expect that the direction of motion of the earth at any moment would enter into the laws of nature, and also that physical systems in their behaviour would be dependent on the orientation in space with respect to the earth. For owing to the alteration in direction of the velocity of revolution of the earth in the course of a year, the earth cannot be at rest relative to the hypothetical system K_0, throughout the whole year. However, the most careful observations have never revealed such anisotropic properties in terrestrial physical space, i.e. a physical non-equivalence of different directions. This is very powerful argument in favour of the principle of relativity.

VI
THE THEOREM OF THE ADDITION OF VELOCITIES EMPLOYED IN CLASSICAL MECHANICS

LET us suppose our old friend the railway carriage to be travelling along the rails with a constant velocity *v*, and that a man traverses the length of the carriage in the direction of travel with a velocity *w*. How quickly or, in other words, with what velocity W does the man advance relative to the embankment during the process? The only possible answer seems to result from the following consideration: If the man were to stand still for a second, he would advance relative to the embankment through a distance *v* equal numerically to the velocity of the carriage. As a consequence of his walking, however, he traverses an additional distance w relative to the carriage, and hence also relative to the embankment, in this second, the distance w being numerically equal to the velocity with which he is walking. Thus in total he covers the distance $W = v + w$ relative to the embankment in the second considered. We shall see later that this result, which expresses the theorem of the addition of velocities employed in classical mechanics, cannot be maintained; in other words, the law that we have just written down does not hold in reality. For the time being, however, we shall assume its correctness.

VII
THE APPARENT INCOMPATIBILITY OF THE LAW OF PROPAGATION OF LIGHT WITH THE PRINCIPLE OF RELATIVITY

THERE is hardly a simpler law in physics than that according to which light is propagated in empty space. Every child at school knows, or believes he knows, that this propagation takes place in straight lines with a velocity $c =$ 300,000 km./sec. At all events we know with great exactness that this velocity is the same for all colours, because if this were not the case, the minimum of emission would not be observed simultaneously for different colours during the eclipse of a fixed star by its dark neighbour. By means of similar considerations based on observations of double stars, the Dutch astronomer De Sitter was also able to show that the velocity of propagation of light cannot depend on the velocity of motion of the body emitting the light. The assumption that this velocity of propagation is dependent on the direction "in space" is in itself improbable.

In short, let us assume that the simple law of the constancy of the velocity of light c (in vacuum) is justifiably believed by the child at school. Who would imagine that this simple law has plunged the conscientiously thoughtful physicist into the greatest intellectual difficulties? Let us consider how these difficulties arise.

Of course we must refer the process of the propagation of light (and indeed every other process) to a rigid reference-body (co-ordinate system). As such a system let us again choose our embankment. We shall imagine the air above it to have been removed. If a ray of light be sent along the embankment, we see from the above that the tip of the ray will be transmitted with the velocity c relative to the embankment. Now let us suppose that our railway carriage is again travelling along the railway lines with the velocity v, and that its direction is the same as that of the ray of light, but its velocity of course much less. Let us inquire about the velocity of propagation of the ray of light relative to the carriage. It is obvious that we can here apply the consideration of the previous section, since the ray of light plays the part of the man walking along relatively to the carriage. The velocity W of the man relative to the embankment is here replaced by the velocity of light relative to the embankment. w is the required velocity of light with respect to the carriage, and we have

$$w = c - v$$

The velocity of propagation of a ray of light relative to the carriage thus comes out smaller than c.

But this result comes into conflict with the principe of relativity set forth in Section V. For, like every other general law of nature, the law of the transmission of light *in vacuo* must, according to the principe of relativity, be the same for the railway carriage as reference-body as when the rails are the body of reference. But, from our above consideration, this would appear to be impossible. If every ray of light is propagated relative to the embankment with the velocity *c*, then for this reason it would appear that another law of propagation of light must necessarily hold with respect to the carriage—a result contradictory to the principle of relativity.

In view of this dilemma there appears to be nothing else for it than to abandon either the principle of relativity or the simple law of the propagation of light *in vacuo*. Those of you who have carefully followed the preceding discussion are almost sure to expect that we should retain the principle of relativity, which appeals so convincingly to the intellect because it is so natural and simple. The law of the propagation of light *in vacuo* would then have to be replaced by a more complicated law conformable to the principle of relativity. The development of theoretical physics shows, however, that we cannot pursue this course. The epoch-making theoretical investigations of H. A. Lorentz on the electrodynamical and optical phenomena connected with moving bodies show that experience in this domain leads conclusively to a theory of electromagnetic phenomena, of which the law of the constancy of the velocity of light *in vacuo* is a necessary consequence. Prominent theoretical physicists were therefore more inclined to reject the principle of relativity, in spite of the fact that no empirical data had been found which were contradictory to this principle.

At this juncture the theory of relativity entered the arena. As a result of an analysis of the physical conceptions of time and space, it became evident that *in reality there is not the least incompatibility between the principle of relativity and the law of propagation of light,* and that by systematically holding fast to both these laws a logically rigid theory could be arrived at. This theory has been called the *special theory of* relativity to distinguish it from the extended theory, with which we shall deal later. In the following pages we shall present the fundamental ideas of the special theory of relativity.

VIII
ON THE IDEA OF TIME IN PHYSICS

LIGHTNING has struck the rails on our railway embankment at two places A and B far distant from each other. I make the additional assertion that these two lightning flashes occurred simultaneously. If I ask you whether there is sense in this statement, you will answer my question with a decided "Yes." But if I now approach you with the request to explain to me the sense of the statement more precisely, you find after some consideration that the answer to this question is not so easy as it appears at first sight.

After some time perhaps the following answer would occur to you: "The significance of the statement is clear in itself and needs no further explanation; of course it would require some consideration if I were to be commissioned to determine by observations whether in the actual case the two events took place simultaneously or not." I cannot be satisfied with this answer for the following reason. Supposing that as a result of ingenious considerations an able meteorologist were to discover that the lightning must always strike the places A and B simultaneously, then we should be faced with the task of testing whether or not this theoretical result is in accordance with the reality. We encounter the same difficulty with all physical statements in which the conception "simultaneous" plays a part The concept does not exist for the physicist until he has the possibility of discovering whether or not it is fulfilled in an actual case. We thus require a definition of simultaneity such that this definition supplies us with the method by means of which, in the present case, he can decide by experiment whether or not both the lightning strokes occurred simultaneously. As long as this requirement is not satisfied, I allow myself to be deceived as a physicist (and of course the same applies if I am not a physicist), when I imagine that I am able to attach a meaning to the statement of simultaneity. (I would ask the reader not to proceed farther until he is fully convinced on this point.)

After thinking the matter over for some time you then offer the following suggestion with which to test simultaneity. By measuring along the rails, the connecting line *AB* should be measured up and an observer placed at the mid-point M of the distance *AB*. This observer should be supplied with an arrangement (c.g. two mirrors inclined at 90°) which allows him visually to observe both places *A* and *B* at the same time. If the observer perceives the two flashes of lightning at the same time, then they are simultaneous.

I am very pleased with this suggestion, but for all that I cannot regard the matter as quite settled, because I feel constrained to raise the following objection: "Your definition would certainly be right, if only I knew that the light by means of which the observer at M perceives the lightning flashes travels along the length A → M with the same velocity as along the length B → M. But an examination of this supposition would only be possible if we already had at our disposal the means of measuring time. It would thus appear as though we were moving here in a logical circle."

After further consideration you cast a somewhat disdainful glance at me and rightly so—and you declare: "I maintain my previous definition nevertheless, because in reality it assumes absolutely nothing about light. There is only *one* demand to be made of the definition of simultaneity, namely, that in every real case it must supply us with an empirical decision as to whether or not the conception that has to be defined is fulfilled. That my definition satisfies this demand is indisputable. That light requires the same time to traverse the path A → M as for the path *B* → *M* is in reality neither a *supposition nor a hypothesis* about the physical nature of light, but a stipulation, which I can make of my own free will in order to arrive at a definition of simultaneity."

It is clear that this definition can be used to give an exact meaning not only to two events, but to as many events as we care to choose, and independently of the positions of the scenes of the events with respect to the body of reference (here the railway embankment). We are thus led also to a definition of "time" in physics. For this purpose we suppose that clocks of identical construction are placed at the points A, B and C of the railway line (co-ordinate system), and that they are set in such a manner that the positions of their pointers are simultaneously (in the above sense) the same. Under these conditions we understand by the "time" of an event the reading (position of the hands) of that one of these clocks which is in the immediate vicinity (in space) of the event. In this manner a time-value is associated with every event which is essentially capable of observation.

This stipulation contains a further physical hypothesis, the validity of which will hardly be doubted without empirical evidence to the contrary. It has been assumed that all these clocks go *at the same rate* if they are of identical construction. Stated more exactly: When two clocks arranged at rest in different places of a reference-body are set in such a manner that a *particular* position of the pointers of the one clock is *simultaneous* (in the above sense) with the same position of the pointers of the other clock, then identical "settings" are always simultaneous (in the sense of the above definition).

IX
THE RELATIVITY OF SIMULTANEITY

UP to now our considerations have been referred to a particular body of reference, which we have styled a "railway embankment." We suppose a very long train travelling along the rails with the constant velocity v and in the direction indicated in Fig. 1.

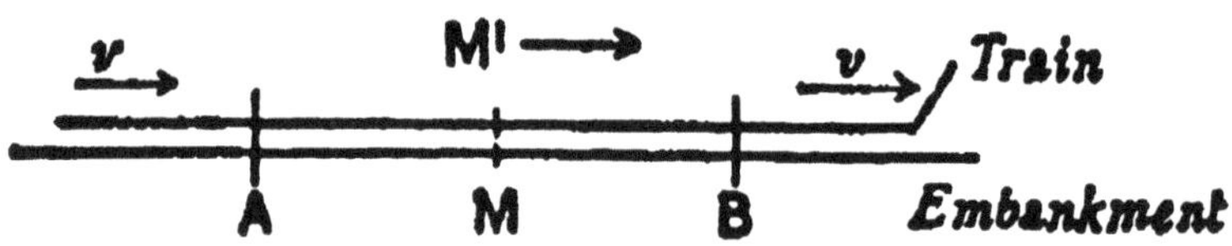

Fig. 1

People travelling in this train will with advantage use the train as a rigid reference-body (co-ordinate system); they regard all events in reference to the train. Then every event which takes place along the line also takes place at a particular point of the train. Also the definition of simultaneity can be given relative to the train in exactly the same way as with respect to the embankment. As a natural consequence, however, the following question arises:

Are two events (e.g. the two strokes of lightning A and B) which are simultaneous *with reference to the railway embankment* also simultaneous *relatively to the train*? We shall show directly that the answer must be in the negative.

When we say that the lightning strokes A and B are simultaneous with respect to the embankment, we mean: the rays of light emitted at the places A and B, where the lightning occurs, meet each other at the mid-point M of the length A → B of the embankment. But the events A and B also correspond to positions A and B on the train. Let M' be the mid-point of the distance A → B on the travelling train. Just when the flashes of lightning occur, this point M' naturally coincides with the point M, but it moves towards the right in the diagram with the velocity *v* of the train. If an observer sitting in the position M' in the train did not possess this velocity, then he would remain permanently at M, and the light rays emitted by the flashes of lightning A and B would reach him simultaneously, *i.e.* they would meet just where he is situated. Now in reality (considered with reference to the railway embankment) he is hastening towards the beam of light coming from B, whilst he is riding on ahead of the beam of light coming from A. Hence the observer will see the beam of light emitted from B earlier than he will see that emitted from A. Observers who take the railway train as their reference-body must therefore come to the conclusion that the lightning flash B took place earlier than the lightning flash A. We thus arrive at the important result:

Events which are simultaneous with reference to the embankment are not simultaneous with respect to the train, and *vice versa* (relativity of simultaneity). Every reference-body (co-ordinate system) has its own particular time; unless we are told the reference-body to which the statement of time refers, there is no meaning in a statement of the time of an event.

Now before the advent of the theory of relativity it had always tacitly been assumed in physics that the statement of time had an absolute significance, *i.e.*, that it is independent of the state of motion of the body of reference. But we have just seen that this assumption is incompatible with the most natural definition of simultaneity; if we discard this assumption, then the conflict between the law of the propagation of light *in vacuo* and the principle of relativity (developed in Section VII) disappears.

We were led to that conflict by the considerations of Section VI, which are now no longer tenable. In that section we concluded that the man in the carriage, who traverses the distance *w per second* relative to the carriage, traverses the same distance also with respect to the embankment in *each second* of time. But, according to the foregoing considerations, the time required by a particular occurrence with respect to the carriage must not be considered equal to the duration of the same occurrence as judged from the embankment (as reference-body). Hence it cannot be contended that the man in walking travels the distance w relative to the railway line in a time which is equal to one second as judged from the embankment.

Moreover, the considerations of Section VI are based on yet a second assumption, which in the light of a strict consideration, appears to be arbitrary, although it was always tacitly made even before the introduction of the theory of relativity.

X
ON THE RELATIVITY OF THE CONCEPTION OF DISTANCE

LET us consider two particular points on the train travelling along the embankment with the velocity v, and inquire as to their distance apart. We already know that it is necessary to have a body of reference for the measurement of a distance, with respect to which body the distance can be measured up. It is the simplest plan to use the train itself as reference-body (co-ordinate system). An observer in the train measures the interval by marking off his measuring-rod in a straight line (*e.g.*, along the floor of the carriage) as many times as is necessary to take him from the one marked point to the other.

Then the number which tells us how often the rod has to be laid down is the required distance.

It is a different matter when the distance has to be judged from the railway line. Here the following method suggests itself. If we call A' and B' the two points on the train whose distance apart is required, then both of these points are moving with the velocity *v* along the embankment. In the first place we require to determine the points A and B of the embankment which are just being passed by the two points A' and B' at a particular time *t*—judged from the embankment. These points A and B of the embankment can be determined by applying the definition of time given in Section VIII. The distance between these points A and B is then measured by repeated application of the measuring-rod along the embankment.

A priori it is by no means certain that this last measurement will supply us with the same result as the first. Thus the length of the train as measured from the embankment may be different from that obtained by measuring in the train itself. This circumstance leads us to a second objection which must be raised against the apparently obvious consideration of Section VI. Namely, if the man in the carriage covers the distance *w* in a unit of time—*measured from the train*,—then this distance—*as measured from the embankment*—is not necessarily also equal to *w*.

* * * * *

POLONIUM

RADIUM

PIERRE AND MARIE CURIE

Marie Curie (1867-1934) was born Marya Sklodowska in Warsaw, Poland. She won a gold medal upon completion of her secondary education at a Russian lycée at the age of 16. When she was 18, she took a position as governess to contribute to her sister's medical education in Paris.

Six years later she herself went to Paris and took up studies at the Sorbonne. In 1893 she received the *license ès physiques*, first in her class, and a year later she received the degree in mathematical sciences, second in the class. It was that year that she met Pierre Curie, supervisor at the École de Physique et de Chimie Industrielle. They were married in 1895.

Determined to continue in advanced work, Marie chose for her doctoral dissertation a follow-up study of Henri Becquerel's discovery of radiation in salts of uranium. She began searching for radiation in other elements. Pitchblende drew her attention as a mineral emitting rays beyond those that could be accounted for only by pure uranium. Now Pierre joined her in research, and they became the greatest husband and wife team in the history of science. Ultimately their efforts led to the discovery of polonium and radium.

Pierre refused a chair at the University of Geneva to continue his work with Marie. He did become a lecturer and then professor at the Sorbonne while Marie accepted an appointment as lecturer in physics at a school for girls in Sèvres.

In 1903 Marie and Pierre, together with Becquerel, received the Nobel Prize in physics.

In December 1904 Marie became chief assistant in a new laboratory directed by Pierre. On April 19, 1906, as he stepped into a street in Paris, Pierre was run over by a dray and was killed.

With determination, Marie continued her work. She was appointed to the professorship vacated by the death of her husband and then became the first woman to teach at the Sorbonne. In 1911 she received her second Nobel Prize, this time in chemistry. She was the only person ever so honored.

During and after World War I she devoted herself to the medical uses of radiation. In 1934, she died of leukemia probably resulting form radiation.

Meanwhile her elder daughter, Irene, and her husband, Frederic Joliot, both of whom had served as Marie's laboratory assistants, had become a notable scientific team in their own right. In 1934 they shared the Nobel Prize in chemistry. Some of their chief contributions were toward the discovery of the neutron, and in nuclear fission and the development of nuclear power.

Eve Curie, author of her mother's biography, was born in Paris in 1904. She became an accomplished pianist and writer.

The Curies' notes on polonium, excerpted here, first appeared in a paper, "On a New Radioactive Substance Contained in Pitchblende," in the *Proceedings of the Academy of Science* for July 18, 1898. The notes on Radium were published in the *Proceedings* for December 26, 1898.

1. What is pitchblende and what was its significance for Marie Curie?

2. What was the method, invented by Pierre and Jacques Curie, that Marie used in her initial investigations of radioactivity?

3. How did the Curies confirm their discoveries to the satisfaction of the scientific community?

4. What differentiated radium from polonium? Why did the Curies think that radium was a different substance from barium?

5. What has been the impact of the discovery of radium?

6. What lessons for scientific research would you draw from the experience of the Curies?

POLONIUM

PIERRE AND MARIE CURIE

Certain minerals containing uranium and thorium (pitchblende, chalcolite, uranite) are very active in emitting Becquerel rays. One of us has already shown that their activity is greater than that of uranium and thorium, and has expressed the opinion that this effect arises from some other very active substance contained in these minerals in small quantity.

The study of the compounds of uranium and of thorium has shown, in fact, that the property of emitting rays which make air conducting and which act on photographic plates, is a specific property of uranium and of thorium, which appears in all the compounds of these metals, being so much the more feeble as the proportion of the active metal in the compound is itself less. The physical state of the substances seems to be of altogether secondary importance. Various experiments have shown that if the substances are mixed with others their condition seems to have no effect except as it varies the proportion of the active body and the absorption produced by the inert substance. Certain causes, such as the presence of impurities, which have so great an effect on phosphorescence or fluorescence, are here altogether without effect. It therefore becomes very probable that if certain minerals are more active than uranium and thorium, it is because they contain a substance more active than these metals.

We have attempted to isolate this substance in pitch blende, and the experiment has confirmed our expectations.

Our chemical investigations have been guided by the tests made of the radiating activity of the products which were separated in each operation. Each product was placed on the plates of a condenser, and the conductibility acquired by the air was measured by an electrometer and a piezoelectric quartz, as in the work already referred to. We thus have not only an indication but a number which gives some measure of the richness of the product in the active substance.

The pitch blende that we analyzed was about two and a half times more active than uranium in our apparatus.

The details of the chemical methods employed are omitted.

By carrying on these different operations we obtained products which were more and more active. Finally, we obtained a substance whose activity is about 400 times greater than that of uranium.

We have attempted to discover among bodies which are already known if there are any which are radioactive. We have examined compounds of almost all the simple bodies; thanks to the kindness of several chemists we have received specimens of very rare substances. Uranium and thorium were the only ones which were evidently active, tantalum perhaps is very feebly so.

We believe, therefore, that the substance which we removed from pitch blende contains a metal which has not yet been known, similar to bismuth in its chemical properties. If the existence of this new metal is confirmed, we propose to call it *polonium,* after the name of the native country of one of us.

* * * * *

RADIUM

PIERRE AND MARIE CURIE

Two of us have shown that by purely chemical processes we may extract from pitch blende a strongly radioactive substance. This substance stands near bismuth in its chemical properties. We have expressed the opinion that pitch blende perhaps contains a new element, for which we proposed the name polonium.

The researches which we have since carried on are in agreement with the first results obtained; but in the course of these researches we encountered a second substance also strongly radioactive and entirely different from the first in its chemical properties. In fact, polonium is precipitated in acid solution by sulphuretted hydrogen; its salts are soluble in acids, and water precipitates them from these solutions; polonium is completely precipitated by ammonia.

The new radioactive substance that we have found presents the chemical aspects of almost pure barium; it is not precipitated either by sulphuretted hydrogen or by ammonium sulphide, or by ammonia; the sulphate is insoluble in water and in acids; the carbonate is insoluble in water; the chloride, very soluble in water, is insoluble in concentrated hydrochloric acid and in alcohol. Finally, it gives the spectrum of barium, which is easy to recognize.

We believe, nevertheless, that this substance, although for the most part consisting of barium, contains in addition a new element which gives it its radioactivity and which furthermore is very near barium in its chemical properties. These are the reasons which speak in favor of this view.

(1) Barium and its compounds are not ordinarily radioactive; now, one of us has shown that radioactivity seems to be an atomic property, persisting in all the chemical and physical states of matter. If we look at the thing this way, the radioactivity of our substance, which does not arise from barium, ought to be attributed to another element.

(2) The first substances which we obtained, in the state of hydrated chlorides, had a radioactivity 60 times greater than that of metallic uranium. (The radioactivity intensity is evaluated by the conductibility of the air in our apparatus.) By dissolving these chlorides in water and precipitating a part of them by alcohol, the precipitated part is much more active than the part which remains dissolved. By starting with this fact we may carry out a series of fractionations, from which we may obtain more and more active chlorides. We have thus obtained chlorides which have an activity 900 times greater than that of uranium. We have been stopped by the lack of material, but from the progress of the operations we may assume that the activity would have been much more increased if we had been able to continue. These facts can be explained by the presence of a radioactive element of which the chloride is less soluble in alcoholic solution than is barium chloride.

(3) M. Demarcay has examined the spectrum of our substance, with a kindness for which we do not know how to thank him enough. The results of his observations are presented in a special note which follows ours. M. Demarcay has found in the spectrum a ray which seems not to belong to any known element. This ray, which is scarcely visible in the chloride that is 60 times more active than uranium, becomes strongly marked in the chloride that was enriched

by fractionation until its activity was 900 times that of uranium. The intensity of this ray increases at the same time as the radioactivity, and this, we think, is a strong reason for attributing it to the radioactive part of our substance.

The various reasons which we have presented lead us to believe that the new radioactive substance contains a new element, to which we propose to give the name *radium*.

MARIE CURIE: A BIOGRAPHY

Eve Curie

CHAPTER XII
The Discovery of Radium

WHILE A YOUNG WIFE kept house, washed her baby daughter and put pans on the fire, in a wretched laboratory at the School of Physics a woman physicist was making the most important discovery of modern science.

At the end of 1897 the balance sheet of Marie's activity showed two university degrees, a fellowship and a monograph on the magnetization of tempered steel. No sooner had she recovered from childbirth than she was back again at the laboratory.

The next stage in the logical development of her career was the doctor's degree. Several weeks of indecision came in here. She had to choose a subject of research which would furnish fertile and original material. Like a writer who hesitates and asks himself questions before settling the subject of his next novel, Marie, reviewing the most recent work in physics with Pierre, was in search of a subject for a thesis.

At this critical moment Pierre's advice had an importance which cannot be neglected. With respect to her husband, the young woman regarded herself as an apprentice: he was an older physicist, much more experienced than she. He was even, to put it exactly, her chief, her "boss."

But without a doubt Marie's character, her intimate nature, had a great part in this all-important choice. From childhood the Polish girl had carried the curiosity and daring of an explorer within her. This was the instinct that had driven her to leave Warsaw for Paris and the Sorbonne, and had made her prefer a solitary room in the Latin Quarter to the Dluskis' downy nest. In her walks in the woods she always chose the wild trail or the unfrequented road.

At this moment she was like a traveler musing on a long voyage. Bent over the globe and pointing out, in some far country, a strange name that excites his imagination, the traveler suddenly decides to go there and nowhere else: so Marie, going through the reports of the latest experimental studies, was attracted by the publication of the French scientist Henri Becquerel of the preceding year. She and Pierre already knew this work; she read it over again and studied it with her usual care.

After Roentgen's discovery of X rays, Henri Poincaré conceived the idea of determining whether rays like the X ray were emitted by "fluorescent" bodies under the action of light. Attracted by the same problem, Henri Becquerel examined the salts of a "rare metal," uranium. Instead of finding the phenomenon he had expected, he observed another, altogether different and incomprehensible: he found that uranium salts *spontaneously* emitted, without exposure to light, some rays of unknown nature. A compound of uranium, placed on a photographic plate surrounded by black paper, made an impression on the plate through the paper. And, like the X ray, these astonishing "uranic" salts discharged an electroscope by rendering the surrounding air a conductor.

Henri Becquerel made sure that these surprising properties were not caused by a preliminary exposure to the sun and that they persisted when the uranium compound had been maintained in darkness for several months. For the first time, a physicist had observed the phenomenon to which Marie Curie was later to give the name of *radioactivity*. But the nature of the radiation and its origin remained an enigma.

Becquerel's discovery fascinated the Curies. They asked themselves whence came the energy—tiny, to be sure—which uranium compounds constantly disengaged in the form of radiation. And what was the nature of this radiation? Here was an engrossing subject of research, a doctor's thesis! The subject tempted Marie most because it was a virgin field: Becquerel's work was very recent and so far as she knew nobody in the laboratories of Europe had yet attempted to make a fundamental study of uranium rays. As a point of departure, and as the only bibliography, there existed some communications presented by Henri Becquerel at the Academy of Science during the year 1896. It was a leap into great adventure, into an unknown realm.

There remained the question of where she was to make her experiments—and here the difficulties began. Pierre made several approaches to the director of the School of Physics with practically no results: Marie was given the free use of a little glassed-in studio on the ground floor of the school. It was a kind of storeroom, sweating with damp, where unused machines and lumber were put away. Its technical equipment was rudimentary and its comfort nil.

Deprived of an adequate electrical installation and of everything that forms material for the beginning of scientific research, she kept her patience, sought and found a means of making her apparatus work in this hole.

It was not easy. Instruments of precision have sneaking enemies: humidity, changes of temperature. Incidentally the climate of this little workroom, fatal to the sensitive electrometer, was not much better for Marie's health. But this had no importance. When she was cold, the young woman took her revenge by noting the degrees of temperature in centigrade in her notebook. On February 6, 1898, we find, among the formulas and figures: "Temperature here 6°25." Six degrees . . .! Marie, to show her disapproval, added ten little exclamation points.

The candidate for the doctor's degree set her first task to be the measurement of the "power of ionization" of uranium rays—that is to say, their power to render the air a conductor of electricity and so to discharge an electroscope. The excellent method she used, which was to be the key to the success of her experiments, had been invented for the study of other phenomena by two physicists well known to her: Pierre and Jacques Curie. Her technical installation consisted of an "ionization chamber," a Curie electrometer and a piezoelectric quartz.

At the end of several weeks the first result appeared: Marie acquired the certainty that the intensity of this surprising radiation was proportional to the quantity of uranium contained in the samples under examination, and that this radiation, which could be measured with precision, was not affected either by the chemical state of combination of the uranium or by external factors such as lighting or temperature.

These observations were perhaps not very sensational to the uninitiated, but they were of passionate interest to the scientist. It often happens in physics that an inexplicable phenomenon can be subjected, after some investigation, to laws already known, and by this very fact loses its interest for the research worker. Thus, in a badly constructed detective story, if we are told in the third chapter that the woman of sinister appearance who might have committed the crime is in reality only an honest little housewife who leads a life without secrets, we feel discouraged and cease to read.

Nothing of the kind happened here. The more Marie penetrated into intimacy with uranium rays, the more they seemed without precedent, essentially unknown. They were like nothing else. Nothing affected them. In spite of their very feeble power, they had an extraordinary individuality.

Turning this mystery over and over in her head, and pointing toward the truth, Marie felt and could soon affirm that the incomprehensible radiation was an *atomic* property. She questioned: Even though the phenomenon had only been observed with uranium, nothing proved that uranium was the only chemical element capable of emitting such radiation. Why should not other bodies possess the same power? Perhaps it was only by chance that this radiation had been observed in uranium first, and had remained attached to uranium in the minds of physicists. Now it must be sought for elsewhere. . . .

No sooner said than done. Abandoning the study of uranium, Marie undertook to examine all *known chemical bodies*, either in the pure state or in compounds. And the result was not long in appearing: compounds of another element, thorium, also emitted spontaneous rays like those of uranium and of similar intensity. The physicist had been right: the surprising phenomenon was by no means the property of uranium alone, and it became necessary to give it a distinct name. Mme Curie suggested the name of *radioactivity*. Chemical substances like uranium and thorium, endowed with this particular "radiance," were called *radio elements*.

Radioactivity so fascinated the young scientist that she never tired of examining the most diverse forms of matter, always by the same method. Curiosity, a marvelous feminine curiosity, the first virtue of a scientist, was developed in Marie to the highest degree. Instead of limiting her observation to simple compounds, salts and oxides, she had the desire to assemble samples of minerals from the collection at the School of Physics, and of making them undergo almost at hazard, for her own amusement, a kind of customs inspection which is an electrometer test. Pierre approved, and chose with her the veined fragments, hard or crumbly, oddly shaped, which she wanted to examine.

Marie's idea was simple—simple as the stroke of genius. At the crossroads where Marie now stood, hundreds of research workers might have remained, nonplussed, for months or even years. After examining all known chemical substances, and discovering—as Marie had done—the radiation of thorium, they would have continued to ask themselves in vain whence came this mysterious radioactivity. Marie, too, questioned and wondered. But her surprise was translated into fruitful acts. She had used up all evident possibilities. Now she turned toward the unplumbed and the unknown.

She knew in advance what she would learn from an examination of the minerals, or rather she thought she knew. The specimens which contained neither uranium nor thorium would be revealed as totally "inactive." The others, containing uranium or thorium, would be radioactive.

Experiment confirmed this prevision. Rejecting the inactive minerals, Marie applied herself to the others and measured their radioactivity. Then came a dramatic revelation: the radioactivity was a *great deal stronger* than could have been normally foreseen by the quantity of uranium or thorium contained in the products examined!

"It must be an error in experiment," the young woman thought; for doubt is the scientist's first response to an unexpected phenomenon.

She started her measurements over again, unmoved, using the same products. She started over again ten times, twenty times. And she was forced to yield to the evidence: the quantities of uranium and of thorium found in these minerals were by no means sufficient to justify the exceptional intensity of the radiation she observed.

Where did this excessive and abnormal radiation come from? Only one explanation was possible: the minerals must contain, in small quantity, a *much more powerfully radioactive substance* than uranium and thorium.

But what substance? In her preceding experiments, Marie had already examined *all known chemical elements.*

The scientist replied to the question with the sure logic and the magnificent audaciousness of a great mind: The minerals certainly contained a radioactive substance, which was at the same time a chemical element unknown until this day: a *new element.*

A new element! It was a fascinating and alluring hypothesis—but still a hypothesis. For the moment this powerfully radioactive substance existed only in the imagination of Marie and of Pierre. But it did exist there. It existed strongly enough to make the young woman go to see Bronya one day and tell her in a restrained, ardent voice:

"You know, Bronya, the radiation that I couldn't explain comes from a new chemical element. The element is there and I've got to find it. We are sure! The physicists we have spoken to believe we have made an error in experiment and advise us to be careful. But I am convinced that I am not mistaken."

These were unique moments in her unique life. The layman forms a theatrical—and wholly false—idea of the research worker and of his discoveries. "The moment of discovery" does not always exist: the scientist's work is too tenuous, too divided, for the certainty of success to crackle out suddenly in the midst of his laborious toil like a stroke of lightning, dazzling him by its fire. Marie, standing in front of her apparatus, perhaps never experienced the sudden intoxication of triumph. This intoxication was spread over several days of decisive labor, made feverish by a magnificent hope. But it must have been an exultant moment when, convinced by the rigorous reasoning of her brain that she was on the trail of new matter, she confided the secret to her elder sister, her ally always. . . . Without exchanging one affectionate word, the two sisters must have lived again, in a dizzying breath of memory, their years of waiting, their mutual sacrifices, their bleak lives as students, full of hope and faith.

It was barely four years before that Marie had written:

> Life is not easy for any of us. But what of that? we must have perseverance and above all confidence in ourselves. We must believe that we are gifted for something, and that this thing, at whatever cost, must be attained.

That "something" was to throw science upon a path hitherto unsuspected.

In a first communication to the Academy, presented by Prof. Lippmann and published in the *Proceedings* on April 12, 1898 "Marie Sklodovska Curie" announced the probable presence in pitchblende ores of a new element endowed with powerful radioactivity. This was the first stage of the discovery of radium.

By the force of her own intuition the physicist had shown to herself that the wonderful substance must exist. She decreed its existence. But its incognito still had to be broken. Now she would have to verify hypothesis by experiment, isolate this material and see it. She must be able to announce with certainty: "It is there."

Pierre Curie had followed the rapid progress of his wife's experiments with passionate interest. Without directly taking part in Marie's work, he had frequently helped her by his remarks and advice. In view of the stupefying character of her results, he did not hesitate to abandon his study of crystals for the time being in order to join his efforts to hers in the search for the new substance.

Thus, when the immensity of a pressing task suggested and exacted collaboration, a great physicist was at Marie's side—a physicist who was the companion of her life. Three years earlier, love had joined this exceptional man and woman together—love, and perhaps some mysterious foreknowledge, some sublime instinct for the work in common.

The available force was now doubled. Two brains, four hands, now sought the unknown element in the damp little workroom in the Rue Lhomond. From this moment onward it is impossible to distinguish each one's part in the work of the Curies. We know that Marie, having chosen to study the radiation of uranium as the subject of her thesis, discovered that other substances were also radioactive. We know that after the examination of minerals she was able to announce the existence of a new chemical element, powerfully radioactive, and that it was the capital importance of this result which decided Pierre Curie to interrupt his very different research in order to try to isolate this element with his wife. At that time—May or June 1898—a collaboration began which was to last for eight years, until it was destroyed by a fatal accident.

We cannot and must not attempt to find out what should be credited to Marie and what to Pierre during these eight years. It would be exactly what the husband and wife did not want. The personal genius of Pierre Curie is known to us by the original work he had accomplished before this collaboration. His wife's genius appears to us in the first intuition of discovery, the brilliant start; and it was to reappear to us again, solitary, when Marie Curie the widow unflinchingly carried the weight of a new science and conducted it, through research, step by step, to its harmonious expansion. We therefore have formal proof that in the fusion of their two efforts, in this superior alliance of man and woman, the exchange was equal.

Let this certainty suffice for our curiosity and admiration. Let us not attempt to separate these creatures full of love, whose handwriting alternates and combines in the working notebooks covered with formulae, these creatures who were to sign nearly all their scientific publications together. They were to write "We found" and "We observed"; and when they were constrained by fact to distinguish between their parts, they were to employ this moving locution:

> Certain minerals containing uranium and thorium (pitchblende, chalcolite, uranite) are very active from the point of view of the emission of Becquerel rays. In a preceding communication, *one of us* showed that their activity was even greater than that of uranium and thorium and stated the opinion that this effect was due to some other very active substance contained in small quantity in these minerals. (Pierre and Marie Curie: *Proceedings of the Academy of Science,* July 18, 1898.)

Marie and Pierre looked for this "very active" substance in an ore of uranium called pitchblende, which in the crude state had shown itself to be four times more radioactive than the pure oxide of uranium that could be extracted from it. But the composition of this ore had been known for a long time with considerable precision. The new element must therefore be present in very small quantity or it would not have escaped the notice of scientists and their chemical analysis.

According to their calculations—"pessimistic" calculations, like those of true physicists, who always take the less attractive of two probabilities—the collaborators thought the ore should contain the new element to a maximum quantity of one per cent. They decided that this was very little. They would have been in consternation if they had known that the radioactive element they were hunting down did not count for more than a millionth part of pitchblende ore.

They began their prospecting patiently, using a method of chemical research invented by themselves, based on radioactivity they separated all the elements in pitchblende by ordinary chemical analysis and then measured the radioactivity of each of the bodies thus obtained. By successive eliminations they saw the "abnormal" radioactivity take refuge in certain parts of the ore. As they went on, the field of investigation was narrowed. It was exactly the technique used by the police when they search the houses of a neighborhood, one by one, to isolate and arrest a malefactor.

But there was more than one malefactor here: the radioactivity was concentrated principally in two different chemical fractions of the pitchblende. For M. and Mme. Curie it indicated the existence of two new elements instead of one. By July 1898 they were able to announce the discovery of one of these substances with certainty.

"You will have to name it," Pierre said to his young wife, in the same tone as if it were a question of choosing a name for little Irene.

The one-time Mlle. Sklodovska reflected in silence for a moment. Then, her heart turning toward her own country which had been erased from the map of the world, she wondered vaguely if the scientific event would be published in Russia, Germany and Austria—the oppressor countries—and answered timidly:

"Could we call it 'polonium'?"

In the *Proceedings of the Academy* for July 1890 we read:

> We believe the substance we have extracted from pitchblende contains a metal not yet observed, related to bismuth by its analytical properties. If the existence of this new metal is confirmed we propose to call it *polonium,* from the name of the original country of one of us.

The choice of this name proves that in becoming a Frenchwoman and a physicist Marie had not disowned her former enthusiasms. Another thing proves it for us: even before the note "On a New Radioactive Substance Contained in Pitchblende" had appeared in the *Proceedings of the Academy,* Marie had sent the manuscript to her native country, to that Joseph Boguski who directed the little laboratory at the Museum of Industry and Agriculture where she had made her first experiments. The communication was published in Warsaw in a monthly photographic review called *Swiatlo* almost as soon as in Paris.

Life was unchanged in the little flat in the Rue de la Glaciere. Marie and Pierre worked even more than usual; that was all. When the heat of summer came, the young wife found time to buy some baskets of fruit in the markets and, as usual, she cooked and put away preserves for the winter, according to the recipes used in the Curie family. Then she locked the shutters on her windows, which gave on burned leaves; she registered their two bicycles at the Orleans station, and, like thousands of other young women in Paris, went off on holiday with her husband and her child.

This year the couple had rented a peasant's house at Auroux, in Auvergne. Happy to breathe good air after the noxious atmosphere of the Rue Lhomond, the Curies made excursions to Mende, Puy, Clermont, Mont-Dore. They climbed hills, visited grottoes, bathed in rivers. Every day, alone in the country, they spoke of what they called their "new metals," polonium and "the other"—the one that remained to be found. In September they would go back to the damp workroom and the dull minerals; with freshened ardor they would take up their search again. . . .

One grief interfered with Marie's intoxication for work: the Dluskis were on the point of leaving Paris. They had decided to settle in Austrian Poland and to build a sanatorium for tubercular sufferers at Zakopane in the Carpathian Mountains. The day of separation arrived: Marie and Bronya exchanged brokenhearted farewells; Marie was losing her friend and protector, and for the first time she had the feeling of exile.

Marie to Bronya, December 2, 1898:

You can't imagine what a hole you have made in my life. With you two, I have lost everything I clung to in Paris except my husband and child. It seems to me that Paris no longer exists, aside from our lodging and the school where we work.

Ask Mme. Dluska if the green plant you left behind should be watered, and how many times a day. Does it need a great deal of heat and sun?

We are well, in spite of the bad weather, the rain and the mud. Irene is getting to be a big girl. She is very difficult about her food, and aside from milk tapioca she will eat hardly anything regularly, not even eggs. Write me what would be a suitable menu for persons of her age. . . .

In spite of their prosaic character—or perhaps because of it—some notes written by Mme Curie in that memorable year 1898 seem to us worth quoting. Some are to be found in the margins of a book called *Family Cooking,* with respect to a recipe for gooseberry jelly:

I took eight pounds of fruit and the same weight in crystallized sugar. After an ebullition of ten minutes, I passed the mixture through a rather fine sieve. I obtained fourteen pots of very good jelly, not transparent, which "took" perfectly.

In a school notebook covered with gray linen, in which the young mother had written little Irene's weight day by day, her diet and the appearance of her first teeth, we read under the date of July 20, 1898, some days after the publication of the discovery of polonium:

Irene says "thanks" with her hand. She can walk very well now on all fours. She says "Gogli, gogli, go." She stays in the garden all day at Sceaux on a carpet. She can roll, pick herself up, and sit down.

On August 15, at Auroux:

Irene has cut her seventh tooth, on the lower left. She can stand for half a minute alone. For the past three days we have bathed her in the river. She cries, but today (fourth bath) she stopped crying and played with her hands in the water.

She plays with the cat and chases him with war cries. She is not afraid of strangers any more. She sings a great deal. She gets up on the table when she is in her chair.

Three months later, on October 17, Marie noted with pride:

Irene can walk very well, and no longer goes on all fours.

On January 5, 1899:

Irene has fifteen teeth!

Between these two notes—that of October 17, 1898, in which Irene no longer goes on all fours, and that of January 5 in which Irene has fifteen teeth—and a few months after the note on the gooseberry preserve, we find another note worthy of remark.

It was drawn up by Marie and Pierre Curie and a collaborator called G. Bemont. Intended for the Academy of Science, and published in the *Proceedings* of the session of December 26, 1898, it announced the existence of a second new chemical element in pitchblende.

Some lines of this communication read as follows:

The various reasons we have just enumerated lead us to believe that the new radioactive substance contains a new element to which we propose to give the name of RADIUM.

The new radioactive substance certainly contains a very strong proportion of barium; in spite of that its radioactivity is considerable. The radioactivity of radium therefore must be enormous.

CHAPTER XIII
Four Years in a Shed

A MAN CHOSEN AT RANDOM from a crowd to read an account of the discovery of radium would not have doubted for one moment that radium existed: beings whose critical sense has not been sharpened and simultaneously deformed by specialized culture keep their imaginations fresh. They are ready to accept an unexpected fact, however extraordinary it may appear, and to wonder at it.

The physicist colleagues of the Curies received the news in slightly different fashion. The special properties of polonium and radium upset fundamental theories in which scientists had believed for centuries. How was one to explain the spontaneous radiation of the radioactive bodies? The discovery upset a world of acquired knowledge and contradicted the most firmly established ideas on the composition of matter. Thus the physicist kept on the reserve. He was violently interested in Pierre and Marie's work, he could perceive its infinite developments, but before being convinced he awaited the acquisition of decisive results.

The attitude of the chemist was even more downright. By definition, a chemist only believes in the existence of a new substance when he has seen the substance, touched it, weighed and examined it, confronted it with acids, bottled it, and when he has determined its "atomic weight."

Now, up to the present, nobody had "seen" radium. Nobody knew the atomic weight of radium. And the chemists, faithful to their principles, concluded: "No atomic weight, no radium. Show us some radium and we will believe you."

To show polonium and radium to the incredulous, to prove to the world the existence of their "children," and to complete their own conviction, M. and Mme Curie were now to labor for four years.

The aim was to obtain pure radium and polonium. In the most strongly radioactive products the scientists had prepared, these substances figured only in imperceptible traces. Pierre and Marie already knew the method by which they could hope to isolate the new metals, but the separation could not be made except by treating very large quantities of crude material.

Here arose three agonizing questions:

How were they to get a sufficient quantity of ore? What premises could they use to effect their treatment? What money was there to pay the inevitable cost of the work?

Pitchblende, in which polonium and radium were hidden, was a costly ore, treated at the St Joachimsthal mines in Bohemia for the extraction of uranium salts used in the manufacture of glass. Tons of pitchblende would cost a great deal: a great deal too much for the Curie household.

Ingenuity was to make up for wealth. According to the expectation of the two scientists, the extraction of uranium should leave, intact in the ore, such traces of polonium and radium as the ore contains. There was no reason why these traces should not be found in the residue. And, whereas crude pitchblende was costly, its residue after treatment had very slight value. By asking an Austrian colleague for a recommendation to the directors of the mine of St Joachimsthal would it not be possible to obtain a considerable quantity of such residue for a reasonable price?

It was simple enough: but somebody had to think of it.

It was necessary, of course, to buy this crude material and pay for its transportation to Paris. Pierre and Marie appropriated the required sum from their very slight savings. They were not so foolish as to ask for official credits. . . . If two physicists on the scent of an immense discovery had asked the University of Paris or the French government for a grant to buy pitchblende residue they would have been laughed at. In any case their letter would have been lost in the files of some office, and they would have had to wait months for a reply, probably unfavorable in the end. Out of the traditions and principles of the French Revolution, which had created the metric system, founded the Normal School, and encouraged science in many circumstances, the State seemed to have retained, after more than a century, only the deplorable words pronounced by Fouquier-Tinville at the trial in which Lavoisier was condemned to the guillotine: "The Republic has no need for scientists."

But at least could there not be found, in the numerous buildings attached to the Sorbonne, some kind of suitable workroom to lend to the Curie couple? Apparently not. After vain attempts, Pierre and Marie staggered back to their point of departure, which is to say to the School of Physics where Pierre taught, to the little room where Marie had done her first experiments. The room gave on a courtyard, and on the other side of the yard there was a wooden shack, an abandoned shed, with a skylight roof in such bad condition that it admitted the rain. The Faculty of Medicine had formerly used the place as a dissecting room, but for a long time now it had not even been considered fit to house the cadavers. No floor: an uncertain layer of bitumen covered the earth. It was furnished with some worn kitchen tables, a blackboard which had landed there for no known reason, and an old cast-iron stove with a rusty pipe.

A workman would not willingly have worked in such a place: Marie and Pierre, nevertheless, resigned themselves to it. The shed had one advantage: it was so untempting, so miserable, that nobody thought of refusing them the use of it. Schutzenberger, the director of the school, had always been very kind to Pierre Curie and no doubt regretted that he had nothing better to offer. However that may be, he offered nothing else; and the couple, very pleased at not being put out into the street with their material, thanked him, saying that "this would do" and that they would "make the best of it."

As they were taking possession of the shed, a reply arrived from Austria. Good news! By extraordinary luck, the residue of recent extractions of uranium had not been scattered. The useless material had been piled up in a no-man's-land planted with pine trees, near the mine of St Joachimsthal. Thanks to the intercession of Professor Suess and the Academy of Science of Vienna, the Austrian government, which was the proprietor of the State factory there, decided to present a ton of residue to the two French lunatics who thought they needed it. If, later on, they wished to be sent a greater quantity of the material, they could obtain it at the mine on the best terms. For the moment the Curies had to pay only the transportation charges on a ton of ore.

One morning a heavy wagon, like those which deliver coal, drew up in the Rue Lhomond before the School of Physics. Pierre and Marie were notified. They hurried bareheaded into the street in their laboratory gowns. Pierre, who was never agitated, kept his calm; but the more exuberant Marie could not contain her joy at the sight of the sacks that were being unloaded. It was pitchblende, *her* pitchblende, for which she had received a notice some days before from the freight station. Full of curiosity and impatience, she wanted to open one of the sacks and contemplate her treasure without further waiting. She cut the strings, undid the coarse sackcloth and plunged her two hands into the dull brown ore, still mixed with pine needles from Bohemia.

There was where radium was hidden. It was from there that Marie must extract it, even if she had to treat a mountain of this inert stuff like dust on the road.

Marya Sklodovska had lived through the most intoxicating moments of her student life in a garret; Marie Curie was to know wonderful joys again in a dilapidated shed. It was a strange sort of beginning over again, in which a sharp subtle happiness (which probably no woman before Marie had ever experienced) twice elected the most miserable setting.

The shed in the Rue Lhomond surpassed the most pessimistic expectations of discomfort. In summer, because of its skylights, it was as stifling as a hothouse. In winter one did not know whether to wish for rain or frost; if it rained, the water fell drop by drop, with a soft, nerve-racking noise, on the ground or on the worktables, in places which the physicists had to mark in order to avoid putting apparatus there. If it froze, one froze. There was no recourse. The stove, even when it was stoked white, was a complete disappointment. If one went near enough to touch it one received a little heat, but two steps away and one was back in the zone of ice.

It was almost better for Marie and Pierre to get used to the cruelty of the outside temperature, since their technical installation—hardly existent—possessed no chimneys to carry off noxious gases, and the greater part of their treatment had to be made in the open air, in the courtyard. When a shower came the physicists hastily moved their apparatus inside: to keep on working without being suffocated they set up draughts between the opened door and windows.

Marie probably did not boast to Dr. Vauthier of this very peculiar cure for attacks of tuberculosis.

> We had no money, no laboratory and no help in the conduct of this important and difficult task [she was to write later]. It was like creating something out of nothing, and if Casimir Dluski once called my student years "the heroic years of my sister-in-law's life," I may say without exaggeration that this period was, for my husband and myself, the heroic period of our common existence.
>
> . . . And yet it was in this miserable old shed that the best and happiest years of our life were spent, entirely consecrated to work. I sometimes passed the whole day stirring a mass in ebullition, with an iron rod nearly as big as myself. In the evening I was broken with fatigue.

In such conditions M. and Mme. Curie worked for four years from 1898 to 1902.

During the first year they busied themselves with the chemical separation of radium and polonium and they studied the radiation of the products (more and more active) thus obtained. Before long they considered it more practical to separate their efforts. Pierre Curie tried to determine the properties of radium, and to know the new metal better. Marie continued those chemical treatments which would permit her to obtain salts of pure radium.

In this division of labor Marie had chosen the "man's job." She accomplished the toil of a day laborer. Inside the shed her husband was absorbed by delicate experiments. In the courtyard, dressed in her old dust-covered and acid-stained smock, her hair blown by the wind, surrounded by smoke which stung her eyes and throat, Marie was a sort of factory all by herself.

> I came to treat as many as twenty kilograms of matter at a time [she writes], which had the effect of filling the shed with great jars full of precipitates and liquids. It was killing work to carry the receivers, to pour off the liquids and to stir, for hours at a stretch, the boiling matter in a smelting basin.

Radium showed no intention of allowing itself to be known by human creatures. Where were the days when Marie naively expected the radium content of pitchblende to be *one per cent?* The radiation of the new substance was so powerful that a tiny quantity of radium, disseminated through the ore, was the source of striking phenomena which could be easily observed and measured. The difficult, the impossible thing, was to isolate this minute quantity, to separate it from the gangue in which it was so intimately mixed.

The days of work became months and years: Pierre and Marie were not discouraged. This material which resisted them, which defended its secrets, fascinated them. United by their tenderness, united by their intellectual passions, they had, in a wooden shack, the "anti-natural" existence for which they had both been made, she as well as he.

> At this period we were entirely absorbed by the new realm that was, thanks to an unhoped-for discovery, opening before us [Marie was to write]. In spite of the difficulties of our working conditions, we felt very happy. Our days were spent at the laboratory. In our poor shed there reigned a great tranquillity: sometimes, as we watched over some operation, we would walk up and down, talking about work in the present and in the future; when we were cold a cup of hot tea taken near the stove comforted us. We lived in our single preoccupation as if in a dream.
>
> . . . We saw only very few persons at the laboratory; among the physicists and chemists there were a few who came from time to time, either to see our experiments or to ask for advice from Pierre Curie, whose competence in several branches of physics was well-known. Then took place some conversations before the blackboard—the sort of conversation one remembers well because it acts as a stimulant for scientific interest and the ardor for work without interrupting the course of reflection and without troubling that atmosphere of peace and meditation which is the true atmosphere of a laboratory.

Whenever Pierre and Marie, alone in this poor place, left their apparatus for a moment and quietly let their tongues run on, their talk about their beloved radium passed from the transcendent to the childish.

"I wonder what *It* will be like, what *It* will look like," Marie said one day with the feverish curiosity of a child who has been promised a toy. "Pierre, what form do you imagine *It* will take?"

"I don't know," the physicist answered gently. "I should like it to have a very beautiful color. . . ."

It is odd to observe that in Marie Curie's correspondence we find, upon this prodigious effort, none of the sensitive comments, decked out with imagery, which used to flash suddenly amid the familiarity of her letters. Was it because the years of exile had somewhat relaxed the young woman's intimacy with her people? Was she too pressed by work to find time?

The essential reason for this reserve is perhaps to be sought elsewhere. It was not by chance that Mme. Curie's letters ceased to be original at the exact moment when the story of her life became exceptional. As student, teacher or young wife, Marie could tell her story. . . . But now she was isolated by all that was secret and inexpressible in her scientific vocation. Among those she loved there was no longer anybody able to understand, to realize her worries and her difficult design. She could share her obsessions with only one person, Pierre Curie, companion. To him alone could she confide rare thoughts and dreams. Marie, from now on, was to present to all others, however near they might be to her heart, an almost commonplace picture of herself. She was to paint for them only the bourgeois side of her life. She was to find sometimes accents full of contained emotion to express her happiness as a woman. But of her work she was to speak only in laconic, inexpressive little phrases: news in three lines, without even attempting to suggest the wonders that work meant to her.

Here we feel an absolute determination not to illustrate the singular profession she had chosen by literature. Through subtle modesty, and also through horror of vain talk and everything superfluous, Marie concealed herself, dug herself in; or rather, she offered only one of her profiles. Shyness, boredom, or reason, whatever it may have been, the scientist of genius effaced and dissimulated herself behind "a woman like all others."

Marie to Bronya, 1899:

> Our life is always the same. We work a lot but we sleep well, so our health does not suffer. The evenings are taken up by caring for the child. In the morning I dress her and give her her food, then I can generally go out at about nine. During the whole of this year we have not been either to the theater or a concert, and we have not paid one visit. For that matter, we feel very well. . . . I miss my family enormously, above all you, my dears, and Father. I often think of my isolation with grief. I cannot complain of anything else, for our health is not bad, the child is growing well, and I have the best husband one could dream of; I could never have imagined finding one like him. He is a true gift of heaven, and the more we live together the more we love each other.

> Our work is progressing. I shall soon have a lecture to deliver on the subject. It should have been last Saturday but I was prevented from giving it, so it will no doubt be this Saturday, or else in a fortnight.

This work, which is so dryly mentioned in passing, was in fact progressing magnificently. In the course of the years 1895 and 1900 Pierre and Marie Curie published a report on the discovery of "induced radioactivity" due to radium, another on the effects of radioactivity, and another on the electric charge carried by the rays. And at last they drew up, for the Congress of Physics of 1900, a general report on the radioactive substances, which aroused immense interest among the scientists of Europe.

The development of the new science of radioactivity was rapid, overwhelming—the Curies needed fellow workers. Up to now they had had only the intermittent help of a laboratory assistant named Petit, an honest man who came to work for them outside his hours of service—working out of personal enthusiasm, almost in secret. But they now required technicians of the first order. Their discovery had important extensions in the domain of chemistry, which demanded attentive study. They wished to associate competent research workers with them.

> Our work on radioactivity began in solitude [Marie was to write]. But before the breadth of the task it became more and more evident that collaboration would be useful. Already in 1898 one of the laboratory chiefs of the school, G. Bemont, had given us some passing help. Toward 1900 Pierre Curie entered into relations with a young chemist, Andre Debierne, assistant in the laboratory of Professor Friedel, who esteemed him highly. Andre Debierne willingly accepted work on radioactivity. He undertook especially the research of a new radio element, the existence of which was suspected in the group of iron and rare clays. He discovered this element, named "actinium." Even though he worked in the physico-chemical laboratory at the Sorbonne directed by Jean Perrin, he frequently came to see us in our shed and soon became a very close friend to us, to Dr. Curie and later on to our children.

Thus, even before radium and polonium were isolated, a French scientist, Andre Debierne, had discovered a "brother," *actinium*.

> At about the same period [Marie tells us], a young physicist, Georges Sagnac, engaged in studying X rays, came frequently to talk to Pierre Curie about the analogies that might exist between these rays, their secondary rays, and the radiation of radioactive bodies. Together they performed a work on the electric charge carried by these secondary rays.

Marie continued to treat, kilogram by kilogram, the tons of pitchblende residue which were sent her on several occasions from St Joachimsthal. With her terrible patience, she was able to be, every day for four years, a physicist, a chemist, a specialized worker, an engineer and a laboring man all at once.

Thanks to her brain and muscle, the old tables in the shed held more and more concentrated products—products more and more rich in radium. Mme Curie was approaching the end: she no longer stood in the courtyard, enveloped in bitter smoke, to watch the heavy basins of material in fusion. She was now at the stage of purification and of the "fractional crystallization" of strongly radioactive solutions. But the poverty of her haphazard equipment hindered her work more than ever. It was now that she needed a spotlessly clean workroom and apparatus perfectly protected against cold, heat and dirt. In this shed, open to every wind, iron and coal dust was afloat which, to Marie's despair, mixed itself into the products purified with so much care. Her heart sometimes constricted before these little daily accidents which took so much of her time and her strength.

Pierre was so tired of the interminable struggle that he would have been quite ready to abandon it. Of course, he did not dream of dropping the study of radium and of radioactivity. But he would willingly have renounced, for the time being, the special operation of preparing pure radium. The obstacles seemed insurmountable. Could they not resume this work later on, under better conditions? More attached to the meaning of natural phenomena than to their material reality, Pierre Curie was exasperated to see the paltry results to which Marie's exhausting effort had led. He advised an armistice.

He counted without his wife's character. Marie wanted to isolate radium and she would isolate it. She scorned fatigue and difficulties, and even the gaps in her own knowledge wnich complicated her task. After all, she was only a very young scientist: she still had not the certainty and great culture Pierre had acquired by twenty years' work, and sometimes she stumbled across phenomena or methods of calculation which she knew very little, and for which she had to make hasty studies.

So much the worse! With stubborn eyes under her great brow, she clung to her apparatus and her test tubes.

In 1902, forty-five months after the day on which the Curies announced the probable existence of radium, Marie finally carried off the victory in this war of attrition: she succeeded in preparing a decigram of pure radium, and made a first determination of the atomic weight of the new substance, which was 225.

The incredulous chemists—of whom there were still a few—could only bow before the facts, before the superhuman obstinacy of a woman.

Radium officially existed.

* * * * *

PHYSICS AND PHILOSOPHY

WERNER HEISENBERG

Werner Heisenberg (1901-1976), atomic physicist, was born at Wurzburg, Germany. He received the Ph.D. both from the University of Munich and the University of Gottingen. In 1937 he married Elisabeth Schumacher. They had seven children.

Heisenberg became professor of theoretical physics at the University of Leipzig in 1927 where he remained until 1941. From 1941 to 1945, he was director of the Kaiser Wilhelm Institute for Physics and professor at the University of Berlin. For twelve years then he was director of the Max Planck Institute for Physics and professor at the University of Gottingen, and then for the next twelve years continued as director of the renamed Max Planck Institute for Physics and Astrophysics, and professor at the University of Munich.

He received special honors for his scientific work from societies and institutions in Spain, Greece, Japan, Denmark, Hungary, Belgium, Yugoslavia, Norway, India, Great Britain, Ireland, and the United States, as well as Germany. In 1932 he received the Nobel prize in physics.

Heisenberg's most notable contribution was in the development of the quantum theory of mechanics. This was another broad step beyond the classical physics of Newton. Heisenberg first presented his quantum theory in 1925 and his celebrated "uncertainty principle" in 1927.

1. The uncertainty principle, first put forward by Heisenberg, seems to place a limit on the precision with which one can observe nature and thus on what one can know from nature. What does this limit imply about the existence of a knowable reality?

2. Quantum mechanics allows one only to calculate the probabilities of various occurrences. Thus the actual outcome of an experiment at the atomic level is not knowable in advance. How does this compare to the known cause, predictable effect of Newtonian physics?

3. Newtonian physics was the basis for a world view known as the clockwork universe. When applied to the human mind, this even seemed to imply that all human thought was predetermined. What does quantum theory do to this view and how does it affect the notion of free will?

4. The so-called romantic movement in literature was in part a rebellion against the mechanistic world view of Newtonian physics. Do you see any current movements in society which might be in part rebellion against the probabilistic world view of Heisenberg?

PHYSICS AND PHILOSOPHY

WERNER HEISENBERG

The History of Quantum Theory

THE origin of quantum theory is connected with a well-known phenomenon, which did not belong to the central parts of atomic physics. Any piece of matter when it is heated starts to glow, gets red hot and white hot at higher temperatures. The color does not depend much on the surface of the material, and for a black body it depends solely on the temperature. Therefore, the radiation emitted by such a black body at high temperatures is a suitable object for physical research; it is a simple phenomenon that should find a simple explanation in terms of the known laws for radiation and heat. The attempt made at the end of the nineteenth century by Lord Rayleigh and Jeans failed, however, and revealed serious difficulties. It would not be possible to describe these difficulties here in simple terms. It must be sufficient to state that the application of the known laws did not lead to sensible results. When Planck, in 1895, entered this line of research he tried to turn the problem from radiation to the radiating atom. This turning did not remove any of the difficulties inherent in the problem, but it simplified the interpretation of the empirical facts. It was just at this time, during the summer of 1900, that Curlbaum and Rubens in Berlin had made very accurate new measurements of the spectrum of heat radiation. When Planck heard of these results he tried to represent them by simple mathematical formulas which looked plausible from his research on the general connection between heat and radiation. One day Planck and Rubens met for tea in Planck's home and compared Rubens' latest results with a new formula suggested by Planck. The comparison showed a complete agreement. This was the discovery of Planck's law of heat radiation.

It was at the same time the beginning of intense theoretical work for Planck. What was the correct physical interpretation of the new formula? Since Planck could, from his earlier work, translate his formula easily into a statement about the radiating atom (the so-called oscillator), he must soon have found that his formula looked as if the oscillator could only contain discrete quanta of energy—a result that was so different from anything known in classical physics that he certainly must have refused to believe it in the beginning. But in a period of most intensive work during the summer of 1900 he finally convinced himself that there was no way of escaping from this conclusion. It was told by Planck's son that his father spoke to him about his new ideas on a long walk through the Grunewald, the wood in the suburbs of Berlin. On this walk he explained that he felt he had possibly made a discovery of the first rank, comparable perhaps only to the discoveries of Newton. So Planck must have realized at this time that his formula had touched the foundations of our description of nature, and that these foundations would one day start to move from their traditional present location toward a new and as yet unknown position of stability. Planck, who was conservative in his whole outlook, did not like this consequence at all, but he published his quantum hypothesis in December of 1900.

The idea that energy could be emitted or absorbed only in discrete energy quanta was so new that it could not be fitted into the traditional framework of physics. An attempt by Planck to reconcile his new hypothesis with the older laws of radiation failed in the essential points. It took five years until the next step could be made in the new direction.

This time it was the young Albert Einstein, a revolutionary genius among the physicists, who was not afraid to go further away from the old concepts. There were two problems in which he could make use of the new ideas. One was the so-called photoelectric effect, the emission of electrons from metals under the influence of light. The experiments, especially those of Lenard, had shown that the energy of the emitted electrons did not depend on the intensity of the light, but only on its color or, more precisely, on its frequency. This could not be understood on the basis of the traditional theory of radiation. Einstein could explain the observations by interpreting Planck's hypothesis as saying that light consists of quanta of energy traveling through space. The energy of one light quantum should, in agreement with Planck's assumptions, be equal to the frequency of the light multiplied by Planck's constant.

The other problem was the specific heat of solid bodies. The traditional theory led to values for the specific heat which fitted the observations at higher temperatures but disagreed with them at low ones. Again Einstein was able to show that one could understand this behavior by applying the quantum hypothesis to the elastic vibrations of the atoms in the solid body. These two results marked a very important advance, since that revealed the presence of Planck's quantum of action—as his constant is called among the physicists—in several phenomena, which had nothing immediately to do with heat radiation. They revealed at the same time the deeply revolutionary character of the new hypothesis, since the first of them led to a description of light completely different from the traditional wave picture. Light could either be interpreted as consisting of electromagnetic waves, according to Maxwell's theory, or as consisting of light quanta, energy packets traveling through space with high velocity. But could it be both? Einstein knew, of course, that the well-known phenomena of diffraction and interference can be explained only on the basis of the wave picture. He was not able to dispute the complete contradiction between this wave picture and the idea of the light quanta; nor did he even attempt to remove the inconsistency of this interpretation. He simply took the contradiction as something which would probably be understood only much later.

In the meantime the experiments of Becquerel, Curie and Rutherford had led to some clarification concerning the structure of the atom. In 1911 Rutherford's observations on the interaction of x-rays penetrating through matter resulted in his famous atomic model. The atom is pictured as consisting of a nucleus, which is positively charged and contains nearly the total mass of the atom, and electrons, which circle around the nucleus like the planets circle around the sun. The chemical bond between atoms of different elements is explained as an interaction between the outer electrons of the neighboring atoms; it has not directly to do with the atomic nucleus. The nucleus determines the chemical behavior of the atom through its charge which in turn fixes the number of electrons in the neutral atom. Initially this model of the atom could not explain the most characteristic feature of the atom, its enormous stability. No planetary system following the laws of Newton's mechanics would ever go back to its original configuration after a collision with another such system. But an atom of the element carbon, for instance, will still remain a carbon atom after any collision or interaction in chemical binding.

The explanation for this unusual stability was given by Bohr in 1913, through the application of Planck's quantum hypothesis. If the atom can change its energy only by discrete energy quanta, this must mean that the atom can exist only in discrete stationary states, the lowest of which is the normal state of the atom. Therefore, after any kind of interaction the atom will finally always fall back into its normal state.

By this application of quantum theory to the atomic model, Bohr could not only explain the stability of the atom but also, in some simple eases, give a theoretical interpretation of the line spectra emitted by the atoms after the excitation through electric discharge or heat. His theory rested upon a combination of classical mechanics for the motion of the electrons with quantum conditions, which were imposed upon the classical motions for defining the discrete stationary states of the system. A consistent mathematical formulation for those conditions was later given by Sommerfeld. Bohr was well aware of the fact that the quantum conditions spoil in some way the consistency of Newtonian mechanics. In the simple case of the hydrogen atom one could calculate from Bohr's theory the frequencies of the light emitted by the atom, and the agreement with the observation was perfect. Yet these frequencies were different from the orbital frequencies and their harmonics of the electrons circling around the nucleus, and this fact showed at once that the theory was still full of contradictions. But it contained an essential part of the truth. It did explain qualitatively the chemical behavior of the atoms and their line spectra; the existence of the discrete stationary states was verified by the experiments of Franck and Hertz, Stern and Gerlach.

Bohr's theory had opened up a new line of research. The great amount of experimental material collected by spectroscopy through several decades was now available for information about the strange quantum laws governing the motions of the electrons in the atom. The many experiments of chemistry could be used for the same purpose. It was from this time on that the physicists learned to ask the right questions; and asking the right question is frequently more than halfway to the solution of the problem.

What were these questions? Practically all of them had to do with the strange apparent contradictions between the results of different experiments. How could it be that the same radiation that produces interference patterns, and therefore must consist of waves, also produces the photoelectric effect, and therefore must consist of moving particles? How could it be that the frequency of the orbital motion of the electron in the atom does not show up in the frequency of the emitted radiation? Does this mean that there is no orbital motion? But if the idea of orbital motion should in incorrect, what happens to the electrons inside the atom? One can see the electrons move through a cloud chamber, and sometimes they are knocked out of an atom; why should they not also move within the atom? It is true that they might be at rest in the normal state of the atom, the state of lowest energy. But there are many states of higher energy, where the electronic shell has an angular momentum. There the electrons cannot possibly be at rest. One could add a number of similar examples. Again and again one found that the attempt to describe atomic events in the traditional terms of physics led to contradictions.

Gradually, during the early twenties, the physicists became accustomed to these difficulties, they acquired a certain vague knowledge about where trouble would occur, and they learned to avoid contradictions. They knew which description of an atomic event would be the correct one for the special experiment under discussion. This was not sufficient to form a consistent general picture of what happens in a quantum process, but it changed the minds of the physicists in such a way that they somehow got into the spirit of quantum theory. Therefore, even some time before one had a consistent formulation of quantum theory one knew more or less what would be the result of any experiment.

One frequently discussed what one called ideal experiments. Such experiments were designed to answer a very critical question irrespective of whether or not they could actually be carried out. Of course it was important that it should be possible in principle to carry out the experiment, but the technique might be extremely complicated. These ideal experiments could be very useful in clarifying certain problems. If there was no agreement among the physicists about the result of such an ideal experiment, it was frequently possible to find a similar but simpler experiment that could be carried out, so that the experimental answer contributed essentially to the clarification of quantum theory.

The strangest experience of those years was that the paradoxes of quantum theory did not disappear during this process of clarification; on the contrary, they became even more marked and more exciting. There was, for instance, the experiment of Compton on the scattering of X-rays. From earlier experiments on the interference of scattered light there could be no doubt that scattering takes place essentially in the following way: The incident light wave makes an electron in the beam vibrate in the frequency of the wave; the oscillating electron then emits a spherical wave with the same frequency and thereby produces the scattered light. However, Compton found in 1923 that the frequency of scattered X-rays was different from the frequency of the incident X-ray. This change of frequency could be formally understood by assuming that scattering is to be described as collision of a light quantum with an electron. The energy of the light quantum is changed during the collision; and since the frequency times Planck's constant should be the energy of the light quantum, the frequency also should be changed. But what happens in this interpretation of the light wave? The two experiments—one on the interference of scattered light and the other on the changes of frequency of the scattered light—seemed to contradict each other without any possibility of compromise.

By this time many physicists were convinced that the apparent contradictions belonged to the intrinsic structure of atomic physics. Therefore, in 1924 de Broglie in France tried to extend the dualism between wave description and particle description to the elementary particles of matter, primarily to the electrons. He showed that a certain matter wave could "correspond" to a moving electron, just as a light wave corresponds to a moving light quantum. It was not clear at the time what the word "correspond" meant in this connection. But de Broglie suggested that the quantum condition in Bohr's theory should be interpreted as a statement about the matter waves. A wave circling around a nucleus can for geometrical reasons only be a stationary wave; and the perimeter of the orbit must be an integral multiple of the wavelength. In this way de Broglie's idea connected the quantum condition, which always had been a foreign element in the mechanics of the electrons, with the dualism between waves and particles.

In Bohr's theory the discrepancy between the calculated orbital frequency of the electrons and the frequency of the emitted radiation had to be interpreted as a limitation to the concept of the electronic orbit. This concept had been somewhat doubtful from the beginning. For the higher orbits, however, the electrons should move at a large distance from the nucleus just as they do when one sees them moving through a cloud chamber. There one should speak about electronic orbits. It was therefore very satisfactory that for these higher orbits the frequencies of the emitted radiation approach the orbital frequency and its higher harmonics. Also Bohr had already suggested in his early papers that the intensities of the emitted spectral lines approach the intensities of the corresponding harmonics. This principle of correspondence had proved very useful for the approximative calculation of the intensities of spectral lines. In this way one had the impression that Bohr's theory gave a qualitative but not a quantitative description of what happens inside the atom; that some new feature of the behavior of matter was qualitatively expressed by the quantum conditions, which in turn were connected with the dualism between waves and particles.

The precise mathematical formulation of quantum theory finally emerged from two different developments. The one started from Bohr's principle of correspondence. One had to give up the concept of the electronic orbit but Still had to maintain it in the limit of high quantum numbers, i.e., for the large orbits. In this latter case the emitted radiation, by means of its frequencies and intensities, gives a picture of the electronic orbit; it represents what the mathematicians call a Fourier expansion of the orbit. The idea suggested itself that one should write down the mechanical laws not as equations for the positions and velocities of the electrons but as equations for the frequencies and amplitudes of their Fourier expansion. Starting from such equations and changing them very little one could hope to come to relations for those quantities which correspond to the frequencies and intensities of the emitted radiation, even for the small orbits and the ground state of the atom. This plan could actually be carried out; in the summer of 1925 it led to a mathematical formalism called matrix mechanics or, more generally, quantum mechanics. The equations of motion in Newtonian mechanics were replaced by similar equations between matrices; it was a strange experience to find that many of the old results of Newtonian mechanics, like conservation of energy, etc., could be derived also in the new scheme. Later the investigations of Born, Jordan and Dirac showed that the matrices representing position and momentum of the electron do not commute. This latter fact demonstrated clearly the essential difference between quantum mechanics and classical mechanics.

The other development followed de Broglie's idea of matter waves. Schrödinger tried to set up a wave equation for de Broglie's stationary waves around the nucleus. Early in 1926 he succeeded in deriving the energy values of the stationary states of the hydrogen atom as "Eigenvalues" of his wave equation and could give a more general prescription for transforming a given set of classical equations of motion into a corresponding wave equation in a space of many dimensions. Later he was able to prove that his formalism of wave mechanics was mathematically equivalent to the earlier formalism of quantum mechanics.

Thus one finally had a consistent mathematical formalism, which could be defined in two equivalent ways, starting either from relations between matrices or from wave equations. This formalism gave the correct energy values for the hydrogen atom; it took less than one year to show that it was also successful for the helium atom and the more complicated problems of the heavier atoms But in what sense did the new formalism describe the atom? The paradoxes of the dualism between wave picture and particle picture were not solved; they were hidden somehow in the mathematical scheme.

A first and very interesting step toward a real understanding of quantum theory was taken by Bohr, Kramers and Slater in 1924. These authors tried to solve the apparent contradiction between the wave picture and the particle picture by the concept of the probability wave. The electromagnetic waves were interpreted not as "real" waves but as probability waves, the intensity of which determines in every point the probability for the absorption (or induced emission) of a light quantum by an atom at this point. This idea led to the conclusion that the laws of conservation of energy and momentum need not be true for the single event, that they are only statistical laws and are true only in the statistical average. This conclusion was not correct, however, and the connections between the wave aspect and the particle aspect of radiation were still more complicated.

But the paper of Bohr, Kramers and Slater revealed one essential feature of the correct interpretation of quantum theory. This concept of the probability wave was something entirely new in theoretical physics since Newton. Probability in mathematics or in statistical mechanics means a statement about our degree of knowledge of the actual situation. In throwing dice we do not know the fine details of the motion of our hands which determine the fall of the dice and therefore we say that the probability for throwing a special number is just one in six. The probability wave of Bohr, Kramers, Slater, however, meant more than that; it meant a tendency for something. It was a quantitative version of the old concept of "potentia" in Aristotelian philosophy. It introduced something standing in the middle between the idea of an event and the actual event, a strange kind of physical reality just in the middle between possibility and reality.

Later when the mathematical framework of quantum theory was fixed, Born took up this idea of the probability wave and gave a clear definition of the mathematical quantity in the formalism, which was to be interpreted as the probability wave. It was not a three-dimensional wave like elastic or radio waves, but a wave in the many-dimensional configuration space, and therefore a rather abstract mathematical quantity.

Even at this time, in the summer of 1926, it was not clear in every case how the mathematical formalism should be used to describe a given experimental situation. One knew how to describe the stationary states of an atom, but one did not know how to describe a much simpler event—as for instance an electron moving through a cloud chamber.

When Schrödinger in that summer had shown that his formalism of wave mechanics was mathematically equivalent to quantum mechanics he tried for some time to abandon the idea of quanta and "quantum jumps" altogether and to replace the electrons in the atoms simply by his three-dimensional matter waves. He was inspired to this attempt by his result, that the energy levels of the hydrogen atom in his theory seemed to be simply the eigenfrequencies of the stationary matter waves. Therefore, he thought it was a mistake to call them energies; they were just frequencies. But in the discussions which took place in the autumn of 1926 in Copenhagen between Bohr and Schrödinger and the

Copenhagen group of physicists it soon became apparent that such an interpretation would not even be sufficient to explain Planck's formula of heat radiation.

During the months following these discussions an intensive study of all questions concerning the interpretation of quantum theory in Copenhagen finally led to a complete and, as many physicists believe, satisfactory clarification of the situation. But it was not a solution which one could easily accept. I remember discussions with Bohr which went through many hours till very late at night and ended almost in despair; and when at the end of the discussion I went alone for a walk in the neighboring park I repeated to myself again and again the question: Can nature possibly be as absurd as it seemed to us in these atomic experiments?

The final solution was approached in two different ways. The one was a turning around of the question. Instead of asking: How can one in the known mathematical scheme express a given experimental situation? the other question was put: Is it true, perhaps, that only such experimental situations can arise in nature as can be expressed in the mathematical formalism? The assumption that this was actually true led to limitations in the use of those concepts that had been the basis of classical physics since Newton. One could speak of the position and of the velocity of an electron as in Newtonian mechanics and one could observe and measure these quantities. But one could not fix both quantities simultaneously with an arbitrarily high accuracy. Actually the product of these two inaccuracies turned out to be not less than Planck's constant divided by the mass of the particle. Similar relations could be formulated for other experimental situations. They are usually called relations of uncertainty or principle of indeterminacy. One had learned that the old concepts fit nature only inaccurately.

The other way of approach was Bohr's concept of complementarity. Schrödinger had described the atom as a system not of a nucleus and electrons but of a nucleus and matter waves. This picture of the matter waves certainly also contained an element of truth. Bohr considered the two pictures—particle picture and wave picture—as two complementary descriptions of the same reality. Any of these descriptions can be only partially true, three must be limitations to the use of the particle concept as well as of the wave concept, else one could not avoid contradictions. If one takes into account those limitations which can be expressed by the uncertainty relations, the contradictions disappear.

In this way since the spring of 1927 one has had a consistent interpretation of quantum theory, which is frequently called the "Copenhagen interpretation." This interpretation received its crucial test in the autumn of 1927 at the Solvay conference in Brussels. Those experiments which had always led to the worst paradoxes were again and again discussed in all details, especially by Einstein. New ideal experiments were invented to trace any possible inconsistency of the theory, but the theory was shown to be consistent and seemed to fit the experiments as far as one could see.

The details of this Copenhagen interpretation will be the subject of the next chapter. It should be emphasized at this point that it has taken more than a quarter of a century to get from the first idea of the existence of energy quanta to a real understanding of the quantum theoretical laws. This indicates the great changes that had to take place in the fundamental concepts concerning reality before one could understand the new situation.

THE ASTRONOMICAL HORIZON

JAMES JEANS

The British mathematician, physicist and astronomer, Sir James Hopwood Jeans (1877-1946) has emerged as one of the premier cosmologists of the twentieth century. In addition to major works in sound, electricity, and radiation, Jeans provided an exceptionally articulate view of the ever-changing cosmos. He not only did significant original research, but also described that research, and the work of others, in a way that won for him enormous popular acclaim. He was the first to propose the theory of continuous creation of matter throughout the universe. He made other major contributions on the energy of stars, the spiral nebulae, binary and multiple star systems, and giant and dwarf stars.

Jeans was born in Ormskirk, Lancashire, England. In 1900 he graduated from Trinity College, Cambridge. He returned to Cambridge as University Lecturer in Mathematics in 1904, and a year later went to the United States to serve as Professor of Applied Mathematics at Princeton University where he remained for four years. He returned to Cambridge then as Stokes Lecturer from 1910 to 1912. Thereafter he devoted most of his time to research and writing. He was honorary secretary of the Royal Society of London from 1919-1929. He was a research associate at the Mount Wilson Observatory, Pasadena, California, from 1923 to 1944. He wrote some sixteen books on physics, astronomy, cosmogony, and philosophy, perhaps the best known of which are *Astronomy* and *Cosmogony* (1928), *The Mysterious Universe* (1931), *Through Space and Time* (1934), and *Physics and Philosophy* (1942).

1. How does Jeans relate the "horizons of knowledge" to the "astronomical horizon"?

2. Why did it take nearly 2,000 years for the idea of Aristarchus that the earth revolves around the sun to be accepted?

3. Why was Bruno burned at the stake?

4. How does Jeans demonstrate "how crowded . . . space is with stars"?

5. How does Einstein's theory of relativity square with Newton's notions of gravitational force and acceleration?

5. What are the implications of the idea of an expanding universe for Einstein's theory of relativity?

6. How would Jeans choose from among differing explanations about the nature of the universe?

7. How do you distinguish between truth and our conceptions of truth?

THE ASTRONOMICAL HORIZON

JAMES JEANS

THE history of astronomy, it has been truly said, is a history of receding horizons. Every increase in telescopic power and—what is hardly less important—every gain in the sensitivity of photographic plates, has opened up new depths of space for exploration, and the centre of astronomical interest has tended to follow the limits of vision. Until the middle of the last century, interest was centred in the planets of the solar system; after that it shifted to the stars of the galactic system; now many astronomers find their main interest in the remote nebulae which lie at the extreme limits of vision of our largest telescopes—the astronomical horizon. Each of these advances—from planets to stars, and from stars to nebulae—has involved a vast march forward into space, for the nearest stars are a million times as distant as the nearest planets, and the nebulae are nearly a million times as distant as the stars.

Knowledge too has its horizons. Every science contains a hard core of well-established fact and, surrounding this, a dim nebulous region in which it is hard to distinguish truth from error, a region of conjecture and doubtful hypotheses. Here is the frontier of knowledge, the region in which progress is being made. In astronomy it is closely associated with the frontier of known space. It is to these vague regions that I shall venture to take you this afternoon, approaching them through regions of well ascertained fact, and pausing here and there to recollect how this nucleus of firm knowledge was gained—for the methods by which the frontier was reached are still of service when we try to make progress beyond.

The ancients thought of the universe as quite a small affair, with our planet as its main ingredient, and the sky as its not very remote ceiling. Then, about 2,200 years ago, Aristarchus of Samos, a teacher in Alexandria, published a book *On the Sizes and Distances of the Sun and Moon.* Astronomers had made observations for many thousands of years before Aristarchus, but he was the first to discuss his observations in a true mathematical spirit. The observations were far from accurate, but he deduced from them by strict reasoning that the sun must be many thousands of times as big as the earth. Then, probably seeing the absurdity of supposing so massive a sun to revolve round so much smaller an earth, he introduced the hypotheses (I) "that the fixed stars and the sun remain motionless, that the earth revolves about the sun, . . . and (2) that the sphere of the fixed stars is so great" that the Earth's orbit "bears the same proportion to the distance of the fixed stars as the centre of the sphere bears to its surface." In brief, he reduces the earth's orbit to a point.

Here is the first known clear statement that the earth revolves round the sun, and also one of the earliest known statements that the stars are immensely distant. The two statements are of course not independent; the latter is a direct consequence of the former. For if the earth revolved round the sun, and the stars were not very distant, they would appear to change their relative positions as the earth moved towards some and away from others in its journey round the sun. Aristarchus, understanding this and noticing no changes, saw that the world of stars must be very great.

His thumb-nail sketch of the universe contained the essence of the truth, but just failed to start astronomy on the right road. The authority of Aristotle, of Plato, and of many others stood against him, with the result that for nearly 2,000 years sound views as to the structure of the universe were driven into hiding. In the second century after Christ we find the Alexandrian astronomer Ptolemy marshalling arguments to prove that the earth could not move in space. Some were quite absurd, as, for instance, that a motion of the earth through space would create such a wind that birds who ventured out into it would never be able to get back to their nests. He repeated the argument that if the earth were moving the stars would appear to change their relative positions, but did not mention Aristarchus' solution that the stars may be too distant for the changes to be noticeable.

In a later age the authority of Aristotle was first reinforced and then superseded by the authority of the Christian Church, but from the Church also came the most damaging criticisms of the current doctrines. From the fourteenth century on we find a succession of ecclesiastics—Oresme, Bishop of Lisieux, Cardinal Nicolas of Cusa, Canon Copernicus of Frauenburg, and the monk Giordano Bruno—all declaring against the Aristotelian cosmology and maintaining that the earth moves, that the stars are other worlds, and that the earth's orbit in space is infinitesimal in comparison with the universe of stars—"as a point is to a clod" was the comparison Copernicus used when he explained, as Aristarchus had done before him, why we notice no changes in the positions of the stars. Bruno went further, maintaining that the infinite goodness of God required that the stars should be infinite in number, and arguing that, as an infinity can have no centre, it is impossible for either the sun or the earth to stand at the centre of the universe. His doctrines probably affected human thought even more profoundly than those of Copernicus, and certainly contained more potential danger to the tenets of orthodoxy. He fell into the hands of the Inquisition, and was condemned and burned at the stake in 1600.

Efforts to explain why the stars showed no changes of position ended in the discovery that there was nothing to explain. In 1718 Halley found that Sirius and Arcturus had moved substantially since Ptolemy had charted their positions some sixteen centuries earlier. From now on stellar positions and their changes were studied systematically, and in time yearly oscillations were found which obviously resulted from the earth's motion round the sun. The more distant a star, the more minute this yearly oscillation would be, so that a measurement of its amount would fix the distance of the star—in terms of course of the dimensions of the earth's orbit. About 106 years ago the distances of three of the nearer stars were measured by this method. Astronomy had now become possessed of a measuring-rod for stellar distances, and only technical difficulties stood in the way of a complete survey of the skies.

Long before this, however, it had become recognized that the stars were suns like our own, so that their comparative faintness showed that they must be very distant. If every star emitted exactly the same amount of light as the sun, a star's faintness would of course provide a precise measure of its distance; Newton had estimated stellar distances in this way, with results that have proved to be not too far from the truth. This method can no longer be used, for we know that its basic supposition is false; the stars shine with very unequal lights, some emitting less than a 300,000th part of the light of the sun, and some more than 300,000 times as much—a range of 90 thousand million to one in luminosity. But the method is still very useful in a modified form. There are special classes of stars—as for instance certain variable stars of assigned periods (Cepheids)—in which all members of any one class are found to be of the same luminosity. These stars provide a system of standard beacons scattered through space, and by their relative faintness and brightness, as seen by us, enable us to measure stellar distances with some accuracy. But even these methods fail for the most distant stars of all, those which form the star-dust of the Milky Way. And these are just the stars that matter when we try to plot out the dimensions and shape of our system of stars. For this problem other methods are needed, and have been found.

Plate I consists of four photographs of the same small area of the sky, an area in which no stars are visible to the naked eye. They show what we should see if we examined this bit of the sky visually through telescopes of apertures 4, 16, 64, and about 160 inches respectively. If space were perfectly transparent to light, pictures (2) and (3) would each probe just four times as far into space as its predecessor, and so would disclose 64 times as big a volume of space. Thus if the stars were uniformly scattered through infinite space, each picture would show 64 times as many stars as the picture before. But a simple count will show that the ratio of increase is nothing like as great as 64, and the same applies, *mutatis mutandis*, to the step from the third picture to the last. If, then, space is perfectly transparent to light, we must conclude that the stars are not uniformly scattered through infinite space, and that the later photographs already penetrate into regions in which the stars have begun to thin out.

Sir William Herschel reached similar conclusions when he used a primitive form of this method. He noted the distances at which the thinning seemed to begin in different parts of the sky, and concluded that the system of stars was shaped like a flat disk, with the sun at, or very near to, its centre.

This description survived until quite recently, but it is unsound. It was based on the supposition that space is perfectly transparent to light, and recent advances in astronomical knowledge and photographic technique have made any such supposition quite untenable. When we examine the sky visually or by a short exposure photograph, we see little beyond stars which appear as mere points of light, as on Plate I. But a long exposure reveals something very different. Plate II shows a long exposure photograph of the Pleiades, and here the stars no longer appear as points of light, but as extensive masses of what we call nebulosity. This and other methods of study have shown that the whole of interstellar space is occupied by a sort of fog of obscuring matter. Sometimes, as in Plate II, the fog may be lighted up by the stars within—much as a street lamp may light up the fog in its vicinity, although the physical process is very different. Perhaps, as in Plate VII, we may find thin, filmy wisps of nebulosity which extend over a whole constellation, and are again raised to luminescence by the hot stars within. Or, as in Plate VIII, the obscuring matter may show itself in its true character, blotting out the light from the stars beyond and so producing an impression of black patches, or even of holes, in the sky.

This obscuring fog prevents our seeing more than a certain distance through space; it imposes a definite range of visibility. When we walk in a forest on a foggy day, we see only those trees which are within a certain distance of us, so that we are at the centre of all the trees we can see. But we must not suppose that we are at the centre of the forest; we are merely at the centre of our individual sphere of visibility. It is the same with the stars.

Nevertheless Herschel, finding that we were at the centre of all the stars we can see, concluded that we were at the centre of the whole system of stars, and in this he was followed by Kapteyn as late as 1922. But again we are only at the centre of our sphere of visibility, the interstellar fog preventing our seeing the more distant stars of our system. Thus the discovery of the dimensions and shape of this system again calls for new methods, and again they have been found.

Although the light from a single star cannot penetrate far through the fog, there are other objects in space—the "globular clusters," close groups of millions of stars—which are brilliant enough to penetrate through almost the whole of the fog. They contain standard stars of known luminosity, so that we can measure their distances and plot out their arrangement. A study by Shapley showed that they occupy a disk-shaped volume of space, somewhat more than 100,000 light-years in radius. It was Herschel's disk-shaped figure repeated on a larger scale, with the difference that the sun did not stand at its centre but about 40,000 light-years away from the centre. It began to seem likely that this figure, rather than Herschel's, must map out the confines of the galactic system of stars.

This conjecture subsequently received confirmation in an unexpected way. A detailed study of the motions of the stars shows that they do not move at random, but after a plan—roughly they move like the planets round the sun. In the solar system the inner planets move faster round the sun than the outer, and so continually overtake the latter in their orbits. The stars of the galactic system are found to move in a similar way, and the centre round which they move proves to be the centre of Shapley's disk, those which are nearest to this centre moving fastest. The sun takes about 250 million years to complete a revolution round this centre and, as the radius of its orbit is about 40,000 light-years, it must move at about 270 km. a second. Actually the situation is not quite so simple as this, since the stars have small random motions of their own superposed on to their regular orbital motions, but these need not concern us here.

We know that the planets would scatter into space if it were not for the restraining force of the sun's gravitational pull. It is the same with the stars; they would scatter into space if it were not for the gravitational pull of the galaxy as a whole. We can estimate the amount of this pull from the particulars of the sun's orbit just given, and so can calculate the total mass of the galaxy from which the pull originates. It is found that this must be of the order of 150,000 million suns like our own. The largest telescope in the world shows only about 1,500 million stars, which thus represent less than one per cent. of the total mass; the remaining 99 per cent. is invisible to us. It will consist of stars which are too distant to be seen through the interstellar fog, stars which are too dark or too faint to be seen anyhow, and the matter of the fog itself.

Large though this mass is, it occupies so great a space that stars are rare occurrences in it. Let us attempt to construct a scale model, taking a large hall as the space available for the galaxy. Let us smoke in it until it contains 150,000 million particles of tobacco smoke, to achieve which we need only smoke a millionth of a gram of tobacco. We now have the right number of particles to represent the stars, but what about their size? Andrade and Parker find that the average diameter of a particle of tobacco smoke is rather less than a millionth of an inch, whence it appears that the particles are far too big—millions of times too big—to represent the stars. Even a single atom is too big to represent a star; its radius represents that of the orbit of Pluto—40 times the radius of the orbit of the earth, 8,000 times the radius of the sun. Thus after we have smoked our millionth of a gram of tobacco, the hall will be millions of times more crowded with tobacco particles than space is with stars.

Galileo's first telescope (1609) had shown nothing more distant than such stars. Three years later Marius observed an object of different type which looked "like a candle-light seen through horn." He had discovered the first of the great extra-galactic nebulae, that which we now call the Great Nebula in Andromeda

Millions of these objects are now known, the majority being of characteristic and regular shapes. As far back as 1755 Kant suggested that they were huge systems of stars, probably very similar to our own galactic system. Herschel too conjectured that sufficient telescopic power would resolve them into clusters of stars. The needed power was not forthcoming until 1924 when Edwin Hubble, working with the great 100-inch telescope at Mount Wilson, succeeded in resolving some of the outer parts of the Andromeda nebula into separate stars. The final resolution of the central parts of this nebula into stars was only effected a few months ago by Baade, also working at Mount Wilson. Many other nebulae also have been studied in detail, and parts of them resolved into stars; they differ greatly in size and shape, but all appear to be stellar systems similar to our own galactic system.

Usually they are found to contain a profusion of the variable stars and other standard objects that are used for measuring distances in our own system. This makes it possible to measure the distances of the nebulae, and also their sizes. The Great Nebula in Andromeda proves to be the nearest of all, at a distance of about 680,000 light-years; its dimensions are found to be comparable with those of the galactic system.

The nebulae are of many different shapes, but it is found to be a general law that nebulae of the same shape are also of the same size (approximately) and also of the same luminosity. Once again, then, nature has provided the astronomer with standard articles—this time, the nebulae themselves. Differences of apparent size and brightness can originate only in differences of distance, so that we can estimate the distances of nebulae which are far too remote for standard stars to be recognized in them—indeed, up to distances of about 500 million light-years, beyond which it is impossible to recognize the nebulae as such. As we pass outwards in space we find no thinning out of the nebulae such as Herschel found with the stars; right up to the limits of our vision, the number visible within any distance remains proportional to the cube of the distance. They are found to be fairly uniformly spaced, at an average distance of about two million light-years. To construct a scale model, we may think of dinner plates scattered through space at an average distance of about twenty feet, and orientated at random. We choose dinner plates to represent the nebulae because most, although by no means all, of them are of the same flat disk shape as our own galaxy. In Plate V, the lower picture is that of such a nebula seen "full face," while the upper is that of a probably very similar nebula seen "edge on." In the latter the dark band round the equator obviously represents a band of obscuring fog which shuts off the light of the stars inside.

Although on the whole the nebulae are scattered uniformly through space, yet here and there they are found to fall into groups and clusters. On Plate VI, the upper picture shows a close group of three in the constellation of Pegasus, while the lower shows a much larger cluster in Coma Berenicis which is at a distance of about 50 million light-years from us. In each of these clusters, the members are all at approximately equal distances from us, and it is of interest to notice that all appear of approximately the same sizes and brightnesses.

Where the nebulae fall into clusters we can estimate their masses or, more briefly, weigh them. For a cluster can only be kept in being by the gravitational attractions of its constituent nebulae. The speeds of motion of the various members can be measured spectroscopically, and from these we can deduce the gravitational pull needed to keep the cluster from disintegrating. This in turn tells us the mass of the whole cluster, and hence, by a simple division, the average mass of all the nebulae of the cluster. Generally this comes out to be of the order of 100,000 million to 200,000 million suns, providing further evidence that these external galaxies are similar systems to our own.

If this were all, the description of the Universe would be very simple--vast aggregations of stars, containing about 150,000 million each, occupying space fairly uniformly as far as our telescopes can probe and we know not how far beyond. Such a description would, we may imagine, have proved very satisfying to an astronomer of a generation ago; he would probably have conjectured that this Chapter of the Book of Nature was closed, except perhaps for elaboration in points of detail and possible explorations into still farther depths of space; little would he have been able to imagine the strange, even fantastic, world he was just about to enter.

The door to the new world was the relativity theory of gravitation which Einstein propounded in 1915. Newton had taught that gravitational acceleration was caused by a force that followed the simple mathematical law of the inverse square of the distance. Einstein found that in his new world of relativity such a law became meaningless, so that Newton's force of gravitation was physically impossible. He regained consistency by attributing acceleration, not to force but to a curvature of space, or, to be more precise, to a curvature of that four-dimensional blend of space and time which forms the background to the whole theory of relativity. The path of a planet or a cricket ball was no longer a curved path in a straight space, but a straight path in a curved space; the curvature was transferred from the path in which the projectile moved to the space in which it moved. This curvature was impressed on any region of space by matter in its proximity. The sun, for instance, crumples up the regions of space in which the planets perform their orbits.

This was found to give a perfect account of the motions of the planets round the sun, but the discussion of the universe as a whole raised more difficult problems. The whole of space contains matter, and the amount of matter in any region of space determines the curvature in that region. With exactly the right amount of matter the total curvature in the whole of space might be just enough to close it up into a finite universe, which would then be in equilibrium. In this case, space would not be like the finite space inside a sphere, but like the finite surface of a sphere, not like the interior of the Earth, but like its surface—finite but unbounded. For a specified density of matter in space, there would be only one size of space for which equilibrium would be possible.

Einstein supposed that astronomical space must be of this kind. Since the average masses and distribution of the nebulae were known from observation, it was possible to calculate what the size and total mass of the universe must be. The particular calculations that were made at this time do not matter much, for they are now superseded; more modern calculations suggest that, in a general way, there must be about as many nebulae in space as there are stars in each nebula—something of the order of 100,000 million—only about 10 million of which are visible in our telescopes. We may visualize the total number in the whole of space as equal to the number of raindrops in a heavy shower—perhaps one giving a tenth of an inch rainfall over the City of Oxford.

This particular picture of a space held in equilibrium by the matter it contained was shattered a few years later when Friedmann and Lemaître showed that the arrangement it portrayed could not be permanent. Space was represented as a sort of coiled spring, its curvature being impressed on it by the matter it contained. If the matter in a particular region of space became less dense, this part of the space would tend to uncoil; if matter moved from one region to another, the curvature would change in both, and the universe would no longer be in equilibrium. The new forces thus brought into play might tend either to restore the original equilibrium or to accentuate the existing disequilibrium. Friedmann and Lemaître showed that they would do the latter. Thus the Einstein arrangement proved to be unstable, so that space left to itself would necessarily start either to expand or to contract. Before the mathematicians had decided which would actually happen, the observers at Mount Wilson announced that, to all appearances, space was actually expanding, and this at no small speed.

Astronomers cannot, of course, observe an expansion of space directly; they can only examine rays of light. In this case they had examined the spectra of the light emitted by distant nebulae and found a systematic displacement towards the red end of the spectrum, a phenomenon which usually indicates that the source of the light is increasing its distance from us, so that fewer waves of light arrive per second. The most distant nebulae of which the spectra can be measured with accuracy show displacements such as would be produced by recession at the terrific speed of about 26,000 miles a second, which is about one-seventh of the speed of light.

The amount of the displacement was found to vary from one nebula to another, being approximately proportional to the distance of the nebula. If, however, a cluster of nebulae was treated as a single unit, so that any random motions of individual nebulae were averaged out, then the displacement for the cluster as a whole was found to be almost exactly proportional to the distance of the cluster. This relation is obeyed so exactly that it obviously must represent something fundamental in the structure of the universe. The simplest interpretation is that it results from a uniform expansion of the whole of space consequent upon the instability of the Einstein configuration, the moving nebulae showing currents in space much as moving particles of dust show currents in air. Yet we may wonder whether an expansion produced in this way would result in so exact a relation between speed and distance as is actually observed. Matter is not distributed in space with perfect uniformity, so that we could hardly expect the expansion resulting from instability to be so perfectly uniform as, on this interpretation, it appears to be. It is at least conceivable that there may be some deeper underlying reason for the law.

There has been a good deal of discussion as to whether these apparent nebular motions really occur, or whether we are interpreting something quite different as motion. It is the kind *of* question that would have seemed outstandingly important to our scientific ancestors of a generation ago, and its answer would have been fundamental in their description of the universe. Today, at least if we accept the doctrines of relativity, it seems to mean very little. For what do we mean if we assert that the apparent motions are illusory? We cannot mean that the nebulae are standing still in absolute space; this is meaningless, since there is no absolute space. Do we then mean that the distance between any pair of nebulae remains unaltered with the passage of time? This again is meaningless. We cannot construct an arch of yardsticks in space from nebula to nebula, and stand by to see whether it breaks up or not—and, *if* it did, we should still not know whether the universe was expanding or the yardsticks contracting. We might measure the angle which one nebula subtended at another, but even if this remained always the same, we could not be sure that the universe and the nebula were not expanding together. In whatever way we try to explain what absence of motion means, there is always some snag to invalidate our explanation; indeed, if the theory of relativity is true, there must be.

The best we can do is to devise an explanation which will "save the phenomena" without making any claim to absolute truth. That of a uniformly expanding universe is certainly one of the simplest, although it may conceivably be open to the objection mentioned above. Another is obtained by supposing that, instead of the universe continually increasing in size, its atoms are continually diminishing in size. When we look at a cluster of nebulae which is 50 million light-years distant, such as that shown in Plate VI, we see it by light which left the nebula 50 million years ago, and so originated in the atoms of 50 million years back in time. We may imagine that these atoms were larger than the corresponding atoms of to-day, and so emitted light of longer wave-length; this explains at once why the observed spectral displacement is so exactly proportional to distance. This last explanation admits of many variants—as, for instance, that the electric attractions and repulsions between charged particles vary with the passage of time; this requires neither the universe nor the atom to change their dimensions. Another and very different description assumes that as a light-quantum ages during its long journey through space, its energy steadily and uniformly decreases; quantum-theory shows that this would result in the spectral lines being displaced by an amount proportional to the distance the light had travelled, which is precisely the relation observed for the light from the nebulae.

These attempts at pictorial descriptions provide us with a first glimpse of the strange new world through which astronomers and mathematicians have been trying to grope their way during the last quarter-century, and to which we shall return in a moment.

We have seen how Einstein's original theory provided a relation between the density of matter in space (which can be estimated from the observational data I have already mentioned) and the size of the universe. This assumed the universe to be at rest. If we picture it as expanding, Einstein's relation must, as Lemaître showed, be replaced by another, which connects the speeds of recession of the nebulae with the radius of the universe, and so—indirectly—with the amount of matter it contains. When the average density of matter and the speeds of recession are assigned, only one value is possible for the total mass of the universe. Using the best observational data available we find that this mass is about that of 10^{79} hydrogen atoms. Or we may avoid any reference to the special substance hydrogen by saying that the universe must consist of about 10^{79} protons and an equal number of electrons. If we had inserted other astronomical data, this number might have come out as 10^{78} or 10^{80}.

Whatever data we assume, the number comes out conspicuously large, and we wonder why the substance of the universe should be broken up into so many separate particles. We wonder, too, what the ultimate interpretation of this number may be—why this particular number rather than any other?

The astronomer of a generation ago might have attempted an answer in terms of a very anthropomorphic Creator manufacturing this number of particles, setting them free in space, and resting on the seventh day because the number seemed good to Him: the number 10^{79} would represent a choice on the part of the Creator.

In recent years Sir Arthur Eddington suggested a much more interesting answer. He believed that the number did not represent any sort of choice, but an inevitable necessity; he thought it was a pure mathematical constant of the same kind as *e* or π; something inherent in the nature of things, which could be discovered by the human intellect working alone—not, it is true, without utilizing any observational knowledge of the world of physics, but without calling upon any observation of the outer world of astronomy. He believed he had shown that its value must be precisely $3/2 \times 136 \times 2^{256}$, of which the value ($1.16 \times 10^{79}$) is certainly of the order suggested by telescopic observation of the astronomical universe.

I have said that this number is not a matter of choice, yet according to Eddington a small element of choice does enter into it. The choice is not on the part of the Creator, but of ourselves. Eddington's contention, in his own words, is "not that there are *N* particles in the universe [*N* denoting the number mentioned above], but that anyone that accepts certain elementary principles of measurement must, if he is consistent, think that there are." In another place he says that "when we find the physicist actually does think there are *N* particles, . . . this is not because of any peculiarity in the external world." In brief, this *N*-ness of the external world appears as a sort of Kantian "category," a contribution of our own minds, not to the universe but to our interpretation of it.

Eddington's primary assumption is that the universe consists of particles of such a primitive kind that they can have no attributes beyond existence and non-existence; in assuming this he elects, as a matter of choice, to describe the universe in terms of what the physicist calls a "particle picture." Until about 1925 the physicist took it for granted that the universe was made up of particles, because he found that this assumption provided the simplest interpretation of the then known phenomena. Actually in depicting matter in this way he was, as we now know, making a choice, and he had to take the consequences of this choice;

one of them, as Planck and Bohr found, was that he had to eject continuity from his scheme of nature. He now finds it more convenient for many purposes to picture matter as waves—waves which are too complex to be represented in space and time, and are too continuous to admit of division into discrete particles or other physical units, waves which are mathematical rather than physical, mental constructs rather than physical entities. The view that the universe consists of particles is only one of many possible views, and one of the consequences to which it commits us, according to Eddington, is that the particles must be N in number. The view that it consists of some other number N' of particles is not a permissible view, since it involves a logical inconsistency.

It is difficult to discuss all this, since Eddington did not claim to be able to give a rigorous proof of it, either logical or mathematical, but only adduced certain considerations which seemed to him to point to its validity, and these I personally find wholly unconvincing. But whether he is right or wrong, he can claim to have revealed to us a suggestive, fascinating and inspiring vision. Even if events should prove his train of thought to be entirely fallacious, the number N is still of great interest. It is no stranger to mathematical thought, especially in the form of its square root of which the value is $\sqrt{N}=3.4 \times 10^{39}$. Mathematical numbers of precisely this order of magnitude have been noticed before; let me mention some.

In ordinary life, our unit of force is the weight of a pound or of a gram, but these units are a matter of human convention, or at best are based on accidental properties of the planet on which we happen to live. But purely observational physics can supply a unit of force which is independent of human conventions, and would be as suitable for use on Mars or on Sirius as it is here, namely, the electrical attraction between the electron and the proton in the hydrogen atom. Astronomy can also provide a universal unit of force—the gravitational attraction between the same two particles. It has often been remarked how unequal these two forces are; their ratio is about 2.3×10^{39}. This ratio, let us notice, is a pure number which must somehow be inherent in the scheme of things—again like e or π. And its value is suspiciously near to Eddington's $\sqrt{N}$.

It is the same with the measurement of length. We terrestrials measure in feet or centimetres, which depend on the human body or the dimensions of our planet, but there is no need to be so parochial. Physics provides a natural unit of length, the so-called "radius" of the electron, which is about 2×10^{-13} cm., while astronomy provides another—the radius of a universe which would contain all the matter of the present universe in equilibrium, or, if we like to adopt a more precise picture, the radius of the universe before expansion began, which, according to Eddington, was about 1,068 million light-years. Again the two units are grossly unequal, their ratio being about 5.0×10^{39}.

It is the same with time. We employ the terrestrial day and year as units, but nature has provided a unit of cosmical significance in the recession of the nebulae. Since their speeds of recession are proportional to their distances, it follows that if the speed of recession of each nebula could be reversed, and the nebula made to move backwards with this reversed speed, all the nebulae would finally meet in some quite small region of space. On inserting the observed values for the speeds of the nebulae, we find that this great concurrence would take place after an interval of about 1,780 million years. We may notice in passing that this is almost exactly equal to the ages of the oldest radioactive rocks found on earth, and so is probably comparable with the age of the universe itself. This has led to the suggestion that the present universe may be merely the debris resulting from the explosion of a sort of big single super-molecule. But without relying on so realistic an interpretation, we must notice that the mere facts of nebular recession, or even of apparent recession, provide us with a natural unit of 1,780 million years which is of cosmical significance; it has a meaning wherever in the universe we happen to live.

Physics also can provide a natural unit of time—the time light takes to travel across an electron, which is about 1.3×10^{-23} sec. The ratio of these two natural units of time is 4.2×10^{39}, which is very near to the ratio of the two natural units of length, this being 5.0×10^{39}.

Actually it is not merely by accident that these two numbers are so nearly equal; the theory of the expanding universe shows that the former must be $\frac{1}{2}\sqrt{3}$ times the latter. But, apart from all precise knowledge, anyone who knew nothing of the mathematical theory of the expanding universe might have conjectured that the approximate equality was not accidental. It is strange that the universe should provide two such unequal units of length, and strange that it should provide two such unequal units of time, but it would be strange beyond belief that the units should stand in such similar ratios through mere accident. If we see two seven-foot men coming out of the same house, we shall probably conjecture that they are brothers. On the same principle, we may conjecture that the two large ratios owe their largeness and approximate equality to some common cause, something inherent in the scheme of nature, in which case it would be natural that they should be of the same order of magnitude.

In the case of the units of length and time, this something can be found in the conception of the expanding universe. But we have also found a similar ratio for the two natural units of force, of which the value is 2.3×10^{39}. As far back as 1920 Hermann Weyl of Göttingen suggested that this ratio and that of the units of length might owe their largeness and approximate equality to some common cause. In 1922 Eddington pointed out that this same cause, if it existed, would also make N great, so that it would not be surprising that each ratio should be approximately equal to $\sqrt{N}$.

If we agree that the approximate equality of these large numbers can be no accident, but must result from something in the fixed order of nature, then it ought to be possible to discover this something, show that it accounts for the approximate equality of the numbers, and also calculate the exact values of the numbers. Eddington claims to have done precisely this, and has deduced values for the fundamental physical constants of nature which agree remarkably well with those determined in the laboratory. Unfortunately the details of the problem are far too intricate for discussion here. But, details apart, Eddington introduces us to a universe which is as it is—or rather, of which our picture is as it is—from sheer mathematical necessity; it is as it is because it could not be otherwise without breaking the laws of mathematics and logic. And as evidence of the truth of his theories, he claims to have calculated the true values of the constants of nature from them. On the other hand it should be remarked that the theory leads to a world consisting solely of electrons and protons—the only two kinds of particles which were known when Eddington formulated the theory. It not only failed to predict the existence of the other fundamental particles which have since been discovered, but (so far as I understand it) also suggests that their existence is impossible. This is particularly true of the positive electron, or positron.

Others have ventured even farther along this road. We have seen how the period of 1,780 million years which we extracted from the apparent motions of the nebulae provides a cosmic unit of time, which is comparable with the age of the universe. An alternative procedure would be to take the age of the universe itself as the unit of time. At present this is of the order of 10^{39} physical units of time, but when the universe is 10 times as old as now, it will be about 10^{40} physical units. On the view we are now considering, the other large numbers which are now of the order of 10^{39} will by then also be of the order of 10^{40}. These numbers are all obtained by combining various of the so-called "constants" of nature, whence it follows that these "constants" cannot be truly constants, but must slowly change their values. As $\sqrt{N}$ is one of these numbers, it would seem to follow, on the view I am now describing, that N itself is not a constant, but must increase with the time—as though creation were still in progress. This conclusion would violate the general principle of the conservation of mass and would be in even more flagrant opposition with Eddington's hypothesis that N is a mathematical constant, for, whatever else changes, the mathematical constants at least must remain eternally the same. A reconciliation can be arranged by making the unit of time continually change its value, so that the age of the universe in terms of this changing unit remains always the same; we must make a world in which "it is always six o'clock." And this is very easily done if we can assume that the constants of nature are themselves capable of change.

Professor Milne has been led, by quite another road, to postulate changes of precisely this kind. Far be it from me to discuss his general work here, on his own native heath, but perhaps I may be permitted to point out its relation to what I have just been saying. Starting from a general hypothetical principle, entirely different from those I have described, he finds it necessary to postulate two distinct kinds of time, one physical and one astronomical. These are measured in different units, the ratio of which changes as the age of the universe increases. At present the ratio of the units is unity because we elect to make it so as a matter of convenience. But it is in process of change through a continual variation of the so-called "constants of nature," such as the constant of gravitation and the charge on the electron. Actually there are more quantities at our disposal than there are equations to be satisfied, so that there are an infinite number of ways of achieving the necessary result. This only amounts to saying that there are many ways in which the observed phenomena can be pictured consistently.

Now it is, I think, generally agreed that Professor Milne's scheme is simply an alternative description to that of the expanding universe, which again is alternative to the shrinking atom, and again, as Milne has recently shown, is substantially similar to that of the ageing light-quanta. All these descriptions predict the same set of phenomena. If they did not, we could appeal to the phenomena to decide between them, and might even be able to reduce the permissible descriptions to a single one. If so, this one would become a working hypothesis, but would still have no claims to represent the truth; all we could say would be that nothing proved it untrue. But so far we know of no way of discriminating by observation.

Before the coming of relativity theory, the final goal which science set before itself was not to find true descriptions of the phenomena, but true explanations. The various descriptions I have mentioned can of course be treated as tentative explanations, and, if absolute space and time still existed, they would all be different. When lengths can be measured against absolute space, an expanding universe cannot be the same thing as a shrinking atom; when we can set our clocks by Newton's "equably flowing stream of time," a varying constant of gravitation cannot be the same thing as one that is fixed. But with absolute space and time banished from the scene, it is meaningless to say that one explanation is more true than another; absolute truth makes its exit in company with absolute time and space.

In this way, it seems to me that astronomy has reached the same state as atomic physics. To the physicist, a shower of electrons is sometimes a stream of particles, and sometimes a train of waves; we are no longer concerned with which is the truer concept, but only with which is the more convenient for the problem in hand; we know, moreover, that neither expresses the whole truth. In the same way, to the astronomer, space is sometimes expanding; sometimes it stands still while the objects in it contract; sometimes it is flat and sometimes curved, sometimes finite and sometimes infinite. None of these pictures is true in any absolute sense, because space is a creation of our own minds, and we can make it what we like, but every one provides a convenient background against which to depict the phenomena. So long as we speak and think in terms of Newtonian space and time, astronomical truth is a meaningless concept; we can aspire only to consistency—to an internal consistency between the different parts of our description, and to an external consistency between the phenomena and our description of them. But if we can adjust our language and concepts to the demands of the theory of relativity—replacing points, distances, and times by events, intervals, and sequences—we may expect that our now diverse pictures will all merge into one. Then it will become permissible to think of a "true" description, although any explanation, except in a purely mathematical form, may be forever beyond our grasp.

LETTERS FROM THE ASTRONOMERS

SIV CEDERING

Siv Cedering was born in Sweden in 1939, moved to the United States in 1953, and became a naturalized citizen in 1958. She was a lecturer at the University of Massachusetts in 1973, but she has devoted herself mostly to the writing of poetry and to editing and translating collections of other poets. She has won numerous awards for her own poetry.

1. What does Cedering's poem have to say about the position of Copernicus?

2. How accurately are the impressions of the respective scientists and their attitudes portrayed in these poems?

3. What does the "letter" of Caroline Herschel have to say about the place of women in science?

4. How does Einstein relate to the astronomers? What is his view of science?

PLATE I

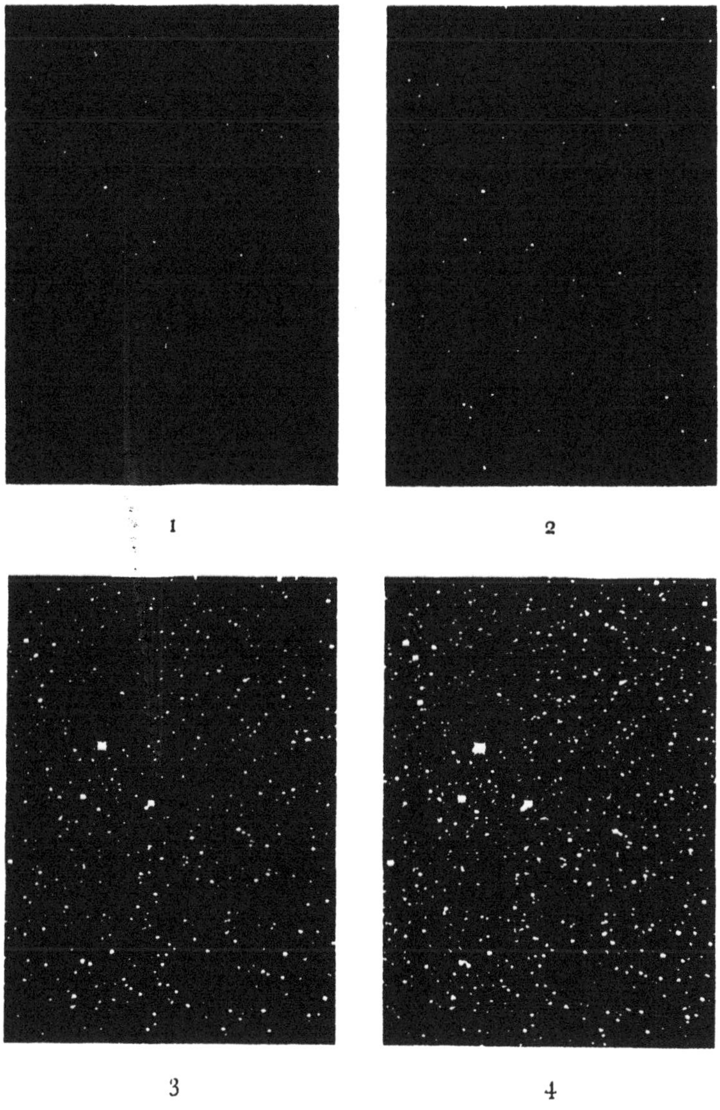

A small region of the sky in the constellation Auriga photographed with different exposures

PLATE II

Kerolyr, Forcalquier

Nebulosity in the Pleiades
(Exposure of photograph, 6 hours)

PLATE III

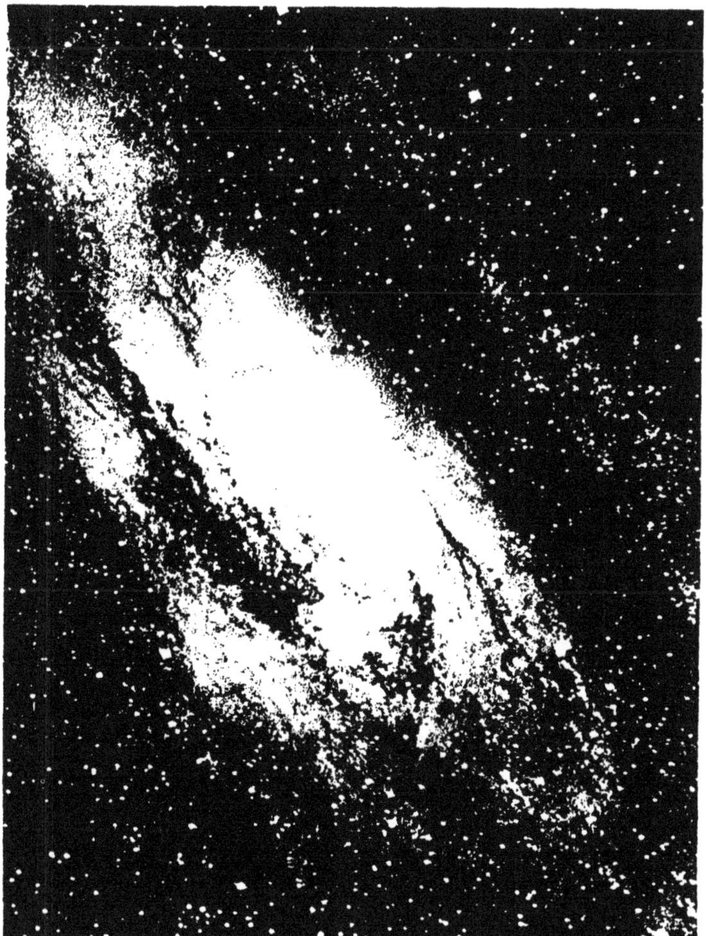

Mount Wilson Observatory

The Central Regions of the Great Nebula in Andromeda

PLATE IV

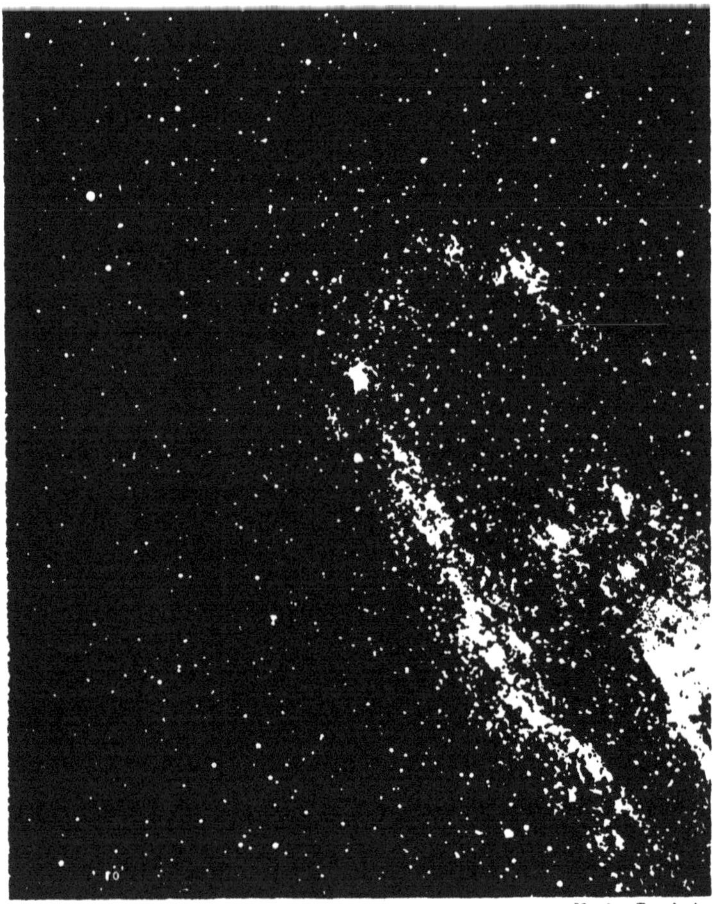

Kerolyr, Forcalquier

The Northern End of the Great Nebula in Andromeda, showing resolution into stars

PLATE V

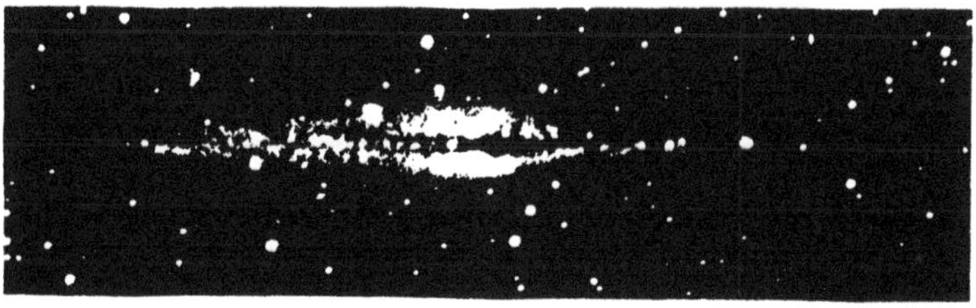

Mount Wilson Observatory

The Nebula N.G.C. 891 in Andromeda

Mount Wilson Observatory

The Nebula N.G.C. 7217

PLATE VI

Mount Wilson Observatory

A close Cluster of Nebulae (N.G.C. 7317–20) in Pegasus

Mount Wilson Observatory

A Cluster of Nebulae in Coma Berenicis, at a distance of about 50 million light-years

PLATE VII

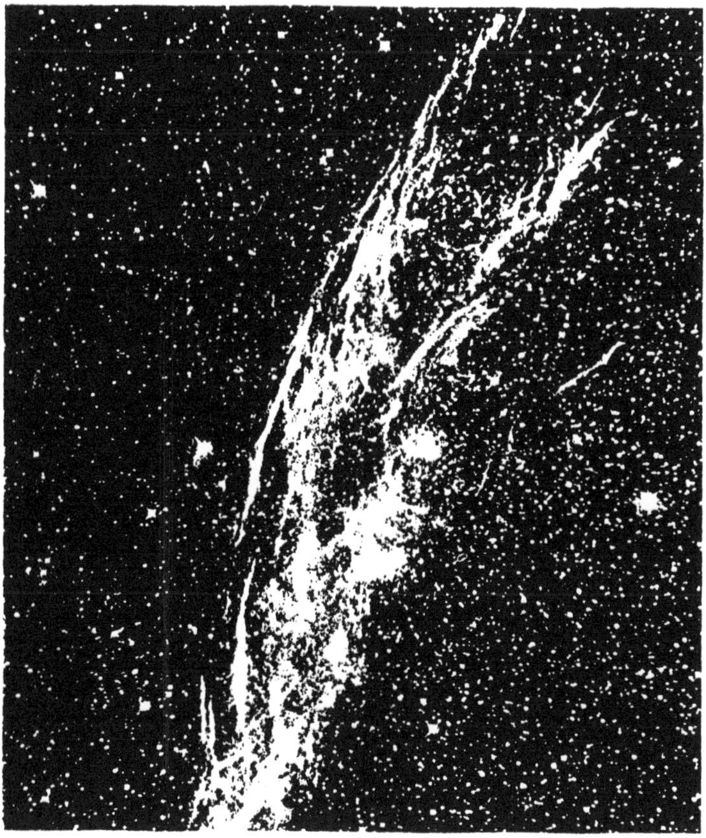

Mount Wilson Observatory

Nebulosity in the Constellation Cygnus
(Exposure of photograph, 12 hours)

PLATE VIII

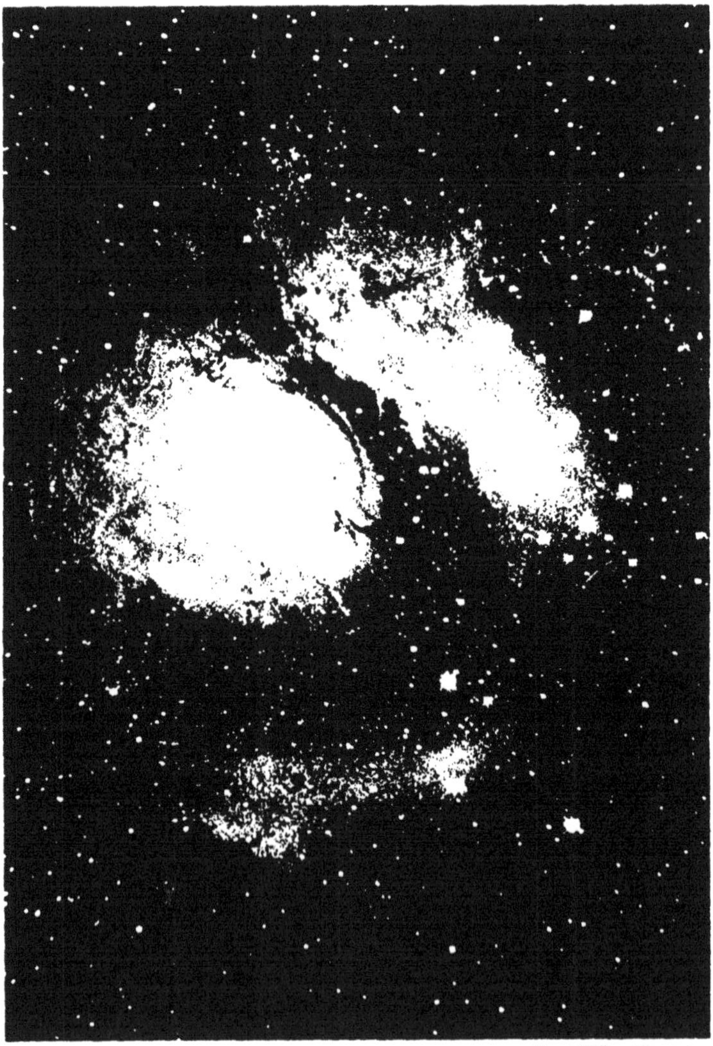

Mount Wilson Observatory

Nebulosity in Sagittarius

LETTERS FROM THE ASTRONOMERS

SIV CEDERING

If then the Astronomers, whereas they spie
A new-found Starre, their Opticks magnifie,
How brave are those, who with their Engine can
Bring man to heaven, and heaven again to man?
—John Donne

I. Nicholas Copernicus (1473-1543)

The sun is the center of the universe.
The planets move around the sun.
Yesterday when I went riding,
it began to snow. The seasons
change. The earth
turns
on its axis.

Share these ideas with your students,
but don't give them my name.
New continents are being discovered.
Books are being printed.
Witches burn.
I am afraid
of the spirit
of the times.
What will they do, if I say
the sun is the center of this cosmic
temple, if I say
its distance from the earth
is infinitesimally small
compared to the distance
between the earth
and the stars?

I carry a pouch
of powdered unicorn's horn
and red sandal wood,
to cure
the ill. I have devised a new
monetary system.

I was asked to Rome to help create
a new calendar, but I declined.
My mathematics are not
sufficient. If you
have been trained in the art,
see if you can find the laws
that would prove

my ideas. Do you have a quadrant?
An armillary sphere? An astrolabe?

II. Johannes Kepler (1571-1630)

If we substitute the word "force"
for the word "soul," we shall have
the basic principle which lies at
the heart of my celestial physics.
—Kepler

They say my mother is a witch.
She was arrested in the rectory.
They dragged her to prison in a trunk.
They want to put her on the rack.
For weeks she has been chained.
I am writing letters
asking them to release her.
My school has been closed.
The Protestant teachers have been burned
at the stake. My youngest child
died, of smallpox.
But I try to continue my exploration
of a celestial science.
I have derived a musical scale
for each planet, from variations
in their daily motions around the sun.
A five-note scale for Jupiter.
Fourteen notes for Mercury, and Venus,
repeating her one long note.
Such harmony. As I picture each planet
floating within the geometric perfections
of space, I think geometry was implanted in man
along with the image of God.
Geometry indeed is God.

III. Galileo Galilei (1564-1642)

They say I was arrogant.
Well, I was.
I had come far
from my father's numbers
and songs.
Before me, no one had seen
the moons of Jupiter,
the phases of Venus,
the mountains of the moon.

Now I am tired of arguing
with Jesuits and Aristotelians.
I have renounced my views
on the Copernican universe.
I have denounced my students
and friends. My book
is condemned. I cannot
put my mind at ease.
I have trouble

sleeping.
My daughter Virginia
reads the prescribed
penitential psalms.
What is Christianity
that it can be threatened?
The sun is turning on its axis.
I am old.
I am going blind.

IV. Caroline Herschel (1750-1848)

for Carol

William is away, and I am minding
the heavens. I have discovered
eight new comets and three nebulae
never before seen by man,
and I am preparing an Index to
Flamsteed's observations, together with
a catalogue of 560 stars omitted from
the British Catalogue, plus a list of errata
in that publication. William says

I have a way with numbers, so I handle
all the necessary reductions and
calculations. I also plan
every night's observation
schedule, for he says my intuition
helps me turn the telescope to discover
star cluster after star cluster.

I have helped him polish the mirrors
and lenses of our new telescope. It is
the largest in existence. Can you imagine
the thrill of turning it to some new
corner of the heavens to see
something never before seen
from earth? I actually like

that he is busy with the Royal Society
and his club, for when I finish my other work
I can spend all night sweeping
the heavens.

Sometimes when I am alone
in the dark, and the universe reveals
yet another secret, I say the names
of my long lost sisters, forgotten
in the books that record
our science—

Aglaonice of Thessaly,
Hyptia,
Hildegard,
Catherina Hevelius,
Maria Agnesi

—as if the stars themselves could

remember. Did you know that Hildegard
proposed a heliocentric universe
300 years before Copernicus?
that she wrote of universal gravitation 500 years
before Newton? But who would listen
to her? She was just a nun, a woman.
What is our age, if that age was dark?

As for my name, it will also be
forgotten, but I am not accused
of being a sorceress, like Aglaonice,
and the Christians do not threaten to
drag me to church, to murder me, like they did
Hyptia of Alexandria, the eloquent, young
woman who devised the instruments
used to accurately measure the position
and motion of

heavenly bodies.
However long we live, life is short, so I
work. And however important man becomes,
he is nothing compared to the stars.
There are secrets, dear sister, and it is
for us to reveal them. Your name, like mine,
is a song. Write soon,

Caroline

V. Albert Einstein (1879-1955)

Yes, I have written
the President. I have told him
that if there is a nuclear war, the Fourth
World War will be fought
with sticks and stones.

Words do hurt me,
and there is no change in my heart
condition, but I am trying to complete
my unified field
theory. I cannot believe
that God is playing roulette
with the world. The mystery
must be locked up
in the elemental infrastructures.

Forgive me for using
scrap paper. The other day
when my wife and I
were being shown
the huge reflecting telescope
at Mount Wilson observatory,
she asked why the instruments
were so large.

On being told
that they were trying to discern
the shape and makeup of the whole
universe, she said:
My husband does that
on the back
of an old
envelope.

SCIENCE AS A WAY OF KNOWING

Peirce

Poincaré

Kuhn

Keller

Hardy

Fisher

Physical space and time are the absolute stupidity of the universe.

José Ortega y Gasset

Wisdom entereth not into a malicious mind, and science without conscience is but the ruin of the soul.

François Rabelais, *Gartantua and Pantagruel,* 1532

The greatest of the changes that science has brought is the acuity of change; the greatest novelty the extent of novelty.

J. Robert Oppenheimer, *Science and the Common Understanding*, 1953

True science teaches, above all, to doubt and be ignorant.

Miguel de Unamuno, *The Tragic Sense of Life*, 1913

The future of Thought, and therefore of History, lies in the hands of the physicists, and. . .the future historian must seek his education in the world of mathematical physics. A new generation must be brought up to think by new methods, and if our historical departments in the Universities cannot enter this next phase, the physical departments will have to assume this task alone.

Henry Adams, *The Degradation of the Democratic Dogme*, 1919

. . .it is more important to have beauty in one's equations than to have them fit experiment.

P. A. M. Dirac, *Scientific American*, 1953

God not only plays dice, he sometimes throws the dice where they cannot be seen.

Stephen Hawking, *Nature*, 1975

I believe the intellectual life of the whole of western society is increasingly being split into two polar groups. When I say the intellectual life, I mean to include also a large part of our practical life, because I should be the last person to suggest the two can at the deepest level be distinguished. . .Two polar groups: at one pole we have the literary intellectuals, who incidentally while no one was looking took to referring to themselves as "intellectuals" as though there were no others. . .Literary intellectuals at one pole--at the other scientists, and as the most representative, the physical scientists. Between the two a gulf of mutual incomprehension--sometimes (particularly among the young) hostility and dislike, but most of all lack of understanding.

C. P. Snow, *The Two Cultures*, 1959

THE FIXATION OF BELIEF

CHARLES PEIRCE

Recognized now as the founder of American pragmatism, Charles Sanders Peirce (1839-1914) was not well known during his lifetime. Although he wrote profusely and with penetrating thought, he was unable to find a publisher for most of his work. His reputation, therefore, rests mainly on manuscripts found after his death, but that was such that a student of American philosophy, Bruce Kublick, has declared that Peirce "is generally recognized as the most important American thinker of the 19th and 20th centuries."

Peirce was born in Cambridge, Massachusetts, son of a leading Harvard mathematician. He graduated from Harvard in 1859 and from the Lawrence Scientific School at Harvard in 1863. After working successively as an astronomer at the Harvard Observatory and as a physicist for the U.S. Coast and Geodetic Survey, he accepted an appointment as lecturer in logic at The Johns Hopkins University in 1879.

Dismissed from Johns Hopkins in 1884, he retired with his second wife, Juliette Froissy, to Milford, Pennsylvania, in 1887, where he devoted himself to his writing. This included some of his best work, though he spent most of his last twenty years in serious illness and abject poverty, relieved only by contributions form William James and other friends.

"The Fixation of Belief" was the first of a series of articles that first appeared in *Popular Science Monthly* in 1877. Peirce wrote the introductory "Concerning the Author" in 1897.

1. How do you account for the different ways in which people arrive at their beliefs?

2. According to Peirce, how does the scientific method differ from other methods in establishing opinion?

3. What is truth?

4. What is Peirce's criticism of Francis Bacon?

5. How do you account for the attempts of some people to impose their beliefs on others, such as in religious proselytizing and heresy trials? Why did the Church object to the findings of Galileo?

THE FIXATION OF BELIEF

CHARLES PEIRCE

I
CONCERNING THE AUTHOR

THE reader has a right to know how the author's opinions were formed. Not, of course, that he is expected to accept any conclusions which are not borne out by argument. But in discussions of extreme difficulty, like these, when good judgment is a factor, and pure ratiocination is not everything, it is prudent to take every element into consideration. From the moment when I could think at all, until now, about forty years, I have been diligently and incessantly occupied with the study of methods [of] inquiry, both those which have been and are pursued and those which ought to be pursued. For ten years before this study began, I had been in training in the chemical laboratory. I was thoroughly grounded not only in all that was then known of physics and chemistry, but also in the way in which those who were successfully advancing knowledge proceeded. I have paid the most attention to the methods of the most exact sciences, have intimately communed with some of the greatest minds of our times in physical science, and have myself made positive contributions—none of them of any very great importance, perhaps—in mathematics, gravitation, optics, chemistry, astronomy, etc. I am saturated, through and through, with the spirit of the physical sciences. I have been a great student of logic, having read everything of any importance on the subject, devoting a great deal of time to medieval thought, without neglecting the works of the Greeks, the English, the Germans, the French, etc., and have produced systems of my own both in deductive and in inductive logic. In metaphysics my training has been less systematic; yet I have read and deeply pondered upon all the main systems, never being satisfied until I was able to think about them as their own advocates thought.

The first strictly philosophical books that I read were of the classical German schools; and I became so deeply imbued with many of their ways of thinking that I have never been able to disabuse myself of them. Yet my attitude was always that of a dweller in a laboratory, eager only to learn what I did not yet know, and not that of philosophers bred in theological seminaries, whose ruling impulse is to teach what they hold to be infallibly true. I devoted two hours a day to the study of Kant's *Critic of the Pure Reason* for more than three years, until I almost knew the whole book by heart, and had critically examined every section of it. For about two years, I had long and almost daily discussions with Chauncey Wright, one of the most acute of the followers of J. S. Mill.

The effect of these studies was that I came to hold the classical German philosophy to be, upon its argumentative side, of little weight; although I esteem it, perhaps am too partial to it, as a rich mine of philosophical suggestions. The English philosophy, meagre and crude, as it is, in its conceptions, proceeds by surer methods and more accurate logic. The doctrine of the association of ideas is, to my thinking, the finest piece of philosophical work of the prescientific ages. Yet I can but pronounce English sensationalism to be entirely destitute of any solid bottom. From the evolutionary philosophers, I have learned little;

although I admit that, however hurriedly their theories have been knocked together, and however antiquated and ignorant Spencer's *First Principles* and general doctrines, yet they are under the guidance of a great and true idea, and are developing it by methods that are in their main features sound and scientific.

The works of Duns Scotus have strongly influenced me. If his logic and metaphysics, not slavishly worshipped, but torn away from its medievalism, be adapted to modern culture, under continual wholesome reminders of nominalistic criticisms, I am convinced that it will go far toward supplying the philosophy which is best to harmonize with physical science. But other conceptions have to be drawn from the history of science and from mathematics.

Thus, in brief, my philosophy may be described as the attempt of a physicist to make such conjecture as to the constitution of the universe as the methods of science may permit, with the aid of all that has been done by previous philosophers. I shall support my propositions by such arguments as I can. Demonstrative proof is not to be thought of. The demonstrations of the metaphysicians are all moonshine. The best that can be done is to supply a hypothesis not devoid of all likelihood, in the general line of growth of scientific ideas, and capable of being verified or refuted by future observers.

Religious infallibilism, caught in the current of the times, shows symptoms of declaring itself to be only practically speaking infallible; and when it has thus once confessed itself subject to gradations, there will remain over no relic of the good old tenth century infallibilism, except that of the infallible scientists, under which head I include, not merely the kind of characters that manufacture scientific catechisms and homilies, churches and creeds, and who are indeed "born missionaries," but all those respectable and cultivated persons who, having acquired their notions of science from reading, and not from research, have the idea that "science" means knowledge, while the truth is, it is a misnomer applied to the pursuit of those who are devoured by a desire to find things out. . . .

Though infallibility in scientific matters seems to me irresistibly comical, I should be in a sad way if I could not retain a high respect for those who lay claim to it, for they comprise the greater part of the people who have any conversation at all. When I say they lay claim to it, I mean they assume the functions of it quite naturally and unconsciously. The full meaning of the adage *Humanum est errare,* they have never waked up to. In those sciences of measurement which are the least subject to error—metrology, geodesy, and metrical astronomy—no man of self-respect ever now states his result, without affixing to it its *probable error;* and if this practice is not followed in other sciences it is because in those the probable errors are too vast to be estimated.

I am a man of whom critics have never found anything good to say. When they could see no opportunity to injure me, they have held their peace. The little laudation I have had has come from such sources, that the only satisfaction I have derived from it, has been from such slices of bread and butter as it might waft my way. Only once, as far as I remember, in all my lifetime have I experienced the pleasure of praise—not for what it might bring but in itself. That pleasure was beatific; and the praise that conferred it was meant for blame. It was that a critic said of me that I did not seem to be *absolutely sure of my own conclusions.* Never, if I can help it, shall that critic's eye ever rest on what I am now writing; for I owe a great pleasure to him; and, such was his evident animus, that should he find that out, I fear the fires of hell would be fed with new fuel in his breast.

My book will have no instruction to impart to anybody. Like a mathematical treatise, it will suggest certain ideas and certain reasons for holding them true; but then, if you accept them, it must be because you like my reasons, and the responsibility lies with you. Man is essentially a social animal: but to be social is one thing, to be gregarious is another: I decline to serve as bellwether. My book is meant for people who *want to find out;* and people who want philosophy ladled out to them can go elsewhere. There are philosophical soup shops at every corner, thank God!

The development of my ideas has been the industry of thirty years. I did not know as I ever should get to publish them, their ripening seemed so slow. But the harvest time has come, at last, and to me that harvest seems a wild one, but of course it is not I who have to pass judgment. It is not quite you, either, individual reader; it is experience and history.

For years in the course of this ripening process, I used for myself to collect my ideas under the designation *fallibilism;* and indeed the first step *toward finding* out is to acknowledge you do not satisfactorily know already: so that no blight can so surely arrest all intellectual growth as the blight of cocksureness; and ninety-nine out of every hundred good heads are reduced to impotence by that malady—of whose inroads they are most strangely unaware!

Indeed, out of a contrite fallibilism, combined with a high faith in the reality of knowledge, and an intense desire to find things out, all my philosophy has always seemed to me to grow. . . .

2
THE FIXATION OF BELIEF

FEW persons care to study logic, because everybody conceives himself to be proficient enough in the art of reasoning already. But I observe that this satisfaction is limited to one's own ratiocination, and does not extend to that of other men.

We come to the full possession of our power of drawing inferences the last of all our faculties; for it is not so much a natural gift as a long and difficult art. The history of its practice would make a grand subject for a book. The medieval schoolmen, following the Romans, made logic the earliest of a boy's studies after grammar, as being very easy. So it was as they understood it. Its fundamental principle, according to them, was, that all knowledge rests either on authority or reason; but that whatever is deduced by reason depends ultimately on a premiss derived from authority. Accordingly, as soon as a boy was perfect in the syllogistic procedure, his intellectual kit of tools was held to be complete.

To Roger Bacon, that remarkable mind who in the middle of the thirteenth century was almost a scientific man, the schoolmen's conception of reasoning appeared only an obstacle to truth. He saw that experience alone teaches anything—a proposition which to us seems easy to understand, because a distinct conception of experience has been handed down to us from former generations; which to him likewise seemed perfectly clear, because its difficulties had not yet unfolded themselves. Of all kinds of experience, the best, he thought, was interior illumination, which teaches many things about Nature which the external senses could never discover, such as the transubstantiation of bread.

Four centuries later, the more celebrated Bacon, in the first book of his *Novum Organum,* gave his clear account of experience as something which must be open to verification and reexamination. But, superior as Lord Bacon's conception is to earlier notions, a modern reader who is not in awe of his grandiloquence is chiefly struck by the inadequacy of his view of scientific procedure. That we have only to make some crude experiments, to draw up briefs of the results in certain blank forms, to go through these by rule, checking off everything disproved and setting down the alternatives, and that thus in a few years physical science would be finished up—what an idea! "He wrote on science like a Lord Chancellor," indeed, as Harvey, a genuine man of science said.

The early scientists, Copernicus, Tycho Brahe, Kepler, Galileo, Harvey, and Gilbert, had methods more like those of their modern brethren. Kepler undertook to draw a curve through the places of Mars, and to state the times occupied by the planet in describing the different parts of that curve; but perhaps his greatest service to science was in impressing on men's minds that this was the thing to be done if they wished to improve astronomy; that they were not to content themselves with inquiring whether one system of epicycles was better than another, but that they were to sit down to the figures and find out what the curve, in truth, was. He accomplished this by his incomparable energy and courage, blundering along in the most inconceivable way (to us), from one irrational hypothesis to another, until, after trying twenty-two of these, he fell, by the mere exhaustion of his invention, upon the orbit which a mind well furnished with the weapons of modern logic would have tried almost at the outset.

In the same way, every work of science great enough to be well remembered for a few generations affords some exemplification of the defective state of the art of reasoning of the time when it was written; and each chief step in science has been a lesson in logic. It was so when Lavoisier and his contemporaries took up the study of Chemistry. The old chemist's maxim had been, "Lege, lege, lege, labora, ora, et relege." Lavoisier's method was not to read and pray, but to dream that some long and complicated chemical process would have a certain effect, to put it into practice with dull patience, after its inevitable failure, to dream that with some modification it would have another result, and to end by publishing the last dream as a fact: his way was to carry his mind into his laboratory, and literally to make of his alembics and cucurbits instruments of thought, giving a new conception of reasoning as something which was to be done with one's eyes open, in manipulating real things instead of words and fancies.

The Darwinian controversy is, in large part, a question of logic. Mr. Darwin proposed to apply the statistical method to biology. The same thing has been done in a widely different branch of science, the theory of gases. Though unable to say what the movements of any particular molecule of gas would be on a certain hypothesis regarding the constitution of this class of bodies, Clausius and Maxwell were yet able, eight years before the publication of Darwin's immortal work, by the application of the doctrine of probabilities, to predict that in the long run such and such a proportion of the molecules would, under given circumstances, acquire such and such velocities; that there would take place, every second, such and such a relative number of collisions, etc.; and from these propositions were able to deduce certain properties of gases, especially in regard to their heat-relations. In like manner, Darwin, while unable to say what the operation of variation and natural selection in any individual case will be, demonstrates that in the long run they will, or would, adapt animals to their circumstances. Whether or not existing animal forms are due to such action, or what position the theory ought to take, forms the subject of a discussion in which questions of fact and questions of logic are curiously interlaced.

The object of reasoning is to find out, from the consideration of what we already know, something else which we do not know. Consequently, reasoning is good if it be such as to give a true conclusion from true premisses, and not otherwise. Thus, the question of validity is purely one of fact and not of thinking. A being the facts stated in the premisses and B being that concluded, the question is, whether these facts are really so related that if A were B would generally be. If so, the inference is valid; if not, not. It is not in the least the question whether, when the premisses are accepted by the mind, we feel an impulse to accept the conclusion also. It is true that we do generally reason correctly by nature. But that is an accident; the true conclusion would remain true if we had no impulse to accept it; and the false one would remain false, though we could not resist the tendency to believe in it.

We are, doubtless, in the main logical animals, but we are not perfectly so. Most of us, for example, are naturally more sanguine and hopeful than logic would justify. We seem to be so constituted that in the absence of any facts to go upon we are happy and self-satisfied; so that the effect of experience is continually to contract our hopes and aspirations. Yet a lifetime of the application of this corrective does not usually eradicate our sanguine disposition. Where hope is unchecked by any experience, it is likely that our optimism is extravagant. Logicality in regard to practical matters (if this be understood, not in the old sense, but as consisting in a wise union of security with fruitfulness of reasoning) is the most useful quality an animal can possess, and might, therefore, result from the action of natural selection; but outside of these it is probably of more advantage to the animal to have his mind filled with pleasing and encouraging visions, independently of their truth; and thus, upon unpractical subjects, natural selection might occasion a fallacious tendency of thought.

That which determines us, from given premisses, to draw one inference rather than another, is some habit of mind, whether it be constitutional or acquired. The habit is good or otherwise, according as it produces true conclusions from true premisses or not; and an inference is regarded as valid or not, without reference to the truth or falsity of its conclusion specially, but according as the habit which determines it is such as to produce true conclusions in general or not. The particular habit of mind which governs this or that inference may be formulated in a proposition whose truth depends on the validity of the inferences which the habit determines, and such a formula is called a *guiding principle* of inference. Suppose, for example, that we observe that a rotating disk of copper quickly comes to rest when placed between the poles of a magnet, and we infer that this will happen with every disk of copper. The guiding principle is, that what is true of one piece of copper is true of another. Such a guiding principle with regard to copper would be much safer than with regard to many other substances—brass, for example.

A book might be written to signalize all the most important of these guiding principles of reasoning. It would probably be, we must confess, of no service to a person whose thought is directed wholly to practical subjects, and whose activity moves along thoroughly-beaten paths. The problems that present themselves to such a mind are matters of routine which he has learned once for all to handle in learning his business. But let a man venture into an unfamiliar field, or where his results are not continually checked by experience, and all history shows that the most masculine intellect will ofttimes lose his orientation and waste his efforts in directions which bring him no nearer to his goal, or even carry him entirely astray. He is like a ship in the open sea, with no one on board who understands the rules of navigation. And in such a case some general study of the guiding principles of reasoning would be sure to be found useful.

The subject could hardly be treated, however, without being first limited; since almost any fact may serve as a guiding principle. But it so happens that there exists a division among facts, such that in one class are all those which are absolutely essential as guiding principles, while in the others are all which have any other interest as objects of research. This division is between those which are necessarily taken for granted in asking why a certain conclusion is thought to follow from certain premisses, and those which are not implied in such a question. A moment's thought will show that a variety of facts are already assumed when the logical question is first asked. It is implied, for instance, that there are such states of mind as doubt and belief—that a passage from one to the other is possible, the object of thought remaining the same, and that this transition is subject to some rules by which all minds are alike bound. As these are facts which we must already know before we can have any clear conception of reasoning at all, it cannot be supposed to be any longer of much interest to inquire into their truth or falsity. On the other hand, it is easy to believe that those rules of reasoning which are deduced from the very idea of the process are the ones which are the most essential; and, indeed, that so long as it conforms to these it will, at least, not lead to false conclusions from true premisses. In point of fact, the importance of what may be deduced from the assumptions involved in the logical question turns out to be greater than might be supposed, and this for reasons which it is difficult to exhibit at the outset. The only one which I shall here mention is, that conceptions which are really products of logical reflection, without being readily seen to be so, mingle with our ordinary thoughts, and are frequently the causes of great confusion. This is the case, for example, with the conception of quality. A quality, as such, is never an object of observation. We can see that a thing is blue or green, but the quality of being blue and the quality of being green are not things which we see; they are products of logical reflections. The truth is, that common-sense, or thought as it first emerges above the level of the narrowly practical, is deeply imbued with that bad logical quality to which the epithet *metaphysical* is commonly applied; and nothing can clear it up but a severe course of logic.

We generally know when we wish to ask a question and when we wish to pronounce a judgment, for there is a dissimilarity between the sensation of doubting and that of believing.

But this is not all which distinguishes doubt from belief. There is a practical difference. Our beliefs guide our desires and shape our actions. The Assassins, or followers of the Old Man of the Mountain, used to rush into death at his least command, because they believed that obedience to him would insure everlasting felicity. Had they doubted this, they would not have acted as they did. So it is with every belief, according to its degree. The feeling of believing is a more or less sure indication of there being established in our nature some habit which will determine our actions. Doubt never has such an effect.

Nor must we overlook a third point of difference. Doubt is an uneasy and dissatisfied state from which we struggle to free ourselves and pass into the state of belief; while the latter is a calm and satisfactory state which we do not wish to avoid, or to change to a belief in anything else. On the contrary, we cling tenaciously, not merely to believing, but to believing just what we do believe.

Thus, both doubt and belief have positive effects upon us, though very different ones. Belief does not make us act at once, but puts us into such a condition that we shall behave in some certain way, when the occasion arises. Doubt has not the least such active effect, but stimulates us to inquiry until it is destroyed. This reminds us of the irritation of a nerve and the reflex action produced thereby; while for the analogue of belief, in the nervous system, we must look to what are called nervous associations—for example, to that habit of the nerves in consequence of which the smell of a peach will make the mouth water.

The irritation of doubt causes a struggle to attain a state of belief. I shall term this struggle *Inquiry*, though it must be admitted that this is sometimes not a very apt designation.

The irritation of doubt is the only immediate motive for the struggle to attain belief. It is certainly best for us that our beliefs should be such as may truly guide our actions so as to satisfy our desires; and this reflection will make us reject every belief which does not seem to have been so formed as to insure this result. But it will only do so by creating a doubt in the place of that belief. With the doubt, therefore, the struggle begins, and with the cessation of doubt it ends. Hence, the sole object of inquiry is the settlement of opinion. We may fancy that this is not enough for us, and that we seek, not merely an opinion, but a true opinion. But put this fancy to the test, and it proves groundless, for as soon as a firm belief is reached we are entirely satisfied, whether the belief be true or false. And it is clear that nothing out of the sphere of our knowledge can be our object, for nothing which does not affect the mind can be the motive for mental effort. The most that can be maintained is, that we seek for a belief that we shall *think* to be true. But we think each one of our beliefs to be true, and indeed, it is mere tautology to say so.

That the settlement of opinion is the sole end of inquiry is a very important proposition. It sweeps away, at once, various vague and erroneous conceptions of proof. A few of these may be noticed here.

1. Some philosophers have imagined that to start an inquiry it was only necessary to utter a question whether orally or by setting it down upon paper, and have even recommended us to begin our studies with questioning everything! But the mere putting of a proposition into the interrogative form does not stimulate the mind to any struggle after belief. There must be a real and living doubt, and without this all discussion is idle.

2. It is a very common idea that a demonstration must rest on some ultimate and absolutely indubitable propositions. These, according to one school, are first principles of a general nature; according to another, are first sensations. But, in point of fact, an inquiry, to have that completely satisfactory result called demonstration, has only to start with propositions perfectly free from all actual doubt. If the premisses are not in fact doubted at all, they cannot be more satisfactory than they are.

3. Some people seem to love to argue a point after all the world is fully convinced of it. But no further advance can be made. When doubt ceases, mental action on the subject comes to an end; and, if it did go on, it would be without a purpose.

If the settlement of opinion is the sole object of inquiry, and if belief is of the nature of a habit, why should we not attain the desired end, by taking as answer to a question any we may fancy, and constantly reiterating it to ourselves, dwelling on all which may conduce to that belief, and learning to turn with contempt and hatred from anything that might disturb it? This simple and direct method is really pursued by many men. I remember once being entreated not to read a certain newspaper lest it might change my opinion upon free-trade. "Lest I might be entrapped by its fallacies and misstatements," was the form of expression. "You are not," my friend said, "a special student of political economy. You might, therefore, easily be deceived by fallacious arguments upon the subject. You might, then, if you read this paper, be led to believe in protection. But you admit that free-trade is the true doctrine; and you do not wish to believe what is not true." I have often known this system to be deliberately adopted. Still oftener, the instinctive dislike of an undecided state of mind, exaggerated into a vague dread of doubt, makes men cling spasmodically to the views they already take. The man feels that if he only holds to his belief without wavering, it will be entirely satisfactory. Nor can it be denied that a steady and immovable faith yields great peace of mind. It may, indeed, give rise to inconveniences, as if a man should resolutely continue to believe that fire would not burn him, or that he would be eternally damned if he received his *ingesta* otherwise than through a stomach-pump. But then the man who adopts this method will not allow that its inconveniences are greater than its advantages. He will say, "I hold steadfastly to the truth, and the truth is always wholesome." And in many cases it may very well be that the pleasure he derives from his calm faith overbalances any inconveniences resulting from its deceptive character. Thus, if it be true that death is annihilation, then the man who believes that he will certainly go straight to heaven when he dies, provided he have fulfilled certain simple observances in this life, has a cheap pleasure which will not be followed by the least disappointment. A similar consideration seems to have weight with many persons in religious topics, for we frequently hear it said, "Oh, I could not believe so-and-so, because I should be wretched if I did." When an ostrich buries its head in the sand as danger approaches, it very likely takes the happiest course. It hides the danger, and then calmly says there is no danger; and, if it feels perfectly sure there is none, why should it raise its head to see? A man may go through life, systematically keeping out of view all that might cause a change in his opinions, and if he only succeeds—basing his method, as he does, on two fundamental psychological laws—I do not see what can be said against his doing so. It would be an egotistical impertinence to object that his procedure is irrational, for that only amounts to saying that his method of settling belief is not ours. He does not propose to himself to be rational, and, indeed, will often talk with scorn of man's weak and illusive reason. So let him think as he pleases.

But this method of fixing belief, which may be called the method of tenacity, will be unable to hold its ground in practice. The social impulse is against it. The man who adopts it will find that other men think differently from him, and it will be apt to occur to him, in some saner moment, that their opinions are quite as good as his own, and this will shake his confidence in his belief. This conception, that another man's thought or sentiment may be equivalent to one's own, is a distinctly new step, and a highly important one. It arises from an impulse too strong in man to be suppressed, without danger of destroying the human species. Unless we make ourselves hermits, we shall necessarily influence each other's opinions; so that the problem becomes how to fix belief, not in the individual merely, but in the community.

Let the will of the state act, then, instead of that of the individual. Let an institution be created which shall have for its object to keep correct doctrines before the attention of the people, to reiterate them perpetually, and to teach them to the young; having at the same time power to prevent contrary doctrines from being taught, advocated, or expressed. Let all possible causes of a change of mind be removed from men's apprehensions. Let them be kept ignorant, lest they should learn of some reason to think otherwise than they do. Let their passions be enlisted, so that they may regard private and unusual opinions with hatred and horror. Then, let all men who reject the established belief be terrified into silence. Let the people turn out and tar-and-feather such men, or let inquisitions be made into the manner of thinking of suspected persons, and when they are found guilty of forbidden beliefs, let them be subjected to some signal punishment. When complete agreement could not otherwise be reached, a general massacre of all who have not thought in a certain way has proved a very effective means of settling opinion in a country. If the power to do this be wanting, let a list of opinions be drawn up, to which no man of the least independence of thought can assent, and let the faithful be required to accept all these propositions, in order to segregate them as radically as possible from the influence of the rest of the world.

This method has, from the earliest times, been one of the chief means of upholding correct theological and political doctrines, and of preserving their universal or catholic character. In Rome, especially, it has been practised from the days of Numa Pompilius to those of Pius Nonus. This is the most perfect example in history; but wherever there is a priesthood—and no religion has been without one—this method has been more or less made use of. Wherever there is an aristocracy, or a guild, or any association of a class of men whose interests depend, or are supposed to depend, on certain propositions, there will be inevitably found some traces of this natural product of social feeling. Cruelties always accompany this system; and when it is consistently carried out, they become atrocities of the most horrible kind in the eyes of any rational man. Nor should this occasion surprise, for the officer of a society does not feel justified in surrendering the interests of that society for the sake of mercy, as he might his own private interests. It is natural, therefore, that sympathy and fellowship should thus produce a most ruthless power.

In judging this method of fixing belief, which may be called the method of authority, we must, in the first place, allow its immeasurable mental and moral superiority to the method of tenacity. Its success is proportionately greater, and, in fact, it has over and over again worked the most majestic results. The mere structures of stone which it has caused to be put together—in Siam, for example, in Egypt, and in Europe—have many of them a sublimity hardly more than rivalled by the greatest works of Nature. And, except the geological epochs, there are no periods of time so vast as those which are measured by some of these organized faiths. If we scrutinize the matter closely, we shall find that there has not been one of their creeds which has remained always the same; yet the change is so slow as to be imperceptible during one person's life, so that individual belief remains sensibly fixed. For the mass of mankind, then, there is perhaps no better method than this. If it is their highest impulse to be intellectual slaves, then slaves they ought to remain.

But no institution can undertake to regulate opinions upon every subject. Only the most important ones can be attended to and on the rest men's minds must be left to the action of natural causes. This imperfection will be no source of weakness so long as men are in such a state of culture that one opinion does not influence another—that is, so long as they cannot put two and two together. But in the most priest-ridden states some individuals will be found who are raised above that condition. These men possess a wider sort of social feeling; they see that men in other countries and in other ages have held to very different doctrines from those which they themselves have been brought up to believe; and they cannot help seeing that it is the mere accident of their having been taught as they have, and of their having been surrounded with the manners and associations they have, that has caused them to believe as they do and not far differently. Nor can their candour resist the reflection that there is no reason to rate their own views at a higher value than those of other nations and other centuries; thus giving rise to doubts in their minds.

They will further perceive that such doubts as these must exist in their minds with reference to every belief which seems to be determined by the caprice either of themselves or of those who originated the popular opinions. The willful adherence to a belief, and the arbitrary forcing of it upon others, must, therefore, both be given up. A different new method of settling opinions must be adopted, that shall not only produce an impulse to believe, but shall also decide what proposition it is which is to be believed. Let the action of natural preferences be unimpeded, then, and under their influence let men, conversing together and regarding matters in different lights, gradually develop beliefs in harmony with natural causes. This method resembles that by which conceptions of art have been brought to maturity. The most perfect example of it is to be found in the history of metaphysical philosophy. Systems of this sort have not usually rested upon any observed facts, at least not in any great degree. They have been chiefly adopted because their fundamental propositions seemed "agreeable to reason." This is an apt expression; it does not mean that which agrees with experience, but that which we find ourselves inclined to believe. Plato, for example, finds it agreeable to reason that the distances of the celestial spheres from one another should be proportional to the different lengths of strings which produce harmonious chords. Many philosophers have been led to their main conclusions by considerations like this*; but this is the lowest and least developed form which the method takes, for it is clear that another man might find Kepler's theory, that the celestial spheres are proportional to the inscribed and circumscribed spheres of the different regular solids, more agreeable to *his* reason. But the shock of opinions will soon lead men to rest on preferences of a far more universal nature. Take, for example, the doctrine that man only acts selfishly—that is, from the consideration that acting in one way will afford him more pleasure than acting in another. This rests on no fact in the world, but it has had a wide acceptance as being the only reasonable theory.

(*Let us see in what manner a few of the greatest philosophers have undertaken to settle opinion, and what their success has been. Descartes, who would have a man begin by doubting everything, remarks that there is one thing he will find himself unable to doubt, and that is, that he does doubt; and when he reflects that he doubts, he can no longer doubt that he exists. Then, because he is all the while doubting whether there are any such things as shape and motion, Descartes thinks he must be persuaded that shape and motion do not belong to his nature, or anything else but consciousness. This is taking it for granted that nothing in his nature lies hidden beneath the surface. Next, Descartes asks the doubter to remark that he has the idea of a Being, in the highest degree intelligent, powerful, and perfect. Now a Being would not have these qualities unless he existed necessarily and eternally. By existing necessarily he means existing by virtue of the existence of the idea. Consequently, all doubt as to the existence of this Being must cease. This plainly supposes that belief is to be fixed by what men find in their minds. He is reasoning like this: I find it written in the volume of my mind that there is something X which is such a sort of thing that the moment it is written down it exists. Plainly, he is aiming at a kind of truth which saying so can make to be so. He gives two further proofs of God's existence. Descartes makes God easier to know than anything else; for whatever we think He is, He is. He fails to remark that this is precisely the definition of a figment. In particular, God cannot be a deceiver; whence it follows, that whatever we quite clearly and distinctly think to be true about any subject, must *be* true. Accordingly, if people will thoroughly discuss a subject, and quite clearly and distinctly make up their minds what they think about it, the desired settlement of the question will be reached. I may remark that the world has pretty thoroughly deliberated upon that theory and has quite distinctly come to the conclusion that it is utter nonsense, whence that judgment is indisputably right.

(Many critics have told me that I misrepresent the *a priori* philosophers when I represent them as adopting whatever opinion there seems to be a natural inclination to adopt. But nobody can say the above does not accurately define the position of Descartes, and upon what does he repose except natural ways of thinking? Perhaps I shall be told, however, that since Kant, that vice has been cured. Kant's great boast is that he critically examines into our natural inclinations toward certain opinions. An opinion that something is *universally* true clearly goes further than experience can warrant. An opinion that something is *necessarily* true (that is, not merely is true in the existing state of things, but would be true in every state of things) equally goes further than experience will warrant. Those remarks had been made by Leibniz and admitted by Hume; and Kant reiterates them. Though they are propositions of a nominalistic cast, they can hardly be denied. I may add that whatever is held to be precisely true goes further than experience can possibly warrant. Accepting those criteria of the origin of ideas, Kant proceeds to reason as follows: Geometrical propositions are held to be universally true. Hence, they are not given by experience. Consequently, it must be owing to an inward necessity of man's nature that he sees everything in space. Ergo, the sum of the angles of a triangle will be equal to two right angles for all the objects of our vision. Just that, and nothing more is Kant's line of thought. But the dry-rot of reason in the seminaries has gone to the point where such stuff is held to be admirable argumentation. I might go through the *Critic of the Pure Reason,* section by section, and show that the thought throughout is precisely of this character. He everywhere shows that ordinary objects, such as trees and gold-pieces, involve elements not contained in the first presentations of sense. But we cannot persuade ourselves to give up the reality of trees and gold-pieces. There is a general inward insistence upon them, and that is the warrant for swallowing the entire bolus of general belief about them. This is merely accepting without question a belief as soon as it is shown to please a great many people very much. When he comes to the ideas of God, Freedom, and Immortality, he hesitates; because people who think only of bread and butter, pleasure and power, are indifferent to those ideas. He subjects these ideas to a different kind of examination, and finally admits them upon grounds which appear to the seminarists more or less suspicious, but which in the eyes of laboratorists are infinitely stronger than the grounds upon which he has accepted space, time, and causality. Those last grounds amount to nothing but this, that what there is a very decided and general inclination to believe must be true. Had Kant merely said, I shall adopt for the present the belief that the three angles of a triangle are equal to two right angles because nobody but brother Lambert and some Italian has ever called it in question, his attitude would be well enough. But on the contrary he and those who today represent his school distinctly maintain the proposition is *proved,* and the Lambertists *refuted*, by what comes merely to general disinclination to think with them.

(As for Hegel, who led Germany for a generation, he recognizes clearly what he is about. He simply launches his boat into the current of thought and allows himself to be carried wherever the current leads. He himself calls his method *dialectic,* meaning that a frank discussion of the difficulties to which any opinion spontaneously gives rise will lead to modification after modification until a tenable position is attained. This is a distinct profession of faith in the method of inclinations.

(Other philosophers appeal to "the test of inconceivability of the opposite," to "presuppositions" (by which they mean *Voraussetzungen,* properly translated, *postulates),* and other devices but all these are but so many systems of rummaging the garret of the skull to find an enduring opinion about the Universe.

(When we pass from the perusal of works upholding the method of authority to those of the philosophers, we not only find ourselves in a vastly higher intellectual atmosphere, but also in a clearer, freer, brighter, and more refreshing moral atmosphere. All this, however, is beside the one significant question of whether the method succeeds in fixing men's opinions. The projects of these authors are most persuasive. One dare swear they should succeed. But in point of fact, up to date they decidedly do not; and the outlook in this direction is most discouraging. The difficulty is that the opinions which today seem most unshakable are found tomorrow to be out of fashion. They are really far more changeable than they appear to a hasty reader to be, since the phrases made to dress out defunct opinions are worn at second hand by their successors.

(We still talk of "cause and effect" although, in the mechanical world, the opinion that phrase was meant to express has been shelved long ago. We now know that the acceleration of a particle at any instant depends upon its position relative to other particles at that same instant; while the old idea was that the past affects the future, while the future does not affect the past. So the "law of demand and supply" has utterly different meanings with different economists.—1893.)

This method is far more intellectual and respectable from the point of view of reason than either of the others which we have noticed. Indeed, as long as no better method can be applied, it ought to be followed, since it is then the expression of instinct which must be the ultimate cause of belief in all cases. But its failure has been the most manifest. It makes of inquiry something similar to the development of taste; but taste, unfortunately, is always more or less a matter of fashion, and accordingly metaphysicians have never come to any fixed agreement, but the pendulum has swung backward and forward between a more material and a more spiritual philosophy, from the earliest times to the latest. And so from this, which has been called the *a priori* method, we are driven, in Lord Bacon's phrase, to a true induction. We have examined into this *a priori* method as something which promised to deliver our opinions from their accidental and capricious element. But development, while it is a process which eliminates the effect of some casual circumstances, only magnifies that of others. This method, therefore does not differ in a very essential way from that of authority. The government may not have lifted its finger to influence my convictions; I may have been left outwardly quite free to choose, we will say, between monogamy and polygamy, and, appealing to my conscience only, I may have concluded that the latter practice is in itself licentious. But when I come to see that the chief obstacle to the spread of Christianity among a people of as high culture as the Hindoos has been a conviction of the immorality of our way of treating women, I cannot help seeing that, though governments do not interfere, sentiments in their development will be very greatly determined by accidental causes. Now, there are some people, among whom I must suppose that my reader is to be found, who, when they see that any belief of theirs is determined by any circumstance extraneous to the facts, will from that moment not merely admit in words that that belief is doubtful, but will experience a real doubt of it, so that it ceases in some degree at least to be a belief.

To satisfy our doubts, therefore, it is necessary that a method should be found by which our beliefs may be determined by nothing human, but by some external permanency—by something upon which our thinking has no effect. Some mystics imagine that they have such a method in a private inspiration from on high. But that is only a form of the method of tenacity, in which the conception of truth as something public is not yet developed. Our external permanency would not be external, in our sense, if it was restricted in its influence to one individual. It must be something which affects, or might affect, every man. And, though these affections are necessarily as various as are individual conditions, yet the method must be such that the ultimate conclusion of every man shall be the same. Such is the method of science. Its fundamental hypothesis restated in more familiar language, is this: There are Real things whose characters are entirely independent of our opinions about them, those Reals affect our senses according to regular laws, and, though our sensations are as different as are our relations to the objects, yet, by taking advantage of the laws of perception. we can ascertain by reasoning how things really and truly are; and any man, if he have sufficient experience and he reason enough about it, will be led to the one True conclusion. The new conception here involved is that of Reality. It may be asked how I know that there are any Reals. If this hypothesis is the sole support of my method of inquiry, my method of inquiry must not be used to support my hypothesis. The reply is this: 1. If investigation cannot be regarded as proving that there are Real things, it at least does not lead to a contrary conclusion; but the method and the conception on which it is based remain ever in harmony. No doubts of the method, therefore, necessarily arise from its practice, as is the case with all the others. 2. The feeling which gives rise to any method of fixing belief is a dissatisfaction at two repugnant propositions. But here already is a vague concession that there is some *one* thing which a proposition should represent. Nobody, therefore, can really doubt that there are Reals, for, if he did, doubt would not be a source of dissatisfaction. The hypothesis, therefore, is one which every mind admits. So that the social impulse does not cause men to doubt it. 3. Everybody uses the scientific method about a great many things, and only ceases to use it when he does not know how to apply it. 4. Experience of the method has not led us to doubt it, but, on the contrary, scientific investigation has had the most wonderful triumphs in the way of settling opinion. These afford the explanation of my not doubting the method or the hypothesis which it supposes; and not having any doubt, nor believing that anybody else whom I could influence has, it would be the merest babble for me to say more about it. If there be anybody with a living doubt upon the subject, let him consider it.

To describe the method of scientific investigation is the object of this series of papers. At present I have only room to notice some points of contrast between it and other methods of fixing belief.

This is the only one of the four methods which presents any distinction of a right and a wrong way. If I adopt the method of tenacity, and shut myself out from all influences, whatever I think necessary to doing this, is necessary according to that method. So with the method of authority: the state may try to put down heresy by means which, from a scientific point of view, seem very ill-calculated to accomplish its purposes; but the only test *on that method* is what the state thinks; so that it cannot pursue the method wrongly. So with the *a priori* method. The very essence of it is to think as one is inclined to think. All metaphysicians will be sure to do that, however they may be inclined to judge each other to be perversely wrong. The Hegelian system recognizes every natural tendency of thought as logical, although it be certain to be abolished by counter-tendencies. Hegel thinks there is a regular system in the succession of these tendencies, in consequence of which, after drifting one way and the other for a long time, opinion will at last go right. And it is true that metaphysicians do get the right ideas at last; Hegel's system of Nature represents tolerably the science of his day; and one may be sure that whatever scientific investigation shall have put out of doubt will presently receive a *priori* demonstration on the part of the metaphysicians. But with the scientific method the case is different. I may start with known and observed facts to proceed to the unknown; and yet the rules which I follow in doing so may not be such as investigation would approve. The test of whether I am truly following the method is not an immediate appeal to my feelings and purposes, but, on the contrary, itself involves the application of the method. Hence it is that bad reasoning as well as good reasoning is possible; and this fact is the foundation of the practical side of logic.

It is not to be supposed that the first three methods of settling opinion present no a vantage whatever over the scientific method. On the contrary, each has some peculiar convenience of its own. The *a priori* method is distinguished for its comfortable conclusions. It is the nature of the process to adopt whatever belief we are inclined to, and there are certain flatteries to the vanity of man which we all believe by nature, until we are awakened from our Pleasing dream by rough facts. The method of authority will always govern the mass of mankind; and those who wield the various forms of organized force in the state will never be convinced that dangerous reasoning ought not to be suppressed in some way. If liberty of speech is to be untrammelled from the grosser forms of constraint, then uniformity of opinion will be secured by a moral terrorism to which the respectability of society will give its thorough approval. Following the method of authority is the path of peace. Certain non-conformities are permitted; certain others (considered unsafe) are forbidden. These are different in different countries and in different ages; but, wherever you are, let it be known that you seriously hold a tabooed belief, and you may be perfectly sure of being treated with a cruelty less brutal but more refined than hunting you like a wolf. Thus, the greatest intellectual benefactors of mankind have never dared, and dare not now, to utter the whole of their thought; and thus a shade of *prima facie* doubt is cast upon every proposition which is considered essential to the security of society. Singularly enough, the persecution does not all come from without; but a man torments himself and is oftentimes most distressed at finding himself believing propositions which he has been brought up to regard with aversion. The peaceful and sympathetic man will, therefore, find it hard to resist the temptation to submit his opinions to authority. But most of all I admire the method of tenacity for its strength. simplicity, and directness. Men who pursue it are distinguished for their decision of character, which becomes very easy with such a mental rule. They do not waste time in trying to make up their minds what they want, but, fastening like lightning upon whatever alternative comes first, they hold it to the end, whatever happens, without an instant's irresolution. This is one of the splendid qualities which generally accompany brilliant, unlasting success. It is impossible not to envy the man who can dismiss reason, although we know how it must turn out at last.

Such are the advantages which the other methods of settling opinion have over scientific investigation. A man should consider well of them; and then he should consider that, after all, he wishes his opinions to coincide with the fact, and that there is no reason why the results of those three first methods should do so. To bring about this effect is the prerogative of the method of science. Upon such considerations he has to make his choice—a choice which is far more than the adoption of any intellectual opinion, which is one of the ruling decisions of his life, to which, when once made, he is bound to adhere. The force of habit will sometimes cause a man to hold on to old beliefs, after he is in a condition to see that they have no sound basis. But reflection upon the state of the case will overcome these habits, and he ought to allow reflection its full weight. People sometimes shrink from doing this, having an idea that beliefs are wholesome which they cannot help feeling rest on nothing. But let such persons suppose an analogous though different case from their own. Let them ask themselves what they would say to a reformed Mussulman who should hesitate to give up his old notions in regard to the relations of the sexes; or to a reformed Catholic who should still shrink from reading the Bible. Would they not say that these persons ought to consider the matter fully, and clearly understand the new doctrine, and then ought to embrace it, in its entirety? But, above all, let it be considered that what is more wholesome than any particular belief is integrity of belief, and that to avoid looking into the support of any belief from a fear that it may turn out rotten is quite as immoral as it is disadvantageous. The person who confesses that there is such a thing as truth, which is distinguished from falsehood simply by this, that if acted on it should, on full consideration, carry us to the point we aim at and not astray, and then, though convinced of this, dares not know the truth and seeks to avoid it, is in a sorry state of mind indeed.

Yes, the other methods do have their merits: a clear logical conscience does cost something—just as any virtue, just as all that we cherish, costs us dear. But we should not desire it to be otherwise. The genius of a man's logical method should be loved and reverenced as his bride, whom he has chosen from all the world. He need not contemn the others; on the contrary, he may honour them deeply, and in doing so he only honours her the more. But she is the one that he has chosen, and he knows that he was right in making that choice. And having made it, he will work and fight for her, and will not complain that there are blows to take, hoping that there may be as many and as hard to give, and will strive to be the worthy knight and champion of her from the blaze of whose splendours he draws his inspiration and his courage.

I often say that when you can measure what you are speaking about, and express it in numbers, you know something about it; but when you cannot measure it, when you cannot express it in numbers, your knowledge is of a meagre and unsatisfactory kind.

William Thomson, Lord Kelvin, 1883

Like a gleam in the darkness, we have appeared for an instant from the black nothingness of the ever-unconscious matter, in order to make good the demands of Reason and create a life worthy of ourselves and the Goal we only dimly perceive.

Andrei Sakharov

So the great edifice of modern physics goes up and the majority of the cleverest people in the western world have about as much insight into it as their neolithic ancestors would have had.

C. P. Snow, *The Two Cultures*, 1959

In mathematics he was greater
Than Tycho Brahe, or Erra Pater:
For he, by geometric scale,
Could take the size of pots of ale;
Resolve, by sines and tangents straight,
If bread or butter wanted weight;
And wisely tell what hour o' the day
The clock does strike, by Algebra.

Samuel Butler, *Hudibras*, 1693

The scientist does not study nature because it is useful; he studies it because he delights in it, and he delights in it because it is beautiful. If nature were not beautiful, it would not be worth knowing, and if nature would not be worth knowing, life would not be worth living.

Henri Poincaré

THE FOUNDATIONS OF SCIENCE

HENRI POINCARÉ

(Translated by George Bruce Halsted)

Jules Henri Poincaré (1854-1912) was a leader in science and mathematics well recognized in his own day. The translator of this volume, George Bruce Halsted, wrote of Poincaré, "His work must be reckoned with the greatest mathematical achievements of mankind." Another wrote, "His right is recognized now, and it is not likely that future generations will revise the judgment, to rank among the greatest mathematicians of all time." Josiah Royce of Harvard wrote, "He is, as geometer, as analyst and as a theoretical physicist, a leader of his age." George Darwin, son of Charles Darwin, said of Poincaré, "He must be regarded as the presiding genius—or, shall I say, my patron saint?"

Poincaré was born in Nancy, France, April 29, 1854, the son of a highly respected member of the medical faculty of the university. As a young lad, he received special instruction from his gifted mother. He excelled in composition, as well as other subjects, in school, though interrupted by the Franco-Prussian War of 1870-71. He finished number one at the École Polytechnique where he followed the courses in mathematics without taking a note and without using the syllabus.

It appeared that he might be heading for a career in engineering when he graduated from the School of Mines in 1879. Instead, he took a doctorate at the University of Paris in 1879, and then joined the science faculty at the University of Caen. After only two years there, he returned to Paris where he remained a member of the University faculty until his death.

Poincaré began serious writing with an important essay in 1878 and over the next forty years, wrote over four hundred articles and books. One of his most brilliant contributions was in non-Euclidian geometry, which began when an idea struck him as he was mounting an omnibus in Coutances in 1881. He revolutionized scientific thought with his notion, "If . . . a phenomenon admits of a complete mechanical explanation, it will admit of an infinity of others which will account equally well for all the peculiarities disclosed by experiment." This demanded a revision of the whole idea of scientific truth.

At the age of thirty-two, Poincaré became a member of the Academy of Sciences and in 1908, he was elected to the French Academy.

1. What makes a *great* mathematician?
2. If "experiment is the sole source of truth," as Poincaré says, what is the role of mathematical physics? What is mathematical physics?
3. What does Poincaré think of the simplicity postulate?
4. What is the role of hypothesis, according to Poincaré?
5. How does it happen that there are people who do not understand mathematics?
6. What is Poincaré's conception of the psychology of mathematical creation or the discovery of new ideas?

THE FOUNDATIONS OF SCIENCE

HENRI POINCARÉ

SCIENCE AND HYPOTHESIS

PART IV
NATURE

CHAPTER IX

HYPOTHESES IN PHYSICS

THE ROLE OF EXPERIMENT AND GENERALIZATION.—Experiment is the sole source of truth. It alone can teach us anything new; it alone can give us certainty. These are two points that can not be questioned.

But then, if experiment is everything, what place will remain for mathematical physics? What has experimental physics to do with such an aid, one which seems useless and perhaps even dangerous?

And yet mathematical physics exists, and has done unquestionable service. We have here a fact that must be explained.

The explanation is that merely to observe is not enough. We must use our observations, and to do that we must generalize. This is what men always have done; only as the memory of past errors has made them more and more careful, they have observed more and more, and generalized less and less.

Every age has ridiculed the one before it, and accused it of having generalized too quickly and too naively. Descartes pitied the Ionians; Descartes, in his turn, makes us smile. No doubt our children will some day laugh at us.

But can we not then pass over immediately to the goal? Is not this the means of escaping the ridicule that we foresee? Can we not be content with just the bare experiment?

No, that is impossible; it would be to mistake utterly the true nature of science. The scientist must set in order. Science is built up with facts, as a house is with stones. But a collection of facts is no more a science than a heap of stones is a house.

And above all the scientist must foresee. Carlyle has somewhere said something like this: "Nothing but facts are of importance. John Lackland passed by here. Here is something that is admirable. Here is a reality for which I would give all the theories in the world." Carlyle was a fellow countryman of Bacon; but Bacon would not have said that. That is the language of the historian. The physicist would say rather: "John Lackland passed by here; that makes no difference to me, for he never will pass this way again."

We all know that there are good experiments and poor ones. The latter will accumulate in vain; though one may have made a hundred or a thousand, a single piece of work by a true master, by a Pasteur, for example, will suffice to tumble them into oblivion. Bacon would have well understood this; it is he who invented the phrase *Experimentum crucis.* But Carlyle would not have understood it. A fact is a fact. A pupil has read a certain number on his thermometer; he has taken no precaution; no matter, he has read it, and if it is only the fact that counts, here is a reality of the same rank as the peregrinations of King John Lackland. Why is the fact that this pupil has made this reading of no interest, while the fact that a skilled physicist had made another reading might be on the contrary very important? It is because from the first reading we could not infer anything. What then is a good experiment? It is that which informs us of something besides an isolated fact; it is that which enables us to foresee, that is, that which enables us to generalize.

For without generalization foreknowledge is impossible. The circumstances under which one has worked will never reproduce themselves all at once. The observed action then will never recur; the only thing that can be affirmed is that under analogous circumstances an analogous action will be produced. In order to foresee, then, it is necessary to invoke at least analogy, that is to say, already then to generalize.

No matter how timid one may be, still it is necessary to interpolate. Experiment gives us only a certain number of isolated points. We must unite these by a continuous line. This is a veritable generalization. But we do more; the curve that we shall trace will pass between the observed points and near these points; it will not pass through these points themselves. Thus one does not restrict himself to generalizing the experiments, but corrects them; and the physicist who should try to abstain from these corrections and really be content with the bare experiment, would be forced to enunciate some very strange laws.

The bare facts, then, would not be enough for us; and that is why we must have science ordered, or rather organized.

It is often said experiments must be made without a preconceived idea. That is impossible. Not only would it make all experiment barren, but that would be attempted which could not be done. Every one carries in his mind his own conception of the world, of which he can not so easily rid himself. We must, for instance, use language; and our language is made up only of preconceived ideas and can not be otherwise. Only these are unconscious preconceived ideas, a thousand times more dangerous than the others.

Shall we say that if we introduce others, of which we are fully conscious, we shall only aggravate the evil? I think not. I believe rather that they will serve as counterbalances to each other—I was going to say as antidotes; they will in general accord ill with one another—they will come into conflict with one another, and thereby force us to regard things under different aspects. This is enough to emancipate us. He is no longer a slave who can choose his master.

Thus, thanks to generalization, each fact observed enables us to foresee a great many others; only we must not forget that the first alone is certain, that all others are merely probable. No matter how solidly founded a prediction may appear to us, we are never *absolutely* sure that experiment will not contradict it, if we undertake to verify it. The probability, however, is often so great that practically we may be content with it. It is far better to foresee even without certainty than not to foresee at all.

One must, then, never disdain to make a verification when opportunity offers. But all experiment is long and difficult; the workers are few; and the number of facts that we need to foresee is immense. Compared with this mass the number of direct verifications that we can make will never be anything but a negligible quantity.

Of this few that we can directly attain, we must make the best use; it is very necessary to get from every experiment the greatest possible number of predictions, and with the highest possible degree of probability. The problem is, so to speak, to increase the yield of the scientific machine.

Let us compare science to a library that ought to grow continually. The librarian has at his disposal for his purchases only insufficient funds. He ought to make an effort not to waste them.

It is experimental physics that is entrusted with the purchases. It alone, then, can enrich the library.

As for mathematical physics, its task will be to make out the catalogue. If the catalogue is well made, the library will not be any richer, but the reader will be helped to use its riches.

And even by showing the librarian the gaps in his collections, it will enable him to make a judicious use of his funds; which is all the more important because these funds are entirely inadequate.

Such, then, is the role of mathematical physics. It must direct generalization in such a manner as to increase what I just now called the yield of science. By what means it can arrive at this, and how it can do it without danger, is what remains for us to investigate.

THE UNITY OF NATURE.—Let us notice, first of all, that every generalization implies in some measure the belief in the unity and simplicity of nature. As to the unity there can be no difficulty. If the different parts of the universe were not like the members of one body, they would not act on one another, they would know nothing of one another; and we in particular would know only one of these parts. We do not ask, then, if nature is one, but how it is one.

As for the second point, that is not such an easy matter. It is not certain that nature is simple. Can we without danger act as if it were?

There was a time when the simplicity of Mariotte's law was an argument invoked in favor of its accuracy; when Fresnel himself, after having said in a conversation with Laplace that nature was not concerned about analytical difficulties, felt himself obliged to make explanations, in order not to strike too hard at prevailing opinion.

To-day ideas have greatly changed; and yet, those who do not believe that natural laws have to be simple, are still often obliged to act as if they did. They could not entirely avoid this necessity without making impossible all generalization, and consequently all science.

It is clear that any fact can be generalized in an infinity of ways, and it is a question of choice. The choice can be guided only by considerations of simplicity. Let us take the most commonplace case, that of interpolation. We pass a continuous line, as regular as possible, between the points given by observation. Why do we avoid points making angles and too abrupt turns? Why do we not make our curve describe the most capricious zigzags? It is because we know beforehand, or believe we know, that the law to be expressed can not be so complicated as all that.

We may calculate the mass of Jupiter from either the movements of its satellites, or the perturbations of the major planets, or those of the minor planets. If we take the averages of the determinations obtained by these three methods, we find three numbers very close together, but different. We might interpret this result by supposing that the coefficient of gravitation is not the same in the three cases. The observations would certainly be much better represented. Why do we reject this interpretation? Not because it is absurd, but because it is needlessly complicated. We shall only accept it when we are forced to, and that is not yet.

To sum up, ordinarily every law is held to be simple till the contrary is proved.

This custom is imposed upon physicists by the causes that I have just explained. But how shall we justify it in the presence of discoveries that show us every day new details that are richer and more complex? How shall we even reconcile it with the belief in the unity of nature? For if everything depends on everything, relationships where so many diverse factors enter can no longer be simple.

If we study the history of science, we see happen two inverse phenomena, so to speak. Sometimes simplicity hides under complex appearances; sometimes it is the simplicity which is apparent, and which disguises extremely complicated realities.

What is more complicated than the confused movements of the planets? What simpler than Newton's law? Here nature, making sport, as Fresnel said, of analytical difficulties, employs only simple means, and by combining them produces I know not what inextricable tangle. Here it is the hidden simplicity which must be discovered.

Examples of the opposite abound. In the kinetic theory of gases, one deals with molecules moving with great velocities, whose paths, altered by incessant collisions, have the most capricious forms and traverse space in every direction. The observable result is Mariotte's simple law. Every individual fact was complicated. The law of great numbers has reestablished simplicity in the average. Here the simplicity is merely apparent, and only the coarseness of our senses prevents our perceiving the complexity.

Many phenomena obey a law of proportionality. But why? Because in these phenomena there is something very small. The simple law observed, then, is only a result of the general analytical rule that the infinitely small increment of a function is proportional to the increment of the variable. As in reality our increments are not infinitely small, but very small, the law of proportionality is only approximate, and the simplicity is only apparent. What I have just said applies to the rule of the superposition of small motions, the use of which is so fruitful, and which is the basis of optics.

And Newton's law itself? Its simplicity, so long undetected, is perhaps only apparent. Who knows whether it is not due to some complicated mechanism, to the impact of some subtile matter animated by irregular movements, and whether it has not become simple only through the action of averages and of great numbers? In any case, it is difficult not to suppose that the true law contains complementary terms, which would become sensible at small distances. If in astronomy they are negligible as modifying Newton's law, and if the law thus regains its simplicity, it would be only because of the immensity of celestial distances.

No doubt, if our means of investigation should become more and more penetrating, we should discover the simple under the complex, then the complex under the simple, then again the simple under the complex, and so on, without our being able to foresee what will be the last term.

We must stop somewhere, and that science may be possible, we must stop when we have found simplicity. This is the only ground on which we can rear the edifice of our generalizations. But this simplicity being only apparent, will the ground be firm enough? This is what must be investigated.

For that purpose, let us see what part is played in our generalizations by the belief in simplicity. We have verified a simple law in a good many particular cases; we refuse to admit that this agreement, so often repeated, is simply the result of chance, and conclude that the law must be true in the general case.

Kepler notices that a planet's positions, as observed by Tycho, are all on one ellipse. Never for a moment does he have the thought that by a strange play of chance Tycho never observed the heavens except at a moment when the real orbit of the planet happened to cut this ellipse.

What does it matter then whether the simplicity be real, or whether it covers a complex reality? Whether it is due to the influence of great numbers, which levels down individual differences, or to the greatness or smallness of certain quantities, which allows us to neglect certain terms, in no case is it due to chance. This simplicity, real or apparent, always has a cause. We can always follow, then, the same course of reasoning, and if a simple law has been observed in several particular cases, we can legitimately suppose that it will still be true in analogous cases. To refuse to do this would be to attribute to chance an inadmissible role.

There is, however, a difference. If the simplicity were real and essential, it would resist the increasing precision of our means of measure. If then we believe nature to be essentially simple, we must, from a simplicity that is approximate, infer a simplicity that is rigorous. This is what was done formerly; and this is what we no longer have a right to do.

The simplicity of Kepler's laws, for example, is only apparent. That does not prevent their being applicable, very nearly, to all systems analogous to the solar system; but it does prevent their being rigorously exact.

THE ROLE OF HYPOTHESIS.—All generalization is a hypothesis. Hypothesis, then, has a necessary role that no one has ever contested. Only, it ought always, as soon as possible and as often as possible, to be subjected to verification. And, of course, if it does not stand this test, it ought to be abandoned without reserve. This is what we generally do, but sometimes with rather an ill humor.

Well, even this ill humor is not justified. The physicist who has just renounced one of his hypotheses ought, on the contrary, to be full of joy; for he has found an unexpected opportunity for discovery. His hypothesis, I imagine, had not been adopted without consideration; it took account of all the known factors that it seemed could enter into the phenomenon. If the test does not support it, it is because there is something unexpected and extraordinary; and because there is going to be something found that is unknown and new.

Has the discarded hypothesis, then, been barren? Far from that, it may be said it has rendered more service than a true hypothesis. Not only has it been the occasion of the decisive experiment, but, without having made the hypothesis, the experiment would have been made by chance, so that nothing would have been derived from it. One would have seen nothing extraordinary; only one fact the more would have been catalogued without deducing from it the least consequence.

Now on what condition is the use of hypothesis without danger?

The firm determination to submit to experiment is not enough; there are still dangerous hypotheses; first, and above all, those which are tacit and unconscious. Since we make them without knowing it, we are powerless to abandon them. Here again, then, is a service that mathematical physics can render us. By the precision that is characteristic of it, it compels us to formulate all the hypotheses that we should make without it, but unconsciously.

Let us notice besides that it is important not to multiply hypotheses beyond measure, and to make them only one after the other. If we construct a theory based on a number of hypotheses, and if experiment condemns it, which of our premises is it necessary to change q It will be impossible to know. And inversely, if the experiment succeeds, shall we believe that we have demonstrated all the hypotheses at once? Shall we believe that with one single equation we have determined several unknowns?

We must equally take care to distinguish between the different kinds of hypotheses. There are first those which are perfectly natural and from which one can scarcely escape. It is difficult not to suppose that the influence of bodies very remote is quite negligible, that small movements follow a linear law, that the effect is a continuous function of its cause. I will say as much of the conditions imposed by symmetry. All these hypotheses form, as it were, the common basis of all the theories of mathematical physics. They are the last that ought to be abandoned.

There is a second class of hypotheses, that I shall term neutral. In most questions the analyst assumes at the beginning of his calculations either that matter is continuous or, on the contrary, that it is formed of atoms. He might have made the opposite assumption without changing his results. He would only have had more trouble to obtain them; that is all. If, then, experiment? confirms his conclusions, will he think that he has demonstrated, for instance, the real existence of atoms?

In optical theories two vectors are introduced, of which one is regarded as a velocity, the other as a vortex. Here again is a neutral hypothesis, since the same conclusions would have been reached by taking precisely the opposite. The success of the experiment, then, can not prove that the first vector is indeed a velocity; it can only prove one thing, that it is a vector. This is the only hypothesis that has really been introduced in the premises. In order to give it that concrete appearance which the weakness of our minds requires, it has been necessary to consider it either as a velocity or as a vortex, in the same way that it has been necessary to represent it by a letter, either x or y. The result, however, whatever it may be, will not prove that it was right or wrong to regard it as a velocity any more than it will prove that it was right or wrong to call it x and not y.

These neutral hypotheses are never dangerous, if only their character is not misunderstood. They may be useful, either as devices for computation, or to aid our understanding by concrete images, to fix our ideas as the saying is. There is, then, no occasion to exclude them.

The hypotheses of the third class are the real generalizations. They are the ones that experiment must confirm or invalidate. Whether verified or condemned, they will always be fruitful But for the reasons that I have set forth, they will only be fruitful if they are not too numerous.

ORIGIN OF MATHEMATICAL PHYSICS.—LET us penetrate further, and study more closely the conditions that have permitted the development of mathematical physics. We observe at once that the efforts of scientists have always aimed to resolve the complex phenomenon directly given by experiment into a very large number of elementary phenomena.

This is done in three different ways: first, in time. Instead of embracing in its entirety the progressive development of a phenomenon, the aim is simply to connect each instant with the instant immediately preceding it. It is admitted that the actual state of the world depends only on the immediate past, without being directly influenced, so to speak, by the memory of a distant past. Thanks to this postulate, instead of studying directly the whole succession of phenomena, it is possible to confine ourselves to writing its 'differential equation.' For Kepler's laws we substitute Newton's law.

Next we try to analyze the phenomenon in space. What experiment gives us is a confused mass of facts presented on a stage of considerable extent. We must try to discover the elementary phenomenon, which will be, on the contrary, localized in a very small region of space.

Some examples will perhaps make my thought better understood. If we wished to study in all its complexity the distribution of temperature in a cooling solid, we should never succeed. Everything becomes simple if we reflect that one point of the solid can not give up its heat directly to a distant point; it will give up its heat only to the points in the immediate neighborhood, and it is by degrees that the flow of heat can reach other parts of the solid. The elementary phenomenon is the exchange of heat between two contiguous points. It is strictly localized, and is relatively simple, if we admit, as is natural, that it is not influenced by the temperature of molecules whose distance is sensible.

I bend a rod. It is going to take a very complicated form, the direct study of which would be impossible. But I shall be able, however, to attack it, if I observe that its flexure is a result only of the deformation of the very small elements of the rod, and that the deformation of each of these elements depends only on the forces that are directly applied to it, and not at all on those which may act on the other elements.

In all these examples, which I might easily multiply, we admit that there is no action at a distance, or at least at a great distance. This is a hypothesis. It is not always true, as the law of gravitation shows us. It must, then, be submitted to verification. If it is confirmed, even approximately, it is precious, for it will enable us to make mathematical physics, at least by successive approximations.

If it does not stand the test, we must look for something else analogous; for there are still other means of arriving at the elementary phenomenon. If several bodies act simultaneously, it may happen that their actions are independent and are simply added to one another, either as vectors or as scalars. The elementary phenomenon is then the action of an isolated body. Or again, we have to deal with small movements, or more generally with small variations, which obey the well-known law of superposition. The observed movement will then be decomposed into simple movements, for example, sound into its harmonics, white light into its monochromatic components.

When we have discovered in what direction it is advisable to look for the elementary phenomenon, by what means can we reach it?

First of all, it will often happen that in order to detect it, or rather to detect the part of it useful to us, it will not be necessary to penetrate the mechanism; the law of great numbers will suffice.

Let us take again the instance of the propagation of heat. Every molecule emits rays toward every neighboring molecule. According to what law, we do not need to know. If we should make any supposition in regard to this, it would be a neutral hypothesis and consequently useless and incapable of verification. And, in fact, by the action of averages and thanks to the symmetry of the medium, all the differences are leveled down, and whatever hypothesis may be made, the result is always the same.

The same circumstance is presented in the theory of electricity and in that of capillarity. The neighboring molecules attract and repel one another. We do not need to know according to what law; it is enough for us that this attraction is sensible only at small distances, that the molecules are very numerous, that the medium is symmetrical, and we shall only have to let the law of great numbers act.

Here again the simplicity of the elementary phenomenon was hidden under the complexity of the resultant observable phenomenon; but, in its turn, this simplicity was only apparent, and concealed a very complex mechanism.

The best means of arriving at the elementary phenomenon would evidently be experiment. We ought by experimental contrivance to dissociate the complex sheaf that nature offers to our researches, and to study with care the elements as much isolated as possible. For example, natural white light would be decomposed into monochromatic lights by the aid of the prism, and into polarized light by the aid of the polarizer.

Unfortunately that is neither always possible nor always sufficient, and sometimes the mind must outstrip experiment. I shall cite only one example, which has always struck me forcibly.

If I decompose white light, I shall be able to isolate a small part of the spectrum, but however small it may be, it will retain a certain breadth. Likewise the natural lights, called *monochromatic,* give us a very narrow line, but not, however, infinitely narrow. It might be supposed that by studying experimentally the properties of these natural lights, by working with finer and finer lines of the spectrum, and by passing at last to the limit, so to speak, we should succeed in learning the properties of a light strictly monochromatic.

That would not be accurate. Suppose that two rays emanate from the same source, that we polarize them first in two perpendicular planes, then bring them back to the same plane of polarization, and try to make them interfere. If the light were *strictly* monochromatic, they would interfere. With our lights, which are nearly monochromatic, there will be no interference, and that no matter how narrow the line. In order to be otherwise it would have to be several million times as narrow as the finest known lines.

Here, then, the passage to the limit would have deceived us. The mind must outstrip the experiment, and if it has done so with success, it is because it has allowed itself to be guided by the instinct of simplicity.

The knowledge of the elementary fact enables us to put the problem in an equation. Nothing remains but to deduce from this by combination the complex fact that can be observed and verified. This is what is called *integration*, and is the business of the mathematician.

It may be asked why, in physical sciences, generalization so readily takes the mathematical form. The reason is now easy to see. It is not only because we have numerical laws to express; it is because the observable phenomenon is due to the superposition of a great number of elementary phenomena *all alike*. Thus quite naturally are introduced differential equations.

It is not enough that each elementary phenomenon obeys simple laws; all those to be combined must obey the same law. Then only can the intervention of mathematics be of use; mathematics teaches us in fact to combine like with like. Its aim is to learn the result of a combination without needing to go over the combination piece by piece. If we have to repeat several times the same operation, it enables us to avoid this repetition by telling us in advance the result of it by a sort of induction. I have explained this above, in the chapter on mathematical reasoning.

But, for this, all the operations must be alike. In the opposite case, it would evidently be necessary to resign ourselves to doing them in reality one after another, and mathematics would become useless.

It is then thanks to the approximate homogeneity of the matter studied by physicists, that mathematical physics could be born.

In the natural sciences, we no longer find these conditions: homogeneity, relative independence of remote parts, simplicity of the elementary fact; and this is why naturalists are obliged to resort to other methods of generalization.

* * * * *

SCIENCE AND METHOD
BOOK I

SCIENCE AND THE SCIENTIST
CHAPTER I

THE CHOICE OF FACTS

TOLSTOI somewhere explains why 'science for its own sake' is in his eyes an absurd conception. We can not know *all* facts, since their number is practically infinite. It is necessary to choose; then we may let this choice depend on the pure caprice of our curiosity; would it not be better to let ourselves be guided by utility, by our practical and above all by our moral needs; have we nothing better to do than to count the number of ladybugs on our planet?

It is clear the word utility has not for him the sense men of affairs give it, and following them most of our contemporaries. Little cares he for industrial applications, for the marvels of electricity or of automobilism, which he regards rather as obstacles to moral progress; utility for him is solely what can make man better.

For my part, it need scarce be said, I could never be content with either the one or the other ideal; I want neither that plutocracy grasping and mean, nor that democracy goody and mediocre, occupied solely in turning the other cheek, where would dwell sages without curiosity, who, shunning excess, would not die of disease, but would surely die of ennui. But that is a matter of taste and is not what I wish to discuss.

The question nevertheless remains and should fix our attention; if our choice can only be determined by caprice or by immediate utility, there can be no science for its own sake, and consequently no science. But is that true? That a choice must be made is incontestable; whatever be our activity, facts go quicker than we, and we can not catch them; while the scientist discovers one fact, there happen milliards of milliards in a cubic millimeter of his body. To wish to comprise nature in science would be to want to put the whole into the part.

But scientists believe there is a hierarchy of facts and that among them may be made a judicious choice. They are right, since otherwise there would be no science, yet science exists. One need only open the eyes to see that the conquests of industry which have enriched so many practical men would never have seen the light, if these practical men alone had existed and if they had not been preceded by unselfish devotees who died poor, who never thought of utility, and yet had a guide far other than caprice.

As Mach says, these devotees have spared their successors the trouble of thinking. Those who might have worked solely in view of an immediate application would have left nothing behind them, and, in face of a new need, all must have been begun over again. Now most men do not love to think, and this is perhaps fortunate when instinct guides them, for most often, when they pursue an aim which is immediate and ever the same, instinct guides them better than reason would guide a pure intelligence. But instinct is routine, and if thought did not fecundate it, it would no more progress in man than in the bee or ant. It is needful then to think for those who love not thinking, and, as they are numerous, it is needful that each of our thoughts be as often useful as possible, and this is why a law will be the more precious the more general it is.

This shows us how we should choose: the most interesting facts are those which may serve many times; these are the facts which have a chance of coming up again. We have been so fortunate as to be born in a world where there are such. Suppose that instead of 60 chemical elements there were 60 milliards of them, that they were not some common, the others rare, but that they were uniformly distributed. Then, every time we picked up a new pebble there would be great probability of its being formed of some unknown substance; all that we knew of other pebbles would be worthless for it; before each new object we should be as the new-born babe; like it we could only obey our caprices or our needs. Biologists would be just as much at a loss if there were only individuals and no species and if heredity did not make sons like their fathers.

In such a world there would be no science; perhaps thought and even life would be impossible, since evolution could not there develop the preservational instincts. Happily it is not so; like all good fortune to which we are accustomed, this is not appreciated at its true worth.

Which then are the facts likely to reappear? They are first the simple facts. It is clear that in a complex fact a thousand circumstances are united by chance, and that only a chance still much less probable could reunite them anew. But are there any simple facts? And if there are, how recognize them? What assurance is there that a thing we think simple does not hide a dreadful complexity? All we can say is that we ought to prefer the facts which *seem* simple to those where our crude eye discerns unlike elements. And then one of two things: either this simplicity is real, or else the elements are so intimately mingled as not to be distinguishable. In the first case there is chance of our meeting anew this same simple fact, either in all its purity or entering itself as element in a complex manifold. In the second case this intimate mixture has likewise more chances of recurring than a heterogeneous assemblage; chance knows how to mix, it knows not how to disentangle, and to make with multiple elements a well-ordered edifice in which something is distinguishable, it must be made expressly. The facts which appear simple, even if they are not so, will therefore be more easily revived by chance. This it is which justifies the method instinctively adopted by the scientist, and what justifies it still better, perhaps, is that oft-recurring facts appear to us simple, precisely because we are used to them.

But where is the simple fact? Scientists have been seeking it in the two extremes, in the infinitely great and in the infinitely small. The astronomer has found it because the distances of the stars are immense, so great that each of them appears but as a point, so great that the qualitative differences are effaced, and because a point is simpler than a body which has form and qualities. The physicist on the other hand has sought the elementary phenomenon in fictively cutting up bodies into infinitesimal cubes, because the conditions of the problem, which undergo slow and continuous variation in passing from one point of the body to another, may be regarded as constant in the interior of each of these little cubes. In the same way the biologist has been instinctively led to regard the cell as more interesting than the whole animal, and the outcome has shown his wisdom, since cells belonging to organisms the most different are more alike, for the one who can recognize their resemblances, than are these organisms themselves. The sociologist is more embarrassed; the elements, which for him are men, are too unlike, too variable, too capricious, in a word, too complex; besides, history never begins over again. How then choose the interesting fact, which is that which begins again? Method is precisely the choice of facts; it is needful then to be occupied first with creating a method, and many have been imagined, since none imposes itself, so that sociology is the science which has the most methods and the fewest results.

Therefore it is by the regular facts that it is proper to begin; but after the rule is well established, after it is beyond all doubt, the facts in full conformity with it are erelong without interest since they no longer teach us anything new. It is then the exception which becomes important. We cease to seek resemblances; we devote ourselves above all to the differences, and among the differences are chosen first the most accentuated, not only because they are the most striking, but because they will be the most instructive. A simple example will make my thought plainer: Suppose one wishes to determine a curve by observing some of its points. The practician who concerns himself only with immediate utility would observe only the points he might need for some special object. These points would be badly distributed on the curve; they would be crowded in certain regions, rare in others, so that it would be impossible to join them by a continuous line, and they would be unavailable for other applications. The scientist will proceed differently; as he wishes to study the curve for itself, he will distribute regularly the points to be observed, and when enough are known he will join them by a regular line and then he will have the entire curve. But for that how does he proceed? If he has determined an extreme point of the curve, he does not stay near this extremity, but goes first to the other end; after the two extremities the most instructive point will be the mid-point, and so on.

* * * * *

CHAPTER III
MATHEMATICAL CREATION

THE genesis of mathematical creation is a problem which should intensely interest the psychologist. It is the activity in which the human mind seems to take least from the outside world, in which it acts or seems to act only of itself and on itself, so that in studying the procedure of geometric thought we may hope to reach what is most essential in man's mind.

This has long been appreciated, and some time back the journal called *L'enseignement mathématique,* edited by Laisant and Fehr, began an investigation of the mental habits and methods of work of different mathematicians. I had finished the main outlines of this article when the results of that inquiry were published, so I have hardly been able to utilize them and shall confine myself to saying that the majority of witnesses confirm my conclusions; I do not say all, for when the appeal is to universal suffrage unanimity is not to be hoped.

A first fact should surprise us, or rather would surprise us if we were not so used to it. HOW does it happen there are people who do not understand mathematics? If mathematics invokes only the rules of logic, such as are accepted by all normal minds; if its evidence is based on principles common to all men, and that none could deny without being mad, how does it come about that so many persons are here refractory?

That not every one can invent is nowise mysterious. That not every one can retain a demonstration once learned may also pass. But that not every one can understand mathematical reasoning when explained appears very surprising when we think of it. And yet those who can follow this reasoning only with difficulty are in the majority: that is undeniable, and will surely not be gainsaid by the experience of secondary-school teachers.

And further: how is error possible in mathematics? A sane mind should not be guilty of a logical fallacy, and yet there are very fine minds who do not trip in brief reasoning such as occurs in the ordinary doings of life, and who are incapable of following or repeating without error the mathematical demonstrations which are longer, but which after all are only an accumulation of brief reasonings wholly analogous to those they make so easily. Need we add that mathematicians themselves are not infallible?

The answer seems to me evident. Imagine a long series of syllogisms, and that the conclusions of the first serve as premises of the following: we shall be able to catch each of these syllogisms, and it is not in passing from premises to conclusion that we are in danger of deceiving ourselves. But between the moment in which we first meet a proposition as conclusion of one syllogism, and that in which we reencounter it as premise of another syllogism occasionally some time will elapse, several links of the chain will have unrolled; so it may happen that we have forgotten it, or worse, that ·re have forgotten its meaning. So it may happen that we replace it by a slightly different proposition, or that, while retaining the same enunciation, we attribute to it a slightly different meaning, and thus it is that we are exposed to error.

Often the mathematician uses a rule. Naturally he begins by demonstrating this rule; and at the time when this proof is fresh in his memory he understands perfectly its meaning and its bearing, and he is in no danger of changing it. But subsequently he trusts his memory and afterward only applies it in a mechanical way; and then if his memory fails him, he may apply it all wrong. Thus it is, to take a simple example, that we sometimes make slips in calculation because we have forgotten our multiplication table.

According to this, the special aptitude for mathematics would be due only to a very sure memory or to a prodigious force of attention. It would be a power like that of the whist-player who remembers the cards played; or, to go up a step, like that of the chess-player who can visualize a great number of combinations and hold them in his memory. Every good mathematician ought to be a good chess-player, and inversely; likewise he should be a good computer. Of course that sometimes happens; thus Gauss was at the same time a geometer of genius and a very precocious and accurate computer.

But there are exceptions; or rather I err; I can not call them exceptions without the exceptions being more than the rule. Gauss it is, on the contrary, who was an exception. As for myself, I must confess, I am absolutely incapable even of adding without mistakes. In the same way I should be but a poor chessplayer; I would perceive that by a certain play I should expose myself to a certain danger; I would pass in review several other plays, rejecting them for other reasons, and then finally I should make the move first examined, having meantime forgotten the danger I had foreseen.

In a word, my memory is not bad, but it would be insufficient to make me a good chess-player. Why then does it not fail me in a difficult piece of mathematical reasoning where most chess-players would lose themselves? Evidently because it is guided by the general march of the reasoning. A mathematical demonstration is not a simple juxtaposition of syllogisms, it is syllogisms *placed in a certain order,* and the order in which these elements are placed is much more important than the elements themselves. If I have the feeling, the intuition, so to speak, of this order, so as to perceive at a glance the reasoning as a whole, I need no longer fear lest I forget one of the elements, for each of them will take its allotted place in the array, and that without any effort of memory on my part.

It seems to me then, in repeating a reasoning learned, that I could have invented it. This is often only an illusion; but even then, even if I am not so gifted as to create it by myself, I myself re-invent it in so far as I repeat it.

We know that this feeling, this intuition of mathematical order, that makes us divine hidden harmonies and relations, can not be possessed by every one. Some will not have either this delicate feeling so difficult to define, or a strength of memory and attention beyond the ordinary, and then they will be absolutely incapable of understanding higher mathematics. Such are the majority. Others will have this feeling only in a slight degree, but they will be gifted with an uncommon memory and a great power of attention. They will learn by heart the details one after another; they can understand mathematics and sometimes make applications, but they cannot create. Others, finally, will possess in a less or greater degree the special intuition referred to, and then not only can they understand mathematics even if their memory is nothing extraordinary, but they may become creators and try to invent with more or less success according as this intuition is more or less developed in them.

In fact, what is mathematical creation? It does not consist in making new combinations with mathematical entities already known. Any one could do that, but the combinations so made would be infinite in number and most of them absolutely without interest. To create consists precisely in not making useless combinations and in making those which are useful and which are only a small minority. Invention is discernment, choice.

How to make this choice I have before explained; the mathematical facts worthy of being studied are those which, by their analogy with other facts, are capable of leading us to the knowledge of a mathematical law just as experimental facts lead us to the knowledge of a physical law. They are those which reveal to us unsuspected kinship between other facts, long known, but wrongly believed to be strangers to one another.

Among chosen combinations the most fertile will often be those formed of elements drawn from domains which are far apart. Not that I mean as sufficing for invention the bringing together of objects as disparate as possible; most combinations so formed would be entirely sterile. But certain among them, very rare, are the most fruitful of all.

To invent, I have said, is to choose; but the word is perhaps not wholly exact. It makes one think of a purchaser before whom are displayed a large number of samples, and who examines them, one after the other, to make a choice. Here the samples would be so numerous that a whole lifetime would not suffice to examine them. This is not the actual state of things. The sterile combinations do not even present themselves to the mind of the inventor. Never in the field of his consciousness do combinations appear that are not really useful, except some that he rejects but which have to some extent the characteristics of useful combinations. All goes on as if the inventor were an examiner for the second degree who would only have to question the candidates who had passed a previous examination.

But what I have hitherto said is what may be observed or inferred in reading the writings of the geometers, reading reflectively.

It is time to penetrate deeper and to see what goes on in the very soul of the mathematician. For this, I believe, I can do best by recalling memories of my own. But I shall limit myself to telling how I wrote my first memoir on Fuchsian functions. I beg the reader's pardon; I am about to use some technical expressions, but they need not frighten him, for he is not obliged to understand them. I shall say, for example, that I have found the demonstration of such a theorem under such circumstances. This theorem will have a barbarous name, unfamiliar to many, but that is unimportant; what is of interest for the psychologist is not the theorem but the circumstances.

For fifteen days I strove to prove that there could not be any functions like those I have since called Fuchsian functions. I was then very ignorant; every day I seated myself at my work table, stayed an hour or two, tried a great number of combinations and reached no results. One evening, contrary to my custom, I drank black coffee and could not sleep. Ideas rose in crowds; I felt them collide until pairs interlocked, so to speak, making a stable combination. By the next morning I had established the existence of a class of Fuchsian functions, those which come from the hypergeometric series; I had only to write out the results, which took but a few hours.

Then I wanted to represent these functions by the quotient of two series; this idea was perfectly conscious and deliberate, the analogy with elliptic functions guided me. I asked myself what properties these series must have if they existed, and I succeeded without difficulty in forming the series I have called theta-Fuchsian.

Just at this time I left Caen, where I was then living, to go on a geologic excursion under the auspices of the school of mines. The changes of travel made me forget my mathematical work. Having reached Coutances, we entered an omnibus to go some place or other. At the moment when I put my foot on the step the idea came to me, without anything in my former thoughts seeming to have paved the way for it, that the transformations I had used to define the Fuchsian functions were identical with those of non-Euclidean geometry. I did not verify the idea; I should not have had time, as, upon taking my seat in the omnibus, I went on with a conversation already commenced, but I felt a perfect certainty. On my return to Caen, for conscience' sake I verified the result at my leisure.

Then I turned my attention to the study of some arithmetical questions apparently without much success and without a suspicion of any connection with my preceding researches. Disgusted with my failure, I went to spend a few days at the seaside, and thought of something else. One morning, walking on the bluff, the idea came to me, with just the same characteristics of brevity, suddenness and immediate certainty, that the arithmetic transformations of indeterminate ternary quadratic forms were identical with those of non-Euclidean geometry.

Returned to Caen, I meditated on this result and deduced the consequences. The example of quadratic forms showed me that there were Fuchsian groups other than those corresponding to the hypergeometric series; I saw that I could apply to them the theory of theta-Fuchsian series and that consequently there existed Uchsian functions other than those from the hypergeometric series, the ones I then knew. Naturally I set myself to form all these functions. I made a systematic attack upon them and carried all the outworks, one after another. There was one however that still held out, whose fall would involve that of the whole place. But all my efforts only served at first the better to show me the difficulty, which indeed was something. All this work was perfectly conscious.

Thereupon I left for Mont-Valérien, where I was to go through my military service; so I was very differently occupied. One day, going along the street, the solution of the difficulty which had stopped me suddenly appeared to me. I did not try to go deep into it immediately, and only after my service did I again take up the question. I had all the elements and had only to arrange them and put them together. So I wrote out my final memoir at a single stroke and without difficulty.

I shall limit myself to this single example; it is useless to multiply them. In regard to my other researches I would have to say analogous things, and the observations of other mathematicians given in *L'enseignement mathématique* would only confirm them.

Most striking at first is this appearance of sudden illumination, a manifest sign of long, unconscious prior work. The role of this unconscious work in mathematical invention appears to me incontestable, and traces of it would be found in other cases where it is less evident. Often when one works at a hard question, nothing good is accomplished at the first attack. Then one takes a rest, longer or shorter, and sits down anew to the work. During the first half-hour, as before, nothing is found, and then all of a sudden the decisive idea presents itself to the mind. It might be said that the conscious work has been more fruitful because it has been interrupted and the rest has given back to the mind its force and freshness. But it is more probable that this rest has been filled out with unconscious work and that the result of this work has afterward revealed itself to the geometer just as in the cases I have cited; only the revelation, instead of coming during a walk or a journey, has happened during a period of conscious work, but independently of this work which plays at most a role of excitant, as if it were the goad stimulating the results already reached during rest, but remaining unconscious, to assume the conscious form.

There is another remark to be made about the conditions of this unconscious work: it is possible, and of a certainty it is only fruitful, if it is on the one hand preceded and on the other hand followed by a period of conscious work. These sudden inspirations (and the examples already cited sufficiently prove this) never happen except after some days of voluntary effort which has appeared absolutely fruitless and whence nothing good seems to have come, where the way taken seems totally astray. These efforts then have not been as sterile as one thinks; they have set agoing the unconscious machine and without them it would not have moved and would have produced nothing.

The need for the second period of conscious work, after the inspiration, is still easier to understand. It is necessary to put in shape the results of this inspiration, to deduce from them the immediate consequences, to arrange them, to word the demonstrations, but above all is verification necessary. I have spoken of the feeling of absolute certitude accompanying the inspiration; in the cases cited this feeling was no deceiver, nor is it usually. But do not think this a rule without exception; often this feeling deceives us without being any the less vivid, and we only find it out when we seek to put on foot the demonstration. I have especially noticed this fact in regard to ideas coming to me in the morning or evening in bed while in a semi-hypnagogic state.

Such are the realities; now for the thoughts they force upon us. The unconscious, or, as we say, the subliminal self plays an important role in mathematical creation; this follows from what we have said. But usually the subliminal self is considered as purely automatic. Now we have seen that mathematical work is not simply mechanical, that it could not be done by a machine, however perfect. It is not merely a question of applying rules, of making the most combinations possible according to certain fixed laws. The combinations so obtained would be exceedingly numerous, useless and cumbersome. The true work of the inventor consists in choosing among these combinations so as to eliminate the useless ones or rather to avoid the trouble of making them, and the rules which must guide this choice are extremely fine and delicate. It is almost impossible to state them precisely; they are felt rather than formulated. Under these conditions, how imagine a sieve capable of applying them mechanically?

A first hypothesis now presents itself: the subliminal self is in no way inferior to the conscious self; it is not purely automatic; it is capable of discernment; it has tact, delicacy; it knows how to choose, to divine. What do I say? It knows better how to divine than the conscious self, since it succeeds where that has failed. In a word, is not the subliminal self superior to the conscious self? You recognize the full importance of this question. Boutroux in a recent lecture has shown how it came up on a very different occasion, and what consequences would follow an affirmative answer.

Is this affirmative answer forced upon us by the facts I have just given? I confess that, for my part, I should hate to accept it. Reexamine the facts then and see if they are not compatible with another explanation.

It is certain that the combinations which present themselves to the mind in a sort of sudden illumination, after an unconscious working somewhat prolonged, are generally useful and fertile combinations, which seem the result of a first impression. Does it follow that the subliminal self, having divined by a delicate intuition that these combinations would be useful, has formed only these, or has it rather formed many others which were lacking in interest and have remained unconscious?

In this second way of looking at it, all the combinations would be formed in consequence of the automatism of the subliminal self, but only the interesting ones would break into the domain of consciousness. And this is still very mysterious. What is the cause that, among the thousand products of our unconscious activity, some are called to pass the threshold, while others remain below? Is it a simple chance which confers this privilege? Evidently not; among all the stimuli of our senses, for example, only the most intense fix our attention, unless it has been drawn to them by other causes. More generally the privileged unconscious phenomena, those susceptible of becoming conscious, are those which, directly or indirectly, affect most profoundly our emotional sensibility.

It may be surprising to see emotional sensibility invoked *à propos* of mathematical demonstrations which, it would seem, can interest only the intellect. This would be to forget the feeling of mathematical beauty, of the harmony of numbers and forms, of geometric elegance. This is a true esthetic feeling that all real mathematicians know, and surely it belongs to emotional sensibility.

Now, what are the mathematic entities to which we attribute this character of beauty and elegance, and which are capable of developing in us a sort of esthetic emotion? They are those whose elements are harmoniously disposed so that the mind without effort can embrace their totality while realizing the details. This harmony is at once a satisfaction of our esthetic needs and an aid to the mind, sustaining and guiding. And at the same time, in putting under our eyes a well-ordered whole, it makes us foresee a mathematical law. Now, as we have said above, the only mathematical facts worthy of fixing our attention and capable of being useful are those which can teach us a mathematical law. So that we reach the following conclusion: The useful combinations are precisely the most beautiful, I mean those best able to charm this special sensibility that all mathematicians know, but of which the profane are so ignorant as often to be tempted to smile at it.

What happens then? Among the great numbers of combinations blindly formed by the subliminal self, almost all are without interest and without utility; but just for that reason they are also without effect upon the esthetic sensibility. Consciousness will never know them; only certain ones are harmonious, and, consequently, at once useful and beautiful. They will be capable of touching this special sensibility of the geometer of which I have just spoken, and which, once aroused, will call our attention to them, and thus give them occasion to become conscious.

This is only a hypothesis, and yet here is an observation which may confirm it: when a sudden illumination seizes upon the mind of the mathematician, it usually happens that it does not deceive him, but it also sometimes happens, as I have said, that it does not stand the test of verification; well, we almost always notice that this false idea, had it been true, would have gratified our natural feeling for mathematical elegance.

Thus it is this special esthetic sensibility which plays the role of the delicate sieve of which I spoke, and that sufflciently explains why the one lacking it will never be a real creator.

Yet all the difficulties have not disappeared. The conscious self is narrowly limited, and as for the subliminal self we know not its limitations, and this is why we are not too reluctant in supposing that it has been able in a short time to make more different combinations than the whole life of a conscious being could encompass. Yet these limitations exist. Is it likely that it is able to form all the possible combinations, whose number would frighten the imagination? Nevertheless that would seem necessary, because if it produces only a small part of these combinations, and if it makes them at random, there would be small chance that the *good,* the one we should choose, would be found among them.

Perhaps we ought to seek the explanation in that preliminary period of conscious work which always precedes all fruitful unconscious labor. Permit me a rough comparison. Figure the future elements of our combinations as something like the hooked atoms of Epicurus. During the complete repose of the mind, these atoms are motionless, they are, so to speak, hooked to the wall; so this complete rest may be indefinitely prolonged without the atoms meeting, and consequently without any combination between them.

On the other hand, during a period of apparent rest and unconscious work, certain of them are detached from the wall and put in motion. They flash in every direction through the space (I was about to say the room) where they are enclosed, as would, for example, a swarm of gnats or, if you prefer a more learned comparison, like the molecules of gas in the kinematic theory of gases. Then their mutual impacts may produce new combinations.

What is the role of the preliminary conscious work? It is evidently to mobilize certain of these atoms, to unhook them from the wall and put them in swing. We think we have done no good, because we have moved these elements a thousand different ways in seelliing to assemble them, and have found no satisfactory aggregate. But, after this shaking up imposed upon them by our will, these atoms do not return to their primitive rest. They freely continue their dance.

Now, our will did not choose them at random; it pursued a perfectly determined aim. The mobilized atoms are therefore not any atoms whatsoever; they are those from which we might reasonably expect the desired solution. Then the mobilized atoms undergo impacts which make them enter into combinations among themselves or with other atoms at rest which they struck against in their course. Again I beg pardon, my comparison is very rough, but I scarcely know how otherwise to make my thought understood.

However it may be, the only combinations that have a chance of forming are those where at least one of the elements is one of those atoms freely chosen by our will. Now, it is evidently among these that is found what I called the *good combination.* Perhaps this is a way of lessening the paradoxical in the original hypothesis.

Another observation. It never happens that the unconscious work gives us the result of a somewhat long calculation *all made,* where we have only to apply fixed rules. We might think the wholly automatic subliminal self particularly apt for this sort of work, which is in a way exclusively mechanical. It seems that thinking in the evening upon the factors of a multiplication we might hope to find the product ready made upon our awakening, or again that an algebraic calculation, for example a verification, would be made unconsciously. Nothing of the sort, as observation proves. All one may hope from these inspirations, fruits of unconscious work, is a point of departure for such calculations. As for the calculations themselves, they must be made in the second period of conscious work, that which follows the inspiration, that in which one verifies the results of this inspiration and deduces their consequences. The rules of these calculations are strict and complicated. They require discipline, attention, will, and therefore consciousness. In the subliminal self, on the contrary, reigns what I should call liberty, if we might give this name to the simple absence of discipline and to the disorder born of chance. Only, this disorder itself permits unexpected combinations.

I shall make a last remark: when above I made certain personal observations, I spoke of a night of excitement when I worked in spite of myself. Such cases are frequent, and it is not necessary that the abnormal cerebral activity be caused by a physical excitant as in that I mentioned. It seems, in such cases, that one is present at his own unconscious work, made partially perceptible to the over-excited consciousness, yet without having changed its nature. Then we vaguely comprehend what distinguishes the two mechanisms or, if you wish, the working methods of the two egos. And the psychologic observations I have been able thus to make seem to me to confirm in their general outlines the views I have given.

Surely they have need of it, for they are and remain in spite of all very hypothetical: the interest of the questions is so great that I do not repent of having submitted them to the reader.

THE STRUCTURE OF SCIENTIFIC REVOLUTIONS

THOMAS S. KUHN

Thomas S. Kuhn was born in Cincinnati in 1922. He graduated summa cum laude in physics from Harvard in 1943. After service during World War II in a radio research laboratory and with the U.S.-British laboratory in the Office of Scientific Research and Development, he returned to Harvard for a master's degree in 1946 and a Ph.D. in 1949. He was elected to both Phi Beta Kappa and Sigma Xi.

After three years as a junior fellow at Harvard, he became a member of the faculty in general education and the history of science there. He served on the faculty of the history of science at the University of California, Berkeley, from 1958 to 1964 when he went to Princeton as professor of the history of science.

A book preceding *The Structure of Scientific Revolutions*, excerpted here, was his *The Copernican Revolution: Planetary Astronomy in the Development of Western Thought* published in 1957.

1. What does Kuhn mean by "normal science" and "paradigms"?

2. How do you account for such varying theories over the years about the nature of such common phenomena as light and electricity?

3. According to Kuhn, how does the emergence of a paradigm affect the structure of the group that practices the field?

4. What is the consequence of what Kuhn sees as "the widening gulf that separates the professional scientist from his colleagues in other fields"?

THE STRUCTURE OF SCIENTIFIC REVOLUTIONS

THOMAS S. KUHN

* * * * *

II. The Route to Normal Science

In this essay, 'normal science' means research firmly based upon one or more past scientific achievements, achievements that some particular scientific community acknowledges for a time as supplying the foundation for its further practice. Today such achievements are recounted, though seldom in their original form, by science textbooks, elementary and advanced. These textbooks expound the body of accepted theory, illustrate many or all of its successful applications, and compare these applications with exemplary observations and experiments. Before such books became popular early in the nineteenth century (and until even more recently in the newly matured sciences), many of the famous classics of science fulfilled a similar function. Aristotle's *Physica*, Ptolemy's *Almagest*, Newton's *Principia* and *Opticks*, Franklin's *Electricity*, Lavoisier's *Chemistry*, and Lyell's *Geology*—these and many other works served for a time implicitly to define the legitimate problems and methods of a research field for succeeding generations of practitioners. They were able to do so because they shared two essential characteristics. Their achievement was sufficiently unprecedented to attract an enduring group of adherents away from competing modes of scientific activity. Simultaneously, it was sufficiently open-ended to leave all sorts of problems for the redefined group of practitioners to resolve.

Achievements that share these two characteristics I shall henceforth refer to as 'paradigms,' a term that relates closely to 'normal science.' By choosing it, I mean to suggest that some accepted examples of actual scientific practice—examples which include law, theory, application, and instrumentation together—provide models from which spring particular coherent traditions of scientific research. These are the traditions which the historian describes under such rubrics as 'Ptolemaic astronomy' (or 'Copernican'), 'Aristotelian dynamics' (or 'Newtonian'), 'corpuscular optics' (or 'wave optics'), and so on. The study of paradigms, including many that are far more specialized than those named illustratively above, is what mainly prepares the student for membership in the particular scientific community with which he will later practice. Because he there joins men who learned the bases of their field from the same concrete models, his subsequent practice will seldom evoke overt disagreement over fundamentals. Men whose research is based on shared paradigms are committed to the same rules and standards for scientific practice. That commitment and the apparent consensus it produces are prerequisites for normal science, i.e., for the genesis and continuation of a particular research tradition.

Because in this essay the concept of a paradigm will often substitute for a variety of familiar notions, more will need to be said about the reasons for its introduction. Why is the concrete scientific achievement, as a locus of professional commitment, prior to the various concepts, laws, theories, and points of view that may be abstracted from it? In what sense is the shared paradigm a fundamental unit for the student of scientific development, a unit that cannot be fully reduced to logically atomic components which might function in its stead? When we encounter them in Section V, answers to these questions and to others like them will prove basic to an understanding both of normal science and of the associated concept of paradigms. That more abstract discussion will depend, however, upon a previous exposure to examples of normal science or of paradigms in operation. In particular, both these related concepts will be clarified by noting that there can be a sort of scientific research without paradigms, or at least without any so unequivocal and so binding as the ones named above. Acquisition of a paradigm and of the more esoteric type of research it permits is a sign of maturity in the development of any given scientific field.

If the historian traces the scientific knowledge of any selected group of related phenomena backward in time, he is likely to encounter some minor variant of a pattern here illustrated from the history of physical optics. Today's physics textbooks tell the student that light is photons, i.e., quantum-mechanical entities that exhibit some characteristics of waves and some of particles. Research proceeds accordingly, or rather according to the more elaborate and mathematical characterization from which this usual verbalization is derived. That characterization of light is, however, scarcely half a century old. Before it was developed by Planck, Einstein, and others early in this century, physics texts taught that light was transverse wave motion, a conception rooted in a paradigm that derived ultimately from the optical writings of Young and Fresnel in the early nineteenth century. Nor was the wave theory the first to be embraced by almost all practitioners of optical science. During the eighteenth century the paradigm for this field was provided by Newton's *Opticks,* which taught that light was material corpuscles. At that time physicists sought evidence, as the early wave theorists had not, of the pressure exerted by light particles impinging on solid bodies.

These transformations of the paradigms of physical optics are scientific revolutions, and the successive transition from one paradigm to another via revolution is the usual developmental pattern of mature science. It is not, however, the pattern characteristic of the period before Newton's work, and that is the contrast that concerns us here. No period between remote antiquity and the end of the seventeenth century exhibited a single generally accepted view about the nature of light. Instead there were a number of competing schools and subschools, most of them espousing one variant or another of Epicurean, Aristotelian, or Platonic theory. One group took light to be particles emanating from material bodies; for another it was a modification of the medium that intervened between the body and the eye; still another explained light in terms of an interaction of the medium with an emanation from the eye; and there were other combinations and modifications besides. Each of the corresponding schools derived strength from its relation to some particular metaphysic, and each emphasized, as paradigmatic observations, the particular cluster of optical phenomena that its own theory could do most to explain. other observations were dealt with by *ad hoc* elaborations, or they remained as outstanding problems for further research.

At various times all these schools made significant contributions to the body of concepts, phenomena, and techniques from which Newton drew the first nearly uniformly accepted paradigm for physical optics. Any definition of the scientist that excludes at least the more creative members of these various schools will exclude their modern successors as well. Those men were scientists. Yet anyone examining a survey of physical optics before Newton may well conclude that, though the field's practitioners were scientists, the net result of their activity was something less than science. Being able to take no common body of belief for granted, each writer on physical optics felt forced to build his field anew from its foundations. In doing so, his choice of supporting observation and experiment was relatively free, for there was no standard set of methods or of phenomena that every optical writer felt forced to employ and explain. Under these circumstances, the dialogue of the resulting books was often directed as much to the members of other schools as it was to nature. That pattern is not unfamiliar in a number of creative fields today, nor is it incompatible with significant discovery and invention. It is not, however, the pattern of development that physical optics acquired after Newton and that other natural sciences make familiar today.

The history of electrical research in the first half of the eighteenth century provides a more concrete and better known example of the way a science develops before it acquires its first universally received paradigm. During that period there were almost as many views about the nature of electricity as there were important electrical experimenters, men like Hauksbee, Gray, Desaguliers, Du Fay, Nollett, Watson, Franklin, and others. All their numerous concepts of electricity had something in common—they were partially derived from one or another version of the mechanico-corpuscular philosophy that guided all scientific research of the day. In addition, all were components of real scientific theories, of theories that had been drawn in part from experiment and observation and that partially determined the choice and interpretation of additional problems undertaken in research. Yet though all the experiments were electrical and though most of the experimenters read each other's works, their theories had no more than a family resemblance.

One early group of theories, following seventeenth-century practice, regarded attraction and frictional generation as the fundamental electrical phenomena. This group tended to treat repulsion as a secondary effect due to some sort of mechanical rebounding and also to postpone for as long as possible both discussion and systematic research on Gray's newly discovered effect, electrical conduction. Other "electricians" (the term is their own) took attraction and repulsion to be equally elementary manifestations of electricity and modified their theories and research accordingly. (Actually, this group is remarkably small—even Franklin's theory never quite accounted for the mutual repulsion of two negatively charged bodies.) But they had as much difficulty as the first group in accounting simultaneously for any but the simplest conduction effects. Those effects, however, provided the starting point for still a third group, one which tended to speak of electricity as a "fluid" that could run through conductors rather than as an "effluvium" that emanated from non-conductors. This group, in its turn, had difficulty reconciling its theory with a number of attractive and repulsive effects. Only through the work of Franklin and his immediate successors did a theory arise that could account with something like equal facility for very nearly all these effects and that therefore could and did provide a subsequent generation of "electricians" with a common paradigm for its research.

Excluding those fields, like mathematics and astronomy, in which the first firm paradigms date from prehistory and also those, like biochemistry, that arose by division and recombination of specialties already matured, the situations outlined above are historically typical. Though it involves my continuing to employ the unfortunate simplification that tags an extended historical episode with a single and somewhat arbitrarily chosen name (e.g., Newton or Franklin), I suggest that similar fundamental disagreements characterized, for example, the study of motion before Aristotle and of statics before Archimedes, the study of heat before Black, of chemistry before Boyle and Boerhaave, and of historical geology before Hutton. In parts of biology—the study of heredity, for example—the first universally received paradigms are still more recent; and it remains an open question what parts of social science have yet acquired such paradigms at all. History suggests that the road to a firm research consensus is extraordinarily arduous.

History also suggests, however, some reasons for the difficulties encountered on that road. In the absence of a paradigm or some candidate for paradigm, all of the facts that could possibly pertain to the development of a given science are likely to seem equally relevant. As a result, early fact-gathering is a far more nearly random activity than the one that subsequent scientific development makes familiar. Furthermore, in the absence of a reason for seeking some particular form of more recondite information, early fact-gathering is usually restricted to the wealth of data that lie ready to hand. The resulting pool of facts contains those accessible to casual observation and experiment together with some of the more esoteric data retrievable from established crafts like medicine, calendar making, and metallurgy. Because the crafts are one readily accessible source of facts that could not have been casually discovered, technology has often played a vital role in the emergence of new sciences. But though this sort of fact-collecting has been essential to the origin of many significant sciences, anyone who examines, for example, Pliny's encyclopedic writings or the Baconian natural histories of the seventeenth century will discover that it produces a morass. One somehow hesitates to call the literature that results scientific. The Baconian "histories" of heat, color, wind, mining, and so on, are filled with information, some of it recondite. But they juxtapose facts that will later prove revealing (e.g., heating by mixture) with others (e.g., the warmth of dung heaps) that will for some time remain too complex to be integrated with theory at all. In addition, since any description must be partial, the typical natural history often omits from its immensely circumstantial accounts just those details that later scientists will find sources of important illumination. Almost none of the early "histories" of electricity, for example, mention that chaff, attracted to a rubbed glass rod, bounces off again. That effect seemed mechanical, not electrical. Moreover, since the casual fact-gatherer seldom possesses the time or the tools to be critical, the natural histories often juxtapose descriptions like the above with others, say, heating by antiperistasis (or by cooling), that we are now quite unable to confirm. Only very occasionally, as in the cases of ancient statics, dynamics, and geometrical optics, do facts collected with so little guidance from pre-established theory speak with sufficient clarity to permit the emergence of a first paradigm.

This is the situation that creates the schools characteristic of the early stages of a science's development. No natural history can be interpreted in the absence of at least some implicit body of intertwined theoretical and methodological belief that permits selection, evaluation, and criticism. If that body of belief is not already implicit in the collection of facts—in which case more than "mere facts" are at hand—it must be externally supplied, perhaps by a current metaphysic, by another science, or by personal and historical accident. No wonder, then, that in the early stages of the development of any science different men confronting the same range of phenomena, but not usually all the same particular phenomena, describe and interpret them in different ways. What is surprising, and perhaps also unique in its degree to the fields we call science, is that such initial divergences should ever largely disappear.

For they do disappear to a very considerable extent and then apparently once and for all. Furthermore, their disappearance is usually caused by the triumph of one of the pre-paradigm schools, which, because of its own characteristic beliefs and preconceptions, emphasized only some special part of the too sizable and inchoate pool of information. Those electricians who thought electricity a fluid and therefore gave particular emphasis to conduction provide an excellent case in point. Led by this belief, which could scarcely cope with the known multiplicity of attractive and repulsive effects, several of them conceived the idea of bottling the electrical fluid. The immediate fruit of their efforts was the Leydeu jar, a device which might never have been discovered by a man exploring nature casually or at random, but which was in fact independently developed by at least two investigators in the early 1740's. Almost from the start of his electrical researches, Franklin was particularly concerned to explain that strange and, in the event, particularly revealing piece of special apparatus. His success in doing so provided the most effective of the arguments that made his theory a paradigm, though one that was still unable to account for quite all the known cases of electrical repulsion. To be accepted as a paradigm, a theory must seem better than its competitors, but it need not, and in fact never does, explain all the facts with which it can be confronted.

What the fluid theory of electricity did for the subgroup that held it, the Franklinian paradigm later did for the entire group of electricians. It suggested which experiments would be worth performing and which, because directed to secondary or to overly complex manifestations of electricity, would not. Only the paradigm did the job far more effectively, partly because the end of interschool debate ended the constant reiteration of fundamentals and partly because the confidence that they were on the right track encouraged scientists to undertake more precise, esoteric, and consuming sorts of work. Freed from the concern with any and all electrical phenomena, the united group of electricians could pursue selected phenomena in far more detail, designing much special equipment for the task and employing it more stubbornly and systematically than electricians had ever done before. Both fact collection and theory articulation became highly directed activities. The effectiveness and efficiency of electrical research increased accordingly, providing evidence for a societal version of Francis Bacon's acute methodological dictum: "Truth emerges more readily from error than from confusion."

We shall be examining the nature of this highly directed or paradigm-based research in the next section, but must first note briefly how the emergence of a paradigm affects the structure of the group that practices the field. When, in the development of a natural science, an individual or group first produces a synthesis able to attract most of the next generation's practitioners, the older schools gradually disappear. In part their disappearance is caused by their members' conversion to the new paradigm. But there are always some men who cling to one or another of the older views, and they are simply read out of the profession, which thereafter ignores their work. The new paradigm implies a new and more rigid definition of the field. Those unwilling or unable to accommodate their work to it must proceed in isolation or attach themselves to some other group. Historically, they have often simply stayed in the departments of philosophy from which so many of the special sciences have been spawned. As these indications hint, it is sometimes just its reception of a paradigm that transforms a group previously interested merely in the study of nature into a profession or, at least, a discipline. In the sciences (though not in fields like medicine, technology, and law, of which the principal *raison d'etre is* an external social need), the formation of specialized journals, the foundation of specialists' societies, and the claim for a special place in the curriculum have usually been associated with a group's first reception of a single paradigm. At least this was the case between the time, a century and a half ago, when the institutional pattern of scientific specialization first developed and the very recent time when the paraphernalia of specialization acquired a prestige of their own.

The more rigid definition of the scientific group has other consequences. When the individual scientist can take a paradigm for granted, he need no longer, in his major works, attempt to build his field anew, starting from first principles and justifying the use of each concept introduced. That can be left to the writer of textbooks. Given a textbook, however, the creative scientist can begin his research where it leaves off and thus concentrate exclusively upon the subtlest and most esoteric aspects of the natural phenomena that concern his group. And as he does this, his research communiques will begin to change in ways whose evolution has been too little studied but whose modern end products are obvious to all and oppressive to many. No longer will his researches usually be embodied in books addressed, like Franklin's *Experiments . . . on Electricity* or Darwin's *Origin of Species,* to anyone who might be interested in the subject matter of the field. Instead they will usually appear as brief articles addressed only to professional colleagues, the men whose knowledge of a shared paradigm can be assumed and who prove to be the only ones able to read the papers addressed to them.

Today in the sciences, books are usually either texts or retrospective reflections upon one aspect or another of the scientific life. The scientist who writes one is more likely to find his professional reputation impaired than enhanced. Only in the earlier, pre-paradigm, stages of the development of the various sciences did the book ordinarily possess the same relation to professional achievement that it still retains in other creative fields. And only in those fields that still retain the book, with or without the article, as a vehicle for research communication are the lines of professionalization still so loosely drawn that the layman may hope to follow progress by reading the practitioners' original reports. Both in mathematics and astronomy, research reports had ceased already in antiquity to be intelligible to a generally educated audience. In dynamics, research became similarly esoteric in the later Middle Ages, and it recaptured general intelligibility only briefly during the early seventeenth century when a new paradigm replaced the one that had guided medieval research. Electrical research began to require translation for the layman before the end of the eighteenth century, and most other fields of physical science ceased to be generally accessible in the nineteenth. During the same two centuries similar transitions can be isolated in the various parts of the biological sciences. In parts of the social sciences they may well be occurring today. Although it has become customary, and is surely proper, to deplore the widening gulf that separates the professional scientist from his colleagues in other fields, too little attention is paid to the essential relationship between that gulf and the mechanisms intrinsic to scientific advance.

* * * * *

REFLECTIONS ON GENDER AND SCIENCE

EVELYN FOX KELLER

Evelyn Fox Keller is a mathematical biologist from New York. She was born in New York City in 1936. She received the B.A. degree from Brandeis University in 1957 and the Ph.D. in physics from Harvard in 1963. After a year as an instructor at New York University, she joined the faculty of the Cornell Graduate School of Medical Science, and then in 1972 became a member of the faculty of Natural Science at the State University of New York College at Purchase. Since 1976, she has also served as a member of the Center for Policy Research.

1. How do you account for the fact that relatively very few women have ever been awarded a Nobel prize in science?

2. According to Keller, what is the significance of Francis Bacon's use of gender and sexual imagery in his writings on science and nature?

3. How does Keller relate the ideas of Freud to those of Bacon?

4. What do you think was Keller's objective in this writing?

REFLECTIONS ON GENDER AND SCIENCE

EVELYN FOX KELLER

CHAPTER TWO

Baconian Science: The Arts of Mastery and Obedience

It has become common for social critics of modern science to assume that control and domination are basic to the scientific impulse. For such critics, the motives of scientific inquiry appear to have been defined by Bacon, for it was he who first and most vividly articulated the equation between scientific knowledge and power, who identified the aims of science as the control and domination of nature. Yet this view represents a particular reading of both science and Bacon.

Bacon's writings lend themselves easily to such a reading. Rejecting Plato's highly abstract and deeply erotic coupling of mind and form—indeed, rejecting all the major traditions that preceded him—Bacon's articulation of his vision was both provocative and aggressive. In viewing science as power, he thought he saw salvation. Indeed, the salvation of mankind was to be found in the very power of science. It thereby became the moral responsibility of men to assume and exercise that power. To his contemporaries, his vision of science as power and as salvation must have seemed fantastic on both counts. To us, he seems to have been prescient in the former and naive in the latter.

With the unfolding of history, Bacon's vision has taken on renewed but different interest. Public confidence and optimism have given way to anxiety, and some critics of science have begun to worry anew about aggression in the scientific impulse. But even as the practice of science has in fact become more explicitly and dramatically involved with power, the writings of scientists have become, if anything, more discreet and more qualifying in their descriptions of the nature of their enterprise. And so, to these critics, Bacon begins to take on the value not only of prescience but also of a bold and revealing honesty.

At the same time, his vision appears to others, particularly to the defenders of science, to be grossly distorted. It speaks more for the technologist than for the scientist, whose search it is said is as much for transcendence as it is for power. Bacon, it is asserted finally, was not really a scientist. Thus science is defended by dissociation from the unwelcome implications of Bacon's vision. That vision may well have been prescient for technology, but it is ultimately not relevant to pure science.

I will argue that Bacon has in fact provided us with a model that is truer to the spirit of the scientific impulse than is generally recognized by the defenders of science, and more complex than is recognized by many of its critics.

One of the ways in which the subtleties of Bacon's model find expression is through the sexual dialectic implicit in his metaphors. Bacon's vision was neither simply prophetic nor simply a misrepresentation of the emerging scientific enterprise. If we search his vivid metaphors with care, as I propose to do, paying particular attention to his use of gender, we find traces of a dialectic that is far more complicated and hence richer in meaning than either the critics or the defenders of science tend to assume. The sexual imagery in Bacon's language is neither as coherent nor as clearly articulated as Plato's, but Bacon nonetheless deserves examination here as a counterpart to Plato; he provided the language from which subsequent generations of scientists extracted a more consistent metaphor of lawful sexual domination.

Bacon's Metaphor

What was Bacon's vision? It was, without a doubt, of a science leading to the sovereignty, dominion, and mastery of man over nature, the "command of nature in action" (Anderson 1960, p. 19). It is in science that "human knowledge and human power meet in one" (ibid., p. 39), where man's native ambition for power finds its constructive, noble, and humane outlet. In distinguishing three kinds, or grades, of ambition, Bacon wrote:

> The first is of those who desire to extend their own power in their native country, a vulgar and degenerate kind. The second is of those who labor to extend the power and dominion of their country among men. This certainly has more dignity, though not less covetousness. But if a man endeavor to establish and extend the power and dominion of the human race itself over the universe, his ambition (if ambition it can be called) is without doubt both a more wholesome and a more noble thing than the other two. Now the empire of man over things depends wholly on the arts and sciences. For we cannot command nature except by obeying her. (ibid., p. 29)

Through science and art (that is, technology, or mechanical art), man can find the power to transform not so much the world as his relation to the world. The goal of science was, for Bacon, "the restitution and reinvesting of man to the sovereignty and power. . . which he had in his first state of creation" (Robertson 1905, p. 188). By what means, we need ask, and from what sources was science to acquire such power? And what was the shape it was to take?

Bacon's answer to these questions is given metaphorically— through his frequent and graphic use of sexual imagery. The invocation of gender and the use of sexual imagery are common enough in descriptions of nature, and perhaps for this reason Bacon's uses have not seemed especially noteworthy. It is important, however, to see how deeply Bacon's use of gender is implicated in his conception of mastery and domination. The fact that mastery and domination are, invariably, exercised over nature as "she" can hardly escape our attention, and indeed has not gone unnoticed (see, for example, Leiss 1972, p. 60). But, under scrutiny, it can be seen that Bacon's use of gender is not so simple. Less conspicuous but still visible is a more complex sexual dialectic—a dialectic whose complexities are not gratuitous but can be interpreted as parts of a description of the scientific impulse. To see the dialectical character of Bacon's imagery, and therefore to perceive the full meaning of his metaphor, it is useful to juxtapose some expressions, even well-worn expressions, of that metaphor. Taken singly they may be familiar enough; taken together they acquire new meaning.

"Let us establish" he wrote, "a chaste and lawful marriage between Mind and Nature" (quoted in Leiss 1972, p. 25); and again, elsewhere, "My dear, dear boy, what I plan for you is to unite you with things themselves in a chaste, holy and legal wedlock. And from this association you will secure an increase beyond all the hopes and prayers of ordinary marriages, to wit, a blessed race of Heroes and Supermen" (Farrington 1951, p. 201). The word *things* here is far from neutral. Elsewhere (even in the same work) he is more explicit. It is Nature herself who is to be the bride, who requires taming, shaping, and subduing by the scientific mind. "I am come in very truth leading to you Nature with all her children to bind her to your service and make her your slave" (ibid., p. 197). Elsewhere, more gently, he writes: "I invite all such to join themselves, as true sons of knowledge, with me, that passing by the outer courts of nature, which numbers have trodden, we may find a way at length into her inner chambers" (Anderson 1960, p. 36). Nature may be coy, but she can be conquered, "For you have but to follow and as it were hound nature in her wanderings, and you will be able, when you like, to lead and drive her afterwards to the same place again" (Spedding et al. 1869, 4: p. 296). The discipline of scientific knowledge, and the mechanical inventions it leads to, do not "merely exert a gentle guidance over nature's course; they have the power to conquer and subdue her, to shake her to her foundations" (Spedding et al. 1869, 5: p. 506). All of this is, however, in the service of truth. In conquering and subduing, in shaking her to her foundations, we do not so much transform Nature as reveal her, for "the nature of things betrays itself more readily under the vexations of art" (meaning practical, or mechanical, art) "than in its natural freedom" (Anderson 1960, p. 25).

But Bacon's formula is not simply aggressive—it is also responsive. He wrote: "For man is but the servant and interpreter of nature: what he does and what he knows is only what he has observed of nature's order in fact or in thought; beyond this he knows nothing and can do nothing. For the chains of causes cannot by any force be loosed or broken, nor can nature be commanded except by being obeyed" (Anderson 1960, p. 29). The aim of science is not to violate but to master nature by following the dictates of the truly natural. That is, it is "natural" to guide, shape, even hound, conquer, and subdue her—only in that way is the true "nature of things" revealed. It is here that the empirical side of Bacon's philosophy finds expression. Experiment expresses the spirit of action, of a "doing" devoted to "finding out." Science controls by following the dictates of nature, but these dictates include the requirement, even demand, for domination.

Not simple violation, or rape, but forceful and aggressive seduction leads to conquest. Alchemy was misguided, not in its aims, but in its methods. Rossi (1968, p. 105) sums up Bacon's interpretation of the myth of Erichthonious as follows: "The basic error of chemical and mechanical productions is not that they strive to dominate nature, but the method by which they attempt to do so; they rape Minerva instead of winning her over."

Yet, the distinction between rape and conquest sometimes seems too subtle. There remains something of a puzzle. Nature is commanded by being obeyed, revealed by being enslaved, hounded, and vexed. The metaphor of seduction, even forceful and aggressive seduction, does not seem adequate to all the ambiguities that are intended. Indeed, in the context of so limited a metaphor, the ambiguities become contradictions. Science is to be aggressive yet responsive, powerful yet benign, masterful yet subservient, shrewd yet innocent, "as if the divine nature enjoyed the kindly innocence in such hide-and-seek, hiding only in order to be found, and with characteristic indulgence desired the human mind to join Him in this sport" (Farrington 1966, p. 92).

Here, as nature becomes divine, the hunt turns to sport—kindly and innocent. At the same time, and surely not coincidentally, "she" becomes "he." The puzzle begins to dissolve as Bacon's metaphor expands. When nature becomes divine, not only does "she" become "he," but by implication as we shall see, the scientific mind becomes more nearly female. The sexual dialectic evident here begins to suggest a larger and richer metaphor for the scientific enterprise, a metaphor which is made generally more explicit in the fragments of a lesser known work of Bacon's entitled *Temporis Partus Masculus,* or The Masculine Birth of Time.

The Masculine Birth of Time

The fragments of this work, written in 1602 or 1603 and never published in Bacon's lifetime, have been translated by Farrington (1951), who says "it adds nothing to our knowledge of Bacon's teachings but it throws much light on his own emotional attitude to his work" (p. 193). The title is interpreted by Farrington as insinuating "that older science represented only a female off-spring, passive, weak, expectant, but now a son was born, active, virile, generative" (p.194).

The key to the birth of a masculine, virile science is to be found in the cleansing of the human mind of "false preconceptions," so as to facilitate its receptivity. In the first chapter he asks, "Do you suppose that men's minds offer flawless polished surfaces for the reception of the true native rays of real things? In fact every avenue to every mind is beset and blocked by the darkest idols, or false preconceptions are deeprooted or burned in" (p. 194).

The crucial concepts in this image of Mind are reception and submission—unobstructed receptivity to the "true native rays of real things." Only then can "the unlocking of the paths of sense and the unkindling of a greater light in Nature" come about. The imagery of submission is in noteworthy contrast to the male imagery of mastery that Bacon typically used to describe the relation of mind to nature.

What is sought in this passage is the proper stance for mind necessary to ensure the reception of truth and the conception of science. To receive God's truth, the mind must be pure and clean, submissive and open. Only then can it give birth to a masculine and virile science. That is, if the mind is pure, receptive, and submissive in its relation to God, it can be transformed by God into a forceful, potent, and virile agent in its relation to nature. Cleansed of contamination, the mind can be impregnated by God and, in that act, virilized: made potent and capable of generating virile offspring in its union with Nature.

The transformation of mind from female to male is made explicit in the structure of this work. The first part is a prayer addressed to God, in the voice of a supplicant. The remainder, and body, of the work is cast in the voice of the mature scientist addressing a son, his virile offspring. Henceforth, Nature becomes indubitably female: the object of actions. It is here that we read, "I am come in very truth leading to you Nature with all her children to bind her to your service and make her your slave." In the same passage, he further bequeathes to this son the heritage of "my only earthly wish, namely to stretch the deplorable narrow limits of man's dominion over the universe to their promised bounds" (p. 197).

Notice the God-like voice in which Bacon now addresses his son and heir. The son, in turn, is invited to assume the posture of supplicant: "Take heart, then, my son, and give yourself to me, so that I may restore you to yourself" (p. 201). Restoration begins with clearing the son's mind of the influence of "sham philosophers," of those who "debauch our minds" (p. 197). To this end the bulk of the work is devoted.

Aristotle, Plato, Galen, Hippocrates successfully come in for a most flagrant debunking. They convey false teachings that block and impede the mind's true receptivity to "the true native rays of real things." Bacon's language is strong. "It was in Aristotle's bosom that were bred and nurtured those crafty triflers . . . the darksome idols from some subterranean cave" (p. 198). Plato is denounced for giving out "the falsehood that truth is, as it were, the native inhabitant of the human mind and need not come in from outside to take up its abode there" (p. 198). Galen is charged with "the malicious intention of lessening human power" (p. 199).

These were the men who gave rise to the older science—the science that Farrington says, in Bacon's view, "represented only a female offspring" in contrast with the "masculine birth" that Bacon now heralds. Yet the terms of their debunking express the same complex duality or dialectic we have already seen. The teachings of the ancients are dangerous not simply because they are unproductive, but also because they discourage receptivity. Plato taught false self-sufficiency; his doctrine left no space for the need of truth to enter the mind from without. The grievous error of the ancients, concludes Farrington, was that "they sought to create the universe out of the human mind. This presumptuous usurpation of the prerogative of the Creator was punished by the curse of barrenness" (p. 202). But not just barrenness, which is a female imperfection—also impotence and the inability to sire virile offspring. The errors of the ancients were more various than this, but they had in common the net effect of breeding impotence, of "lessening human power," of retarding the birth of the true and masculine science that, Bacon says, "is to be sought from the light of nature, not the darkness of antiquity"(p. 200). The causes of past impotence and feminization in Bacon's vision are as complex as the sources of future potency.

Concluding the work, Bacon warns the neophyte scientist that the task of cleansing the mind is a difficult one, and requires a long and faithful apprenticeship. Bacon offers himself as guide and leader, promising thereby to "restore" the youth to himself, to lead him into a marriage that will yield "Heroes and Supermen," to prepare him for his rightful ascendancy.

Interpretation and Conclusion

The metaphor that Bacon employs to articulate his vision for the birth of science is now seen to be suffused with overtones of considerable subtlety. Behind the overt insistence on the virility and masculinity of the scientific mind lies a covert assumption and acknowledgment of the dialectical, even hermaphroditic, nature of the "marriage between Mind and Nature." This recognition can stand on its own, for Bacon's imagery reveals itself. It is, however, possible to make a further interpretation.

The two-sided character of Bacon's metaphor can be seen as expressing not only the dual nature of the scientific venture, simultaneously receptive and potent, but also a ubiquitous fantasy of childhood sexuality. The virilization of the scientist, and of the child as well, is acquired as a gift received from the father. This gift permits the scientist to father "Heroes and Supermen"; it provides him with the capacity, as it were, to give birth to himself. It is this aspect of Bacon's metaphor, and of the child's fantasy, that reveals the implicitly bisexual character of the oedipal project. Freud wrote, in *The Ego and the Id:*

> The simple Oedipus complex is by no means its commonest form, but rather represents a simplification or schematization which, to be sure, is often enough adequate for practical purposes. Closer study usually discloses the more complete Oedipus complex, which is twofold, positive and negative, and is due to the bisexuality originally present in children: that is to say, a boy has not merely an ambivalent attitude towards his father and an affectionate object-relation towards his mother, but at the same time he also behaves like a girl and displays an affectionate feminine attitude to his father and a corresponding hostility and jealousy towards his mother. (1927, p. 220)

The impulse to identify oneself simultaneously with the mother and the father finds exquisitely face-saving and economical expression in what Freud described as the oedipal boy's "wish to be the father of himself" (1950, 4, p. 201). Its economy is achieved by condensation, and budding masculine pride is sustained by elision. In fact that fantasy, by virtue of its compression of so many conflicting wishes, reveals the poetic genius of the young boy's unconscious. Through condensation and elision, it allows for the subterranean survival of wishes no longer deemed acceptable, wishes regarded as feminine. Thus, in identifying himself with the race of fathers who can give birth, the young boy can simultaneously assert his independence and safeguard the earlier and conflicting wish for identification with the mother; in presuming to father himself, he satisfies his wish for omnipotent self-sufficiency.

The Baconian metaphor, in condensing the dual impulse to appropriate and deny the maternal, seems to resemble the boy's oedipal ambitions. As such it represents a bypass, a compensation for and a way of doing without the mother. Omnipotence is secured by an identification with the father which allows simultaneously for the appropriation and denial of the feminine. Both the child and science can now enter the man's world.

In the context of this interpretation, the sexual aggressivity of Bacon's imagery begins to assume a somewhat defensive quality. What is most immediately conspicuous in that imagery is its denial of the feminine as subject—a denial often taken to be generally characteristic of the scientific endeavor. But, when we examine the imagery more carefully, we find, behind that simple denial, a prior cooptation of the female mode—a cooptation which, given the initial impulse toward denial, necessitates an ever more urgent and aggressive repudiation. That is to say, the aggressively male stance of Bacon's scientist could, and perhaps now should, be seen as driven by the need to deny what all scientists, including Bacon, privately have known, namely, that the scientific mind must be, on some level, a hermaphroditic mind.

It is in this sense that Bacon's metaphor, taken in its entirety, becomes richer and more complete, and, as such, provides a more accurate description of the scientific impulse. This description may well represent a version of the oedipal project, but one that retains all the bisexual implications of that project intact. In contemporary times, the explicit role of God has disappeared, and the scientific fantasy has become more self-contained. Where Bacon was able to parcel out the dual aspects of the scientific mind by assigning one of these functions to the relation of mind to God, or to divine nature, and the other to Nature, contemporary scientists cannot. For most scientists today, there is only one Nature, and only one mind. The scientist himself has assumed the procreative function that Bacon reserved for God: his mind is now a single entity, both phallus and womb. However, his kinship with Bacon survives in his simultaneous appropriation and denial of the feminine.

and dream of masculine
filiation, dream of God the father
emerging from himself
in his son,—and
no mother then.

HELENE CIXOUS. Sorties (1981)

A MATHEMATICIAN'S APOLOGY

G.H. HARDY

Godfrey Harold Hardy (1877-1947) showed a proclivity for numbers from his earliest years, and especially at Winchester School. Both of his parents were educators with talents for mathematics and they encouraged his intellectual activity.

After graduation from Cambridge University, he returned to Trinity College as a Fellow in 1900 and was lecturer in mathematics there from 1906 to 1919, and Cayley lecturer in Cambridge from 1914 to 1919.

In 1908 he, concurrently with Wilhelm Weinberg, a German physician, framed the Hardy-Weinberg law of genetics having to do with the proportions of dominant and recessive traits propagated in a large mixed population.

Most of Hardy's major contributions in mathematics were in collaboration with John E. Littlewood in perhaps the most famous collaboration in the history of mathematics, and in collaboration with a poorly educated mathematical genius from India, Ramanujan.

In 1919, Hardy became a Fellow of New College and Savilian Professor of Geometry at Oxford. He returned to Cambridge in 1931 as Sadleiran Professor of Pure Mathematics and Fellow of Trinity, where he remained until his death.

Hardy liked to play tennis and other sports, and he was an avid cricket fan throughout his life. John Maynard Keynes once told him that if he (Hardy) had read the stock market reports for half an hour each day with the same concentration with which he read the cricket scores, he surely would have been a rich man.

1. Why, according to Hardy, should one devote a lifetime career to mathematics?
2. Why should "any genuine mathematician. . . feel . . . that the popular reputation of mathematics is based largely on ignorance and confusion"?
3. What is the difference between "pure" and "applied" mathematics?
4. What does Hardy mean in saying that, "the mathematician is in much more direct contact with reality" than is the physicist?
5. What is the difference between the "real mathematics of the real mathematicians," and the "trivial mathematics"?
6. What about Hardy's statement made in 1940 that, "No one has yet discovered any warlike purpose to be served by the theory of . . . relativity"?

A MATHEMATICIAN'S APOLOGY

G. H. HARDY

2

I PROPOSE to put forward an apology for mathematics; and I may be told that it needs none, since there are now few studies more generally recognized, for good reasons or bad, as profitable and praiseworthy. This may be true; indeed it is probable, since the sensational triumphs of Einstein, that stellar astronomy and atomic physics are the only sciences which stand higher in popular estimation. A mathematician need not now consider himself on the defensive. He does not have to meet the sort of opposition described by Bradley in the admirable defence of metaphysics which forms the introduction to *Appearance and Reality*.

A metaphysician, says Bradley, will be told that 'metaphysical knowledge is wholly impossible', or that 'even if possible to a certain degree, it is practically no knowledge worth the name'. 'The same problems,' he will hear, 'the same disputes, the same sheer failure. Why not abandon it and come out? Is there nothing else more worth your labour?' There is no one so stupid as to use this sort of language about mathematics. The mass of mathematical truth is obvious and imposing; its practical applications, the bridges and steam-engines and dynamos, obtrude themselves on the dullest imagination. The public does not need to be convinced that there is something in mathematics.

All this is in its way very comforting to mathematicians, but it is hardly possible for a genuine mathematician to be content with it. Any genuine mathematician must feel that it is not on these crude achievements that the real case for mathematics rests, that the popular reputation of mathematics is based largely on ignorance and confusion, and that there is room for a more rational defence. At any rate, I am disposed to try to make one. It should be a simpler task, at any rate, than Bradley's difficult apology.

I shall ask, then, why is it really worth while to make a serious study of mathematics? What is the proper justification of a mathematician's life? And my answers will be, for the most part, such as are to be expected from a mathematician: I think that it is worth while, that there is ample justification. But I should say at once that, in defending mathematics, I shall be defending myself, and that my apology is bound to be to some extent egotistical. I should not think it worth while to apologize for my subject if I regarded myself as one of its failures.

Some egotism of this sort is inevitable, and I do not feel that it really needs justification. Good work is not done by 'humble' men. It is one of the first duties of a professor, for example, in any subject, to exaggerate a little both the importance of his subject and his own importance in it. A man who is always asking 'Is what I do worth while?' and 'Am I the right person to do it?' will always be ineffective himself and a discouragement to others. He must shut his eyes a little and think a little more of his subject and himself than they deserve. This is not too difficult: it is harder not to make his subject and himself ridiculous by shutting his eyes too tightly.

3

A MAN who sets out to justify his existence and his activities has to distinguish two different questions. The first is whether the work which he does is worth doing; and the second is why he does it (whatever its value may be).

The first question is often very difficult, and the answer very discouraging, but most people will find the second easy enough even then. Their answers, if they are honest, will usually take one or other of two forms; and the second form is merely a humbler variation of the first, which is the only answer which we need consider seriously.

(I) 'I do what I do because it is the one and only thing that I can do at all well. I am a lawyer, or a stockbroker, or a professional cricketer, because I have some real talent for that particular job. I am a lawyer because I have a fluent tongue, and am interested in legal subtleties; I am a stockbroker because my judgement of the markets is quick and sound; I am a professional cricketer because. I can bat unusually well. I agree that it might be better to be a poet or a mathematician, but unfortunately I have no talent for such pursuits.'

I am not suggesting that this is a defence which can be made by most people, since most people can do nothing at all well. But it is impregnable when it can be made without absurdity, as it can by a substantial minority: perhaps five or even ten per cent of men can do something rather well. It is a tiny minority who can do anything *really* well, and the number of men who can do two things well is negligible. If a man has any genuine talent, he should be ready to make almost any sacrifice in order to cultivate it to the full.

This view was endorsed by Dr Johnson—

> When I told him that I had been to see [his namesake] Johnson ride upon three horses, he said 'Such a man, sir, should be encouraged, for his performances show the extent of the human powers. . . '—

and similarly he would have applauded mountain climbers, channel swimmers, and blindfold chess-players. For my own part, I am entirely in sympathy with all such attempts at remarkable achievement. I feel some sympathy even with conjurors and ventriloquists; and when Alekhine and Bradman set out to beat records, I am quite bitterly disappointed if they fail. And here both Dr Johnson and I find ourselves in agreement with the public. As W.J. Turner has said so truly, it is only the 'highbrows' (in the unpleasant sense) who do not admire the 'real swells'.

We have of course to take account of the differences in value between different activities. I would rather be a novelist or a painter than a statesman of similar rank; and there are many roads to fame which most of us would reject as actively pernicious. Yet it is seldom that such differences of value will turn the scale in a man's choice of a career, which will almost always be dictated by the limitations of his natural abilities. Poetry is more valuable than cricket, but Bradman would be a fool if he sacrificed his cricket in order to write second-rate minor poetry (and I suppose that it is unlikely that he could do better). If the cricket were a little less supreme, and the poetry better, then the choice might be more difficult: I do not know whether I would rather have been

Victor Trumper or Rupert Brooke. It is fortunate that such dilemmas occur so seldom.

I may add that they are particularly unlikely to present themselves to a mathematician. It is usual to exaggerate rather grossly the differences between the mental processes of mathematicians and other people, but it is undeniable that a gift for mathematics is one of the most specialized talents, and that mathematicians as a class are not particularly distinguished for general ability or versatility. If a man is in any sense a real mathematician, then it is a hundred to one that his mathematics will be far better than anything else he can do, and that he would be silly if he surrendered any decent opportunity of exercising his one talent in order to do undistinguished work in other fields. Such a sacrifice could be justified only by economic necessity or age.

* * * * *

10

A MATHEMATICIAN, like a painter or a poet, is a maker of patterns. If his patterns are more permanent than theirs, it is because they are made with *ideas*. A painter makes patterns with shapes and colours, a poet with words. A painting may embody an 'idea', but the idea is usually commonplace and unimportant. In poetry, ideas count for a good deal more; but, as Housman insisted, the importance of ideas in poetry is habitually exaggerated: 'I cannot satisfy myself that there are any such things as poetical ideas. . . . Poetry is not the thing said but a way of saying it.'

> Not all the water in the rough rude sea
> Can wash the balm from an anointed King.

Could lines be better, and could ideas be at once more trite and more false? The poverty of the ideas seems hardly to affect the beauty of the verbal pattern. A mathematician, on the other hand, has no material to work with but ideas, and so his patterns are likely to last longer, since ideas wear less with time than words.

The mathematician's patterns, like the painter's or the poet's, must be *beautiful;* the ideas, like the colours or the words, must fit together in a harmonious way. Beauty is the first test: there is no permanent place in the world for ugly mathematics. And here I must deal with a misconception which is still widespread (though probably much less so now than it was twenty years ago), what Whitehead has called the 'literary superstition' that love of and aesthetic appreciation of mathematics is 'a monomania confined to a few eccentrics in each generation.'

It would be difficult now to find an educated man quite insensitive to the aesthetic appeal of mathematics. It may be very hard to *define* mathematical beauty, but that is just as true of beauty of any kind—we may not know quite what we mean by a beautiful poem, but that does not prevent us from recognizing one when we read it. Even Professor Hogben, who is out to minimize at all costs the importance of the aesthetic element in mathematics, does not venture to deny its reality. 'There are, to be sure, individuals for whom mathematics exercises a coldly impersonal attraction. . . .The aesthetic appeal of mathematics may be very real for a chosen few.' But they are 'few', he suggests, and they feel 'coldly' (and are really rather ridiculous people, who live in silly little university towns sheltered from the fresh breezes of the wide open spaces) . In this he is merely echoing Whitehead's 'literary superstition'.

The fact is that there are few more 'popular' subjects than mathematics. Most people have some appreciation of mathematics, just as most people can enjoy a pleasant tune; and there are probably more people really interested in mathematics than in music. Appearances may suggest the contrary, but there are easy explanations. Music can be used to stimulate mass emotion, while mathematics cannot; and musical incapacity is recognized (no doubt rightly) as mildly discreditable, whereas most people are so frightened of the name of mathematics that they are ready, quite unaffectedly, to exaggerate their own mathematical stupidity.

A very little reflection is enough to expose the absurdity of the 'literary superstition'. There are masses of chess-players in every civilized country—in Russia, almost the whole educated population; and every chess-player can recognize and appreciate a 'beautiful' game or problem. Yet a chess problem is *simply* an exercise in pure mathematics (a game not entirely, since psychology also plays a part), and everyone who calls a problem 'beautiful' is applauding mathematical beauty, even if it is beauty of a comparatively lowly kind. Chess problems are the hymn-tunes of mathematics.

We may learn the same lesson, at a lower level but for a wider public, from bridge, or descending further, from the puzzle columns of the popular newspapers. Nearly all their immense popularity is a tribute to the drawing power of rudimentary mathematics, and the better makers of puzzles, such as Dudeney or 'Caliban', use very little else. They know their business; what the public wants is a little intellectual 'kick', and nothing else has quite the kick of mathematics.

I might add that there is nothing in the world which pleases even famous men (and men who have used disparaging language about mathematics) quite so much as to discover, or rediscover, a genuine mathematical theorem. Herbert Spencer republished in his autobiography a theorem about circles which he proved when he was twenty (not knowing that it had been proved over two thousand years before by Plato). Professor Soddy is a more recent and a more striking example (but *his* theorem really is his own).

* * * * *

22

THERE is another misconception against which we must guard. It is quite natural to suppose that there is a great difference in utility between 'pure' and 'applied' mathematics. This is a delusion: there is a sharp distinction between the two kinds of mathematics, which I will explain in a moment, but it hardly affects their utility.

How do pure and applied mathematics differ from one another? This is a question which can be answered definitely and about which there is general agreement among mathematicians. There will be nothing in the least unorthodox about my answer, but it needs a little preface.

My next two sections will have a mildly philosophical flavour. The philosophy will not cut deep, or be in any way vital to my main theses; but I shall use words which are used very frequently with definite philosophical implications, and a reader might well become confused if I did not explain how I shall use them.

I have often used the adjective 'real', and as we use it commonly in conversation. I have spoken of 'real mathematics' and 'real mathematicians', as I might have spoken of 'real poetry' or 'real poets', and I shall continue to do so. But I shall also use the word 'reality', and with two different connotations.

In the first place, I shall speak of 'physical reality', and here again I shall be using the word in the ordinary sense. By physical reality I mean the material world, the world of day and night, earthquakes and eclipses, the world which physical science tries to describe.

I hardly suppose that, up to this point, any reader is likely to find trouble with my language, but now I am near to more difficult ground. For me, and I suppose for most mathematicians, there is another reality, which I will call 'mathematical reality'; and there is no sort of agreement about the nature of mathematical reality among either mathematicians or philosophers. Some hold that it is 'mental' and that in some sense we construct it, others that it is outside and independent of us. A man who could give a convincing account of mathematical reality would have solved very many of the most difficult problems of metaphysics. If he could include physical reality in his account, he would have solved them all.

I should not wish to argue any of these questions here even if I were competent to do so, but I will state my own position dogmatically in order to avoid minor misapprehensions. I believe that mathematical reality lies outside us, that our function is to discover or *observe* it, and that the theorems which we prove, and which we describe grandiloquently as our 'creations', are simply our notes of our observations. This view has been held, in one form or another, by many philosophers of high reputation from Plato onwards, and I shall use the language which is natural to a man who holds it. A reader who does not like the philosophy can alter the language: it will make very little difference to my conclusions.

23

THE contrast between pure and applied mathematics stands out most clearly, perhaps, in geometry. There is the science of pure geometry, in which there are many geometries, projective geometry, Euclidean geometry, non-Euclidean geometry, and so forth. Each of these geometries is a *model,* a pattern of ideas, and is to be judged by the interest and beauty of its particular pattern. It is a *map* or *picture,* the joint product of many hands, a partial and imperfect copy (yet exact so far as it extends) of a section of mathematical reality. But the point which is important to us now is this, that there is one thing at any rate of which pure geometries are *not* pictures, and that is the spatio-temporal reality of the physical world. It is obvious, surely, that they cannot be, since earthquakes and eclipses are not mathematical concepts.

This may sound a little paradoxical to an outsider, but it is a truism to a geometer; and I may perhaps be able to make it clearer by an illustration. Let us suppose that I am giving a lecture on some system of geometry, such as ordinary Euclidean geometry, and that I draw figures on the blackboard to stimulate the imagination of my audience, rough drawings of straight lines or circles or ellipses. It is plain, first, that the truth of the theorems which I prove is in no way affected by the quality of my drawings. Their function is merely to bring home my meaning to my hearers, and, if I can do that, there would be no gain in having them redrawn by the most skilful draughtsman. They are pedagogical illustrations, not part of the real subject-matter of the lecture.

Now let us go a stage further. The room in which I am lecturing is part of the physical world, and has itself a certain pattern. The study of that pattern, and of the general pattern of physical reality, is a science in itself, which we may call 'physical geometry'. Suppose now that a violent dynamo, or a massive gravitating body, is introduced into the room. Then the physicists tell us that the geometry of the room is changed, its whole physical pattern slightly but definitely distorted. Do the theorems which I have proved become false? Surely it would be nonsense to suppose that the proofs of them which I have given are affected in any way. It would be like supposing that a play of Shakespeare is changed when a reader spills his tea over a page. The play is independent of the pages on which it is printed, and 'pure geometries' are independent of lecture rooms, or of any other detail of the physical world.

This is the point of view of a pure mathematician. Applied mathematicians, mathematical physicists, naturally take a different view, since they are preoccupied with the physical world itself, which also has its structure or pattern. We cannot describe this pattern exactly, as we can that of a pure geometry, but we can say something significant about it. We can describe, sometimes fairly accurately, sometimes very roughly, the relations which hold between some of its constituents, and compare them with the exact relations holding between constituents of some system of pure geometry. We may be able to trace a certain resemblance between the two sets of relations, and then the pure geometry will become interesting to physicists; it will give us, to that extent, a map which 'fits the facts' of the physical world. The geometer offers to the physicist a whole set of maps from which to choose. One map, perhaps, will fit the facts better than others, and then the geometry which provides that particular map will be the geometry most important for applied mathematics. I may add that even a pure mathematician may find his appreciation of this geometry quickened, since there is no mathematician so pure that he feels no interest at all in the physical world; but, in so far as he succumbs to this temptation, he will be abandoning his purely mathematical position.

24

THERE is another remark which suggests itself here and which physicists may find paradoxical, though the paradox will probably seem a good deal less than it did eighteen years ago. I will express it in much the same words which I used in 1922 in an address to Section A of the British Association. My audience then was composed almost entirely of physicists, and I may have spoken a little provocatively on that account; but I would still stand by the substance of what I said.

I began by saying that there is probably less difference between the positions of a mathematician and of a physicist than is generally supposed, and that the most important seems to me to be this, that the mathematician is in much more direct contact with reality. This may seem a paradox, since it is the physicist who deals with the subject-matter usually described as 'real'; but a very little reflection is enough to show that the physicist's reality, whatever it may be, has few or none of the attributes which common sense ascribes instinctively to reality. A chair may be a collection of whirling electrons, or an idea in the mind of God: each of these accounts of it may have its merits, but neither conforms at all closely to the suggestions of common sense.

I went on to say that neither physicists nor philosophers have ever given any convincing account of what 'physical reality' is, or of how the physicist passes, from the confused mass of fact or sensation with which he starts, to the construction of the objects which he calls 'real'. Thus we cannot be said to know what the subject-matter of physics is; but this need not prevent us from understanding roughly what a physicist is trying to do. It is plain that he is trying to correlate the incoherent body of crude fact confronting him with some definite and orderly scheme of abstract relations, the kind of scheme which he can borrow only from mathematics.

A mathematician, on the other hand, is working with his own mathematical reality. Of this reality, as I explained in § 22, I take a 'realistic' and not an 'idealistic' view. At any rate (and this was my main point) this realistic view is much more plausible of mathematical than of physical reality, because mathematical objects are so much more what they seem. A chair or a star is not in the least like what it seems to be; the more we think of it, the fuzzier its outlines become in the haze of sensation which surrounds it; but '2' or '317' has nothing to do with sensation, and its properties stand out the more clearly the more closely we scrutinize it. It may be that modern physics fits best into some framework of idealistic philosophy—I do not believe it, but there are eminent physicists who say so. Pure mathematics, on the other hand, seems to me a rock on which all idealism founders: 317 is a prime, not because we think so, or because our minds are shaped in one way rather than another, but *because it is so,* because mathematical reality is built that way.

* * * * *

28

THERE are then two mathematics. There is the real mathematics of the real mathematicians, and there is what I will call the 'trivial' mathematics, for want of a better word. The trivial mathematics may be justified by arguments which would appeal to Hogben, or other writers of his school, but there is no such defence for the real mathematics, which must be justified as art if it can be justified at all. There is nothing in the least paradoxical or unusual in this view, which is that held commonly by mathematicians.

We have still one more question to consider. We have concluded that the trivial mathematics is, on the whole, useful, and that the real mathematics, on the whole, is not; that the trivial mathematics does, and the real mathematics does not, 'do good' in a certain sense; but we have still to ask whether either sort of mathematics does *harm.* It would be paradoxical to suggest that mathematics of any sort does much harm in time of peace, so that we are driven to the consideration of the effects of mathematics on war. It is very difficult to argue such questions at all dispassionately now, and I should have preferred to avoid them; but some sort of discussion seems inevitable. Fortunately, it need not be a long one.

There is one comforting conclusion which is easy for a real mathematician. Real mathematics has no effects on war. No one has yet discovered any warlike purpose to be served by the theory of numbers or relativity, and it seems very unlikely that anyone will do so for many years. It is true that there are branches of applied mathematics, such as ballistics and aerodynamics, which have been developed deliberately for war and demand a quite elaborate technique: it is perhaps hard to call them 'trivial', but none of them has any claim to rank as 'real'. They are indeed repulsively ugly and intolerably dull; even Littlewood could not make ballistics respectable, and if he could not who can? So a real mathematician has his conscience clear; there is nothing to be set against any value his work may have; mathematics is, as I said at Oxford, a 'harmless and innocent' occupation.

The trivial mathematics, on the other hand, has many applications in war. The gunnery experts and aeroplane designers, for example, could not do their work without it. And the general effect of these applications is plain: mathematics facilitates (if not so obviously as physics or chemistry) modern, scientific, 'total' war.

It is not so clear as it might seem that this is to be regretted, since there are two sharply contrasted views about modern scientific war. The first and the most obvious is that the effect of science on war is merely to magnify its horror, both by increasing the sufferings of the minority who have to fight and by extending them to other classes. This is the most natural and the orthodox view. But there is a very different view which seems also quite tenable, and which has been stated with great force by Haldane in *Callinicus*. It can be maintained that modern warfare is *less* horrible than the warfare of pre-scientific times; that bombs are probably more merciful than bayonets; that lachrymatory gas and mustard gas are perhaps the most humane weapons yet devised by military science; and that the orthodox view rests solely on loose-thinking sentimentalism. It may also be urged (though this was not one of Haldane's theses) that the equalization of risks which science was expected to bring would be in the long run salutary; that a civilian's life is not worth more than a soldier's, nor a woman's than a man's; that anything is better than the concentration of savagery on one particular class; and that, in short, the sooner war comes 'all out' the better.

I do not know which of these views is nearer to the truth. It is an urgent and a moving question, but I need not argue it here. It concerns only the 'trivial' mathematics, which it would be Hogben's business to defend rather than mine. The case for his mathematics may be rather more than a little soiled; the case for mine is unaffected.

Indeed, there is more to be said, since there is one purpose at any rate which the real mathematics may serve in war. When the world is mad, a mathematician may find in mathematics an incomparable anodyne. For mathematics is, of all the arts and sciences, the most austere and the most remote, and a mathematician should be of all men the one who can most easily take refuge where, as Bertrand Russell says, 'one at least of our nobler impulses can best escape from the dreary exile of the actual world'. It is a pity that it should be necessary to make one very serious reservation—he must not be too old. Mathematics is not a contemplative but a creative subject; no one can draw much consolation from it when he has lost the power or the desire to create; and that is apt to happen to a mathematician rather soon. It is a pity, but in that case he does not matter a great deal anyhow, and it would be silly to bother about him.

29

I WILL end with a summary of my conclusions, but putting them in a more personal way. I said at the beginning that anyone who defends his subject will find that he is defending himself; and my justification of the life of a professional mathematician is bound to be, at bottom, a justification of my own. Thus this concluding section will be in its substance a fragment of autobiography.

I cannot remember ever having wanted to be anything but a mathematician. I suppose that it was always clear that my specific abilities lay that way, and it never occurred to me to question the verdict of my elders. I do not remember having felt, as a boy, any *passion* for mathematics, and such notions as I may have had of the career of a mathematician were far from noble. I thought of mathematics in terms of examinations and scholarships: I wanted to beat other boys, and this seemed to be the way in which I could do so most decisively.

I was about fifteen when (in a rather odd way) my ambitions took a sharper turn. There is a book by 'Alan St Aubyn' called *A Fellow of Trinity,* one of a series dealing with what is supposed to be Cambridge college life. I suppose that it is a worse book than most of Marie Corelli's; but a book can hardly be entirely bad if it fires a clever boy's imagination. There are two heroes, a primary hero called Flowers, who is almost wholly good, and a secondary hero, a much weaker vessel, called Brown. Flowers and Brown find many dangers in university life, but the worst is a gambling saloon in Chesterton run by the Misses Bellenden, two fascinating but extremely wicked young ladies. Flowers survives all these troubles, is Second Wrangler and Senior Classic, and succeeds automatically to a Fellowship (as I suppose he would have done then). Brown succumbs, ruins his parents, takes to drink, is saved from delirium tremens during a thunderstorm only by the prayers of the Junior Dean, has much difficulty in obtaining even an Ordinary Degree, and ultimately becomes a missionary. The friendship is not shattered by these unhappy events, and Flowers's thoughts stray to Brown, with affectionate pity, as he drinks port and eats walnuts for the first time in Senior Combination Room.

Now Flowers was a decent enough fellow (so far as 'Alan St Aubyn' could draw one), but even my unsophisticated mind refused to accept him as clever. If he could do these things, why not I? In particular, the final scene in Combination Room fascinated me completely, and from that time, until I obtained one, mathematics meant to me primarily a Fellowship of Trinity.

I found at once, when I came to Cambridge, that a Fellowship implied 'original work', but it was a long time before I formed any definite idea of research. I had of course found at school, as every future mathematician does, that I could often do things much better than my teachers; and even at Cambridge I found, though naturally much less frequently, that I could sometimes do things better than the College lecturers. But I was really quite ignorant, even when I took the Tripos, of the subjects on which I have spent the rest of my life; and I still thought of mathematics as essentially a 'competitive' subject. My eyes were first opened by Professor Love, who taught me for a few terms and gave me my first serious conception of analysis. But the great debt which I owe to him—he was, after all, primarily an applied mathematician—was his advice to read Jordan's famous *Cours d'analyse;* and I shall never forget the astonishment with which I read that remarkable work, the first inspiration for so many mathematicians of my generation, and learnt for the first time as I read it what mathematics really meant. From that time onwards I was in my way a real mathematician, with sound mathematical ambitions and a genuine passion for mathematics.

I wrote a great deal during the next ten years, but very little of any importance; there are not more than four or five papers which I can still remember with some satisfaction. The real crises of my career came ten or twelve years later, in 1911, when I began my long collaboration with Littlewood, and in 1913, when I discovered Ramanujan. All my best work since then has been bound up with theirs, and it is obvious that my association with them was the decisive event of my life. I still say to myself when I am depressed, and find myself forced to listen to pompous and tiresome people, 'Well, I have done one thing you could never have done, and that is to have collaborated with both Littlewood and Ramanujan on something like equal terms.' It is to them that I owe an unusually late maturity: I was at my best at a little past forty, when I was a professor at Oxford. Since then I have suffered from that steady deterioration which is the common fate of elderly men and particularly of elderly mathematicians. A mathematician may still be competent enough at sixty, but it is useless to expect him to have original ideas.

It is plain now that my life, for what it is worth, is finished, and that nothing I can do can perceptibly increase or diminish its value. It is very difficult to be dispassionate, but I count it a 'success'; I have had more reward and not less than was due to a man of my particular grade of ability. I have held a series of comfortable and 'dignified' positions. I have had very little trouble with the duller routine of universities. I hate 'teaching', and have had to do very little, such teaching as I have done having been almost entirely supervision of research; I love lecturing, and have lectured a great deal to extremely able classes; and I have always had plenty of leisure for the researches which have been the one great permanent happiness of my life. I have found it easy to work with others, and have collaborated on a large scale with two exceptional mathematicians; and this has enabled me to add to mathematics a good deal more than I could reasonably have expected. I have had my disappointments, like any other mathematician, but none of them has been too serious or has made me particularly unhappy. If I had been offered a life neither better nor worse when I was twenty, I would have accepted without hesitation.

It seems absurd to suppose that I could have 'done better'. I have no linguistic or artistic ability, and very little interest in experimental science. I might have been a tolerable philosopher, but not one of a very original kind. I think that I might have made a good lawyer; but journalism is the only profession, outside academic life, in which I should have felt really confident of my chances. There is no doubt that I was right to be a mathematician, if the criterion is to be what is commonly called success.

My choice was right, then, if what I wanted was a reasonably comfortable and happy life. But solicitors and stockbrokers and bookmakers often lead comfortable and happy lives, and it is very difficult to see how the world is the richer for their existence. Is there any sense in which I can claim that my life has been less futile than theirs? It seems to me again that there is only one possible answer: yes, perhaps, but, if so, for one reason only.

I have never done anything 'useful'. No discovery of mine has made, or is likely to make, directly or indirectly, for good or ill, the least difference to the amenity of the world. I have helped to train other mathematicians, but mathematicians of the same kind as myself, and their work has been, so far at any rate as I have helped them to it, as useless as my own. Judged by all practical standards, the value of my mathematical life is nil; and outside mathematics it is trivial anyhow. I have just one chance of escaping a verdict of complete triviality, that I may be judged to have created something worth creating. And that I have created something is undeniable: the question is about its value.

The case for my life, then, or for that of any one else who has been a mathematician in the same sense in which I have been one, is this: that I have added something to knowledge, and helped others to add more; and that these somethings have a value which differs in degree only, and not in kind, from that of the creations of the great mathematicians, or of any of the other artists, great or small, who have left some kind of memorial behind them.

MATHEMATICS OF A LADY TASTING TEA

RONALD A. FISHER

Educated at Harrow and at Gonville and Caius College, Cambridge, Ronald A. Fisher (1890-1962) began his professional career as statistician to the Mercantile and General Investment Company from 1913 to 1915. Then, after teaching in English public schools (i.e., private schools) for four years, he joined the statistical department of Rothamsted Experimental Station where he remained until 1933, when he became Galton Professor of Eugenics, University College, London. From 1943 to 1957 he was Arthur Balfour Professor of Genetics at the University of Cambridge, and in 1956 he became President of Gonville and Caius College.

He was a recipient of the Darwin Medal of the Royal Society and many other honors in science. He was knighted in 1952.

Fisher was the author of *The Genetical Theory of Natural Selection* that was reprinted twenty-eight years after its publication in 1930, and he authored five other books and numerous articles on statistics or genetics. Five of his books went through five to twelve editions.

1. How would Fisher determine whether or not a lady by tasting a cup of tea was able to discriminate whether the milk or the tea infusion was first added to the cup?

2. What does Fisher consider as "statistically significant" evidence?

3. What is the "null hypothesis" and what is its significance?

4. How may the value of a statistical experiment be improved?

MATHEMATICS OF A LADY TASTING TEA

RONALD A. FISHER

Statement of Experiment

A LADY declares that by tasting a cup of tea made with milk she can discriminate whether the milk or the tea infusion was first added to the cup. We will consider the problem of designing an experiment by means of which this assertion can be tested. For this purpose let us first lay down a simple form of experiment with a view to studying its limitations and its characteristics, both those which appear to be essential to the experimental method, when well developed, and those which are not essential but auxiliary.

Our experiment consists in mixing eight cups of tea, four in one way and four in the other, and presenting them to the subject for judgment in a random order. The subject has been told in advance of what the test will consist, namely that she will be asked to taste eight cups, that these shall be four of each kind, and that they shall be presented to her in a random order, that is in an order not determined arbitrarily by human choice, but by the actual manipulation of the physical apparatus used in games of chance, cards, dice, roulettes, etc., or, more expeditiously, from a published collection of random sampling numbers purporting to give the actual results of such manipulation. Her task is to divide the 8 cups into two sets of 4, agreeing, if possible, with the treatments received.

Interpretation and its Reasoned Basis

In considering the appropriateness of any proposed experimental design, it is always needful to forecast all possible results of the experiment, and to have decided without ambiguity what interpretation shall be placed upon each one of them. Further, we must know by what argument this interpretation is to be sustained. In the present instance we may argue as follows. There are 70 ways of choosing a group of 4 objects out of 8. This may be demonstrated by an argument familiar to students of "permutations and combinations," namely, that if we were to choose the 4 objects in succession we should have successively 8, 7, 6, 5 objects to choose from, and could make our succession of choices in 8 x 7 x 6 x 5, or 1680 ways. But in doing this we have not only chosen every possible set of 4, but every possible set in every possible order; and since 4 objects can be arranged in order in 4 x 3 x 2 x 1, or 24 ways, we may find the number of possible choices by dividing 1680 by 24. The result, 70, is essential to our interpretation of the experiment. At best the subject can judge rightly with every cup and, knowing that 4 are of each kind, this amounts to choosing, out of the 70 sets of 4 which might be chosen, that particular one which is correct. A subject without any faculty of discrimination would in fact divide the 8 cups correctly into two sets of 4 in one trial out of 70, or, more properly, with a frequency which would approach 1 in 70 more and more nearly the more often the test were repeated. Evidently this frequency, with which unfailing success would be achieved by a person lacking altogether the faculty under test, is calculable from the number of cups used. The odds could be made much higher by enlarging the experiment, while, if the experiment were much smaller even the greatest possible success would give odds so low that the result might, with considerable probability, be ascribed to chance.

The Test Of Significance

It is open to the experimenter to be more or less exacting in respect of the smallness of the probability he would require before he would be willing to admit that his observations have demonstrated a positive result. It is obvious that an experiment would be useless of which no possible result would satisfy him. Thus, if he wishes to ignore results having probabilities as high as 1 in 20—the probabilities being of course reckoned from the hypothesis that the phenomenon to be demonstrated is in fact absent—then it would be useless for him to experiment with only 3 cups of tea of each kind. For 3 objects can be chosen out of 6 in only 20 ways, and therefore complete success in the test would be achieved without sensory discrimination, i.e., by "pure chance," in an average of 5 trials out of 100. It is usual and convenient for experimenters to take 5 per cent. as a standard level of significance, in the sense that they are prepared to ignore all results which fail to reach this standard, and, by this means, to eliminate from further discussion the greater part of the fluctuations which chance causes have introduced into their experimental results. No such selection can eliminate the whole of the possible effects of chance coincidence, and if we accept this convenient convention, and agree that an event which would occur by chance only once in 70 trials is decidedly "significant," in the statistical sense, we thereby admit that no isolated experiment, however significant in itself, can suffice for the experimental demonstration of any natural phenomenon; for the "one chance in a million" will undoubtedly occur, with no less and no more than its appropriate frequency, however surprised we may be that it should occur to us. In order to assert that a natural phenomenon is experimentally demonstrable we need, not an isolated record, but a reliable method of procedure. In relation to the test of significance, we may say that a phenomenon is experimentally demonstrable when we know how to conduct an experiment which will rarely fail to give us a statistically significant result.

Returning to the possible results of the psycho-physical experiment, having decided that if every cup were rightly classified a significant positive result would be recorded, or, in other words, that we should admit that the lady had made good her claim, what should be our conclusion if, for each kind of cup, her judgments are 3 right and 1 wrong? We may take it, in the present discussion, that any error in one set of judgments will be compensated by an error in the other, since it is known to the subject that there are 4 cups of each kind. In enumerating the number of ways of choosing 4 things out of 8, such that 3 are right and 1 wrong, we may note that the 3 right may be chosen, out of the 4 available, in 4 ways and, independently of this choice, that the 1 wrong may be chosen, out of the 4 available, also in 4 ways. So that in all we could make a selection of the kind supposed in 16 different ways. A similar argument shows that, in each kind of judgment, 2 may be right and 2 wrong in 36 ways, 1 right and 3 wrong in 16 ways and none right and 4 wrong in 1 way only. It should be noted that the frequencies of these five possible results of the experiment make up together, as it is obvious they should, the 70 cases out of 70.

It is obvious, too, that 3 successes to 1 failure, although showing a bias, or deviation, in the right direction, could not be judged as statistically significant evidence of a real sensory discrimination. For its frequency of chance occurrence is 16 in 70, or more than 20 per cent. Moreover, it is not the best possible result, and in judging of its significance we must take account not only of its own frequency, but also of the frequency for any better result. In the present instance "3 right and 1 wrong" occurs 16 times, and "4 right" occurs once in 70 trials, making 17 cases out of 70 as good as or better than that observed. The reason for including cases better than that observed becomes obvious on considering what our conclusions would have been had the case of 3 right and 1 wrong only 1 chance, and the case of 4 right 16 chances of occurrence out of 70. The rare case of 3 right and 1 wrong could not be judged significant merely because it was rare, seeing that a higher degree of success would frequently have been scored by mere chance.

The Null Hypothesis

Our examination of the possible results of the experiment has therefore led us to a statistical test of significance, by which these results are divided into two classes with opposed interpretations. Tests of significance are of many different kinds, which need not be considered here. Here we are only concerned with the fact that the easy calculation in permutations which we encountered, and which gave us our test of significance, stands for something present in every possible experimental arrangement; or, at least, for something required in its interpretation. The two classes of results which are distinguished by our test of significance are, on the one hand, those which show a significant discrepancy from a certain hypothesis; namely, in this case, the hypothesis that the judgments given are in no way influenced by the order in which the ingredients have been added; and on the other hand, results which show no significant discrepancy from this hypothesis. This hypothesis, which may or may not be impugned by the result of an experiment, is again characteristic of all experimentation. Much confusion would often be avoided if it were explicitly formulated when the experiment is designed. In relation to any experiment we may speak of this hypothesis as the "null hypothesis," and it should be noted that the null hypothesis is never proved or established, but is possibly disproved, in the course of experimentation. Every experiment may be said to exist only in order to give the facts a chance of disproving the null hypothesis.

It might be argued that if an experiment can disprove the hypothesis that the subject possesses no sensory discrimination between two different sorts of object, it must therefore be able to prove the opposite hypothesis, that she can make some such discrimination. But this last hypothesis, however reasonable or true it may be, is ineligible, as a null hypothesis to be tested by experiment, because it is inexact. If it were asserted that the subject would never be wrong in her judgments we should again have an exact hypothesis, and it is easy to see that this hypothesis could be disproved by a single failure, but could never be proved by any finite amount of experimentation. It is evident that the null hypothesis must be exact, that is free from vagueness and ambiguity, because it must supply the basis of the "problem of distribution," of which the test of significance is the solution. A null hypothesis may, indeed, contain arbitrary elements, and in more complicated cases often does so: as, for example, if it should assert that the death-rates of two groups of animals are equal, without specifying what these death-rates usually are. In such cases it is evidently the equality rather than any particular values of the death-rates that the experiment is designed to test, and possibly to disprove.

In cases involving statistical "estimation" these ideas may be extended to the simultaneous consideration of a series of hypothetical possibilities. The notion of an error of the so-called "second kind," due to accepting the null hypothesis "when it is false" may then be given a meaning in reference to the quantity to be estimated. It has no meaning with respect to simple tests of significance, in which the only available expectations are those which flow from the null hypothesis being true.

Randomisation; the Physical Basis of the Validity of the Test

We have spoken of the experiment as testing a certain null hypothesis, namely, in this case, that the subject possesses no sensory discrimination whatever of the kind claimed; we have, too, assigned as appropriate to this hypothesis a certain frequency distribution of occurrences, based on the equal frequency of the 70 possible ways of assigning 8 objects to two classes of 4 each; in other words, the frequency distribution appropriate to a classification by pure chance. We have now to examine the physical conditions of the experimental technique needed to justify the assumption that, if discrimination of the kind under test is absent, the result of the experiment will be wholly governed by the laws of chance. It is easy to see that it might well be otherwise. If all those cups made with the milk first had sugar added, while those made with the tea first had none, a very obvious difference in flavour would have been introduced which might well ensure that all those made with sugar should be classed alike. These groups might either be classified all right or all wrong, but in such a case the frequency of the critical event in which all cups are classified correctly would not be 1 in 70, but 35 in 70 trials, and the test of significance would be wholly vitiated. Errors equivalent in principle to this are very frequently incorporated in otherwise well-designed experiments.

It is no sufficient remedy to insist that "all the cups must be exactly alike" in every respect except that to be tested. For this is a totally impossible requirement in our example, and equally in all other forms of experimentation. In practice it is probable that the cups will differ perceptibly in the thickness or smoothness of their material, that the quantities of milk added to the different cups will not be exactly equal, that the strength of the infusion of tea may change between pouring the first and the last cup, and that the temperature also at which the tea is tasted will change during the course of the experiment. These are only examples of the differences probably present; it would be impossible to present an exhaustive list of such possible differences appropriate to any one kind of experiment, because the uncontrolled causes which may influence the result are always strictly innumerable. When any such cause is named, it is usually perceived that, by increased labour and expense, it could be largely eliminated. Too frequently it is assumed that such refinements constitute improvements to the experiment. Our view, which will be much more fully exemplified in later sections, is that it is an essential characteristic of experimentation that it is carried out with limited resources, and an essential part of the subject of experimental design to ascertain how these should be best applied; or, in particular, to which causes of disturbance care should be given, and which *ought* to be deliberately ignored. To ascertain, too, for those which are not to be ignored, to what *extent* it is worth while to take the trouble to diminish their magnitude. For our present purpose, however, it is only necessary to recognise that, whatever degree of care and experimental skill is expended in equalising the conditions, other than the one under test, which are liable to affect the result, this equalisation must always be to a greater or less extent incomplete, and in many important practical cases will certainly be grossly defective. We are concerned, therefore, that this inequality, whether it be great or small, shall not impugn the exactitude of the frequency distribution, on the basis of which the result of the experiment is to be appraised.

The Effectiveness of Randomisation

The element in the experimental procedure which contains the essential safeguard is that the two modifications of the test beverage are to be prepared "in random order." This, in fact, is the only point in the experimental procedure in which the laws of chance, which are to be in exclusive control of our frequency distribution, have been explicitly introduced. The phrase "random order" itself, however, must be regarded as an incomplete instruction, standing as a kind of shorthand symbol for the full procedure of randomisation, by which the validity of the test of significance may be guaranteed against corruption by the causes of disturbance which have not been eliminated. To demonstrate that, with satisfactory randomisation, its validity is, indeed, wholly unimpaired, let us imagine all causes of disturbance—the strength of the infusion, the quantity of milk, the temperature at which it is tasted, etc.—to be predetermined for each cup; then since these, on the null hypothesis, are the only causes influencing classification, we may say that the probabilities of each of the 70 possible choices or classifications which the subject can make are also predetermined. If, now, after the disturbing causes are fixed, we assign, strictly at random, 4 out of the 8 cups to each of our experimental treatments, then every set of 4, whatever its probability of being so classified, will certainly have a probability of exactly 1 in 70 of *being* the 4, for example, to which the milk is added first. However important the causes of disturbance may be, even if they were to make it certain that one particular set of 4 should receive this classification. the probability that the 4 so classified and the 4 which ought to have been so classified should be the same, must be rigorously in accordance with our test of significance.

It is apparent, therefore, that the random choice of the objects to be treated in different ways would be a complete guarantee of the validity of the test of significance, if these treatments were the last in time of the stages in the physical history of the objects which might affect their experimental reaction. The circumstance that the experimental treatments cannot always be applied last, and may come relatively early in their history, causes no practical inconvenience; for subsequent causes of differentiation, if under the experimenter's control, as, for example, the choice of different pipettes to be used with different flasks, can either be predetermined before the treatments have been randomised, or, if this has not been done, can be randomised on their own account; and other causes of differentiation will be either (a) consequences of differences already randomised, or (b) natural consequences of the difference in treatment to be tested, of which on the null hypothesis there will be none, by definition, or (c) effects supervening by chance independently from the treatments applied. Apart, therefore, from the avoidable error of the experimenter himself introducing with his test treatments, or subsequently, other differences in treatment, the effects of which the experiment is not intended to study, it may be said that the simple precaution of randomisation will suffice to guarantee the validity of the test of significance, by which the result of the experiment is to be judged.

The Sensitiveness of an Experiment. Effects of Enlargement and Repetition

A probable objection, which the subject might well make to the experiment so far described, is that only if every cup is classified correctly will she be judged successful. A single mistake will reduce her performance below the level of significance. Her claim, however, might be, not that she could draw the distinction with invariable certainty, but that, though sometimes mistaken, she would be right more often than not; and that the experiment should be enlarged sufficiently, or repeated sufficiently often, for her to be able to demonstrate the predominance of correct classifications in spite of occasional errors.

An extension of the calculation upon which the test of significance was based shows that an experiment with 12 cups, six of each kind, gives, on the null hypothesis, 1 chance in 924 for complete success, and 36 chances for 5 of each kind classified right and 1 wrong. As 37 is less than a twentieth of 924, such a test could be counted as significant, although a pair of cups have been wrongly classified; and it is easy to verify that, using larger numbers still, a significant result could be obtained with a still higher proportion of errors. By increasing the size of the experiment, we can render it more sensitive, meaning by this that it will allow of the detection of a lower degree of sensory discrimination, or, in other words, of a quantitatively smaller departure from the null hypothesis. Since in every case the experiment is capable of disproving, but never of proving this hypothesis, we may say that the value of the experiment is increased whenever it permits the null hypothesis to be more readily disproved.

The same result could be achieved by repeating the experiment, as originally designed, upon a number of different occasions, counting as a success all those occasions on which 8 cups are correctly classified. The chance of success on each occasion being 1 in 70, a simple application of the theory of probability shows that 2 or more successes in 10 trials would occur, by chance, with a frequency below the standard chosen for testing significance; so that the sensory discrimination would be demonstrated, although, in 8 attempts out of 10, the subject made one or more mistakes. This procedure may be regarded as merely a second way of enlarging the experiment and, thereby, increasing its sensitiveness, since in our final calculation we take account of the aggregate of the entire series of results, whether successful or unsuccessful. It would clearly be illegitimate, and would rob our calculation of its basis, if the unsuccessful results were not all brought into the account.

Qualitative Methods of Increasing Sensitiveness

Instead of enlarging the experiment we may attempt to increase its sensitiveness by qualitative improvements; and these are, generally speaking, of two kinds: (a) the reorganisation of its structure, and (b) refinements of technique. To illustrate a change of structure we might consider that, instead of fixing in advance that 4 cups should be of each kind, determining by a random process how the subdivision should be effected, we might have allowed the treatment of each cup to be determined independently by chance, as by the toss of a coin, so that each treatment has an equal chance of being chosen. The chance of classifying correctly 8 cups randomised in this way, without the aid of sensory discrimination, is 1 in 2^8, or 1 in 256 chances, and there are only 8 chances of classifying 7 right and 1 wrong; consequently the sensitiveness of the experiment has been increased, while still using only 8 cups, and it is possible to score a significant success, even if one is classified wrongly. In many types of experiment, therefore, the suggested change in structure would be evidently advantageous. For the special requirements of a psycho-physical experiment, however, we should probably prefer to forego this advantage, since it would occasionally occur that all the cups would be treated alike, and this, besides bewildering the subject by an unexpected occurrence, would deny her the real advantage of judging by comparison.

Another possible alteration to the structure of the experiment, which would, however, decrease its sensitiveness, would be to present determined, but unequal, numbers of the two treatments. Thus we might arrange that 5 cups should be of the one kind and 3 of the other, choosing them properly by chance, and informing the subject how many of each to expect. But since the number of ways of choosing 3 things out of 8 is only 56, there is now, on the null hypothesis, a probability of a completely correct classification of 1 in 56. It appears in fact that we cannot by these means do better than by presenting the two treatments in equal numbers, and the choice of this equality is now seen to be justified by its giving to the experiment its maximal sensitiveness.

With respect to the refinements of technique, we have seen above that these contribute nothing to the validity of the experiment, and of the test of significance by which we determine its result. They may, however, be important, and even essential, in permitting the phenomenon under test to manifest itself. Though the test of significance remains valid, it may be that without special precautions even a definite sensory discrimination would have little chance of scoring a significant success. If some cups were made with India and some with China tea, even though the treatments were properly randomised, the subject might not be able to discriminate the relatively small difference in flavour under investigation, when it was confused with the greater differences between leaves of different origin. Obviously, a similar difficulty could be introduced by using in some cups raw milk and in others boiled, or even condensed milk, or by adding sugar in unequal quantities. The subject has a right to claim, and it is in the interests of the sensitiveness of the experiment, that gross differences of these kinds should be excluded, and that the cups should, not as far as *possible*, but as far as is practically convenient, be made alike in all respects except that under test.

How far such experimental refinements should be carried is entirely a matter of judgment, based on experience. The validity of the experiment is not affected by them. Their sole purpose is to increase its sensitiveness, and this object can usually be achieved in many other ways, and particularly by increasing the size of the experiment. If, therefore, it is decided that the sensitiveness of the experiment should be increased, the experimenter has the choice between different methods of obtaining equivalent results; and will be wise to choose whichever method is easiest to him, irrespective of the fact that previous experimenters may have tried, and recommended as very important, or even essential, various ingenious and troublesome precautions.

Beauty is the first test; there is no permanent place in the world for ugly mathematics.

G. H. Hardy, *A Mathematician's Apology*, 1941

The outcome of any serious research can only be to make two questions where there had been one before.

Thorstein Veblen, *The Place of Science in Modern Civilization and Other Essays*, 1919

The universe is not only queerer than we imagine, it is queerer than we *can* imagine.

J. B. S. Haldane

Indeed we find ourselves here on the very path taken by Einstein of adapting our modes of perception borrowed from the sensations to the gradually deeping knowledge of the laws of Nature. the hindrances met on this path originate above all in the fact that, so to say, every word in the language refers to our ordinary perception.

Neils Bohr, 1928

I remember once going to see him [Srinivasa Ramanujan] when he was lying ill at Putney. I had ridden in taxi-cab No. 1729 and remarked that the number seemed to me to be a rather dull one, and that I hoped it was not an unfavourable omen. "No," he replied, "it is a very interesting number; it is the smallest number expressible as the sum of two cubes in two different ways."

G. H. Hardy, *Ramanujan*, 1940

I have never done anything "useful." No discovery of mine has made, or is likely to make, directly or indirectly, for good or ill, the least difference to the amenity of the world.

G. H. Hardy, *A Mathematician's Apology*, 1941

SCIENCE AND RELIGION

Galileo Galilei

Albert Einstein

What terrible questions we are learning to ask! The former men believed in magic, by which temples, cities and men were swallowed up, and all trace of them gone. We are coming on the secret of a magic which sweeps out of men's minds all vestige of theism and beliefs which they and their fathers held and were framed upon.

Ralph Waldo Emerson, *The Conduct of Life*, 1860

The God whom science recognizes must be a God of universal laws exclusively, a God who does a wholesale, not a retail business. He cannot accomodate his processes to the convenience of indificuals.

William James, *The Varieties of Religious Experience*, 1902

Religion is a way of life and an attitude to the universe. It brings man in closer touch with the inner nature of reality. Statements of fact made in its name are untrue in detail, but often contain some truth at their core. Science is also a way of life and a attitude to the universe. It is concerned with everything but the nature of reality. Statements of fact made in its name are generally right in detail, but can only reveal the form and not the real nature of existence. The wise man regulates his conduct by the theories both of religion and science. But he regards these theories not as statements of ultimate fact, but as art forms.

J. B. S. Haldane, *Possible Worlds*, 1927

LETTER TO THE GRAND DUCHESS CHRISTINA

GALILEO GALILEI

For a biographical sketch of Galileo Galilei (1564-1642), see page 41. The *Letter to the Grand Duchess Christina* carried little weight at Galileo's trial for heresy in 1633, though it did impress those who read it at the time, and it had a much greater impact in the years to come.

Galileo's *Dialogue Concerning the Two Chief World Systems* remained on the Catholic Index of Prohibited Books for two hundred years. It was dropped from the list in 1875.

Then in 1893, in the encyclical *Providentissimus Deus*, Pope Leo XIII expressed a view of the relationship between biblical interpretation and scientific investigation that closely corresponded to the arguments of Galileo in his *Letter to the Grand Duchess Christina.*

Finally, in 1979, Pope John Paul II declared that the Church might have been mistaken in condemning Galileo, and he asked a church commission to restudy the case. Four years later the commission reported that Galileo should not have been condemned, but it would not go so far as to say that the conviction should be overturned. But on its recommendation the Pontifical Academy of Sciences published the proceedings of the Inquisition relating to Galileo's trial.

Note: In 1950, Pope Pius XII said that the Darwinian theory of evolution was a serious hypothesis worthy of study. In 1996, Pope John Paul II, in a letter to the Pontifical Academy of Sciences, declared that evolution is "more than just a theory," and that a "convergence" of scientific evidence in the last fifty years makes "a significant argument in favor of this theory." He saw no incompatibility between evolution and the Christian faith.

1. Do you see a fundamental conflict between science and religion?

2. What were Galileo's arguments on the relationship between religion and scientific inquiry?

3. Do you think the Index of Forbidden Books serves a useful purpose? Why do you think Galileo's *Dialogue* remained on the list so long?

4. Why did Galileo's views become acceptable to religious leaders in the late nineteenth and twentieth centuries when they were not acceptable in the seventeenth century?

5. Has resistance to new scientific ideas had any lasting effect? If so, in what way?

LETTER TO THE GRAND DUCHESS CHRISTINA

GALILEO GALILEI

LETTER
TO
MADAME CHRISTINA OF LORRAINE

GRAND DUCHESS OF TUSCANY

GALILEO GALILEI
TO
THE MOST SERENE
GRAND DUCHESS MOTHER:

Some years ago, as Your Serene Highness well knows, I discovered in the heavens many things that had not been seen before our own age. The novelty of these things, as well as some consequences which followed from them in contradiction to the physical notions commonly held among academic philosophers, stirred up against me no small number of professors—as if I had placed these things in the sky with my own hands in order to upset nature and overturn the sciences. They seemed to forget that the increase of known truths stimulates the investigation, establishment, and growth of the arts; not their diminution or destruction.

Showing a greater fondness for their own opinions than for truth, they sought to deny and disprove the new things which, if they had cared to look for themselves, their own senses would have demonstrated to them. To this end they hurled various charges and published numerous writings filled with vain arguments, and they made the grave mistake of sprinkling these with passages taken from places in the Bible which they had failed to understand properly, and which were ill suited to their purposes.

These men would perhaps not have fallen into such error had they but paid attention to a most useful doctrine of St. Augustine's, relative to our making positive statements about things which are obscure and hard to understand by means of reason alone. Speaking of a certain physical conclusion about the heavenly bodies, he wrote: "Now keeping always our respect for moderation in grave piety, we ought not to believe anything inadvisedly on a dubious point, lest in favor to our error we conceive a prejudice against .something that truth hereafter may reveal to be not contrary in any way to the sacred books of either the Old or the New Testament."

Well, the passage of time has revealed to everyone the truths that I previously set forth; and, together with the truth of the facts, there has come to light the great difference in attitude between those who simply and dispassionately refused to admit the discoveries to be true, and those who combined with their incredulity some reckless passion of their own. Men who were well grounded in astronomical and physical science were persuaded as soon as they received my first message. There were others who denied them or remained in doubt only because of their novel and unexpected character, and because they had not yet had the opportunity to see for themselves. These men have by degrees come to be satisfied. But some, besides allegiance to their original error, possess I know not what fanciful interest in remaining hostile not so much toward the things in question as toward their discoverer. No longer being able to deny them, these men now take refuge in obstinate silence, but being more than ever exasperated by that which has pacified and quieted other men, they divert their thoughts to other fancies and seek new ways to damage me.

I should pay no more attention to them than to those who previously contradicted me—at whom I always laugh, being assured of the eventual outcome—were it not that in their new calumnies and persecutions I perceive that they do not stop at proving themselves more learned than I am (a claim which I scarcely contest), but go so far as to cast against me imputations of crimes which must be, and are, more abhorrent to me than death itself. I cannot remain satisfied merely to know that the injustice of this is recognized by those who are acquainted with these men and with me, as perhaps it is not known to others.

Persisting in their original resolve to destroy me and everything mine by any mean they can think of, these men are aware of my views in astronomy and philosophy. They know that as to the arrangement of the parts of the universe, I hold the sun to be situated motionless in the center of the revolution of the celestial orbs while the earth rotates on its axis and revolves about the sun. They know also that I support this position not only by refuting the arguments of Ptolemy and Aristotle, but by producing many counterarguments; in particular, some which relate to physical effects whose causes can perhaps be assigned in no other way. In addition there are astronomical arguments derived from many things in my new celestial discoveries that plainly confute the Ptolemaic system while admirably agreeing with and confirming the contrary hypothesis. Possibly because they are disturbed by the known truth of other propositions of mine which differ from those commonly held, and therefore mistrusting their defense so long as they confine themselves to the field of philosophy, these men have resolved to fabricate a shield for their fallacies out of the mantle of pretended religion and the authority of the Bible. These they apply, with little judgment, to the refutation of arguments that they do not understand and have not even listened to.

First they have endeavored to spread the opinion that such propositions in general are contrary to the Bible and are consequently damnable and heretical. They know that it is human nature to take up causes whereby a man may oppress his neighbor, no matter how unjustly, rather than those from which a man may receive some just encouragement. Hence they have had no trouble in finding men who would preach the damnability and heresy of the new doctrine from their very pulpits with unwonted confidence, thus doing impious and inconsiderate injury not only to that doctrine and its followers but to all mathematics and mathematicians in general. Next, becoming bolder, and hoping (though vainly) that this seed which first took root in their hypocritical minds would send out branches and ascend to heaven, they began scattering rumors among the people that before long this doctrine would be condemned by the supreme authority. They know, too, that official condemnation would not only suppress the two propositions which I have mentioned, but would render damnable all other astronomical and physical statements and observations that have any necessary relation or connection with these.

In order to facilitate their designs, they seek so far as possible (at least among the common people) to make this opinion seem new and to belong to me alone. They pretend not to know that its author, or rattler its restorer and confirmer, was Nicholas Copernicus; and that he was not only a Catholic, but a priest and a canon. He was in fact so esteemed by the church that when the Lateran Council under Leo X took up the correction of the church calendar, Copernicus was called to Rome from the most remote parts of Germany to undertake its reform. At that time the calendar was defective because the true measures of the year and the lunar month were not exactly known. The Bishop of Culm, then superintendent of this matter, assigned Copernicus to seek more light and greater certainty concerning the celestial motions by means of constant study and labor. With Herculean toil he set his admirable mind to this task, and he made such great progress in this science and brought our knowledge of the heavenly motions to such precision that he became celebrated as an astronomer. Since that time not only has the calendar been regulated by his teachings, but tables of all the motions of the planets have been calculated as well.

Having reduced his system into six books, he published these at the instance of the Cardinal of Capua and the Bishop of Culm. And since he had assumed his laborious enterprise by order of the supreme pontiff, he dedicated this book *On the celestial revolutions* to Pope Paul III. When printed, the book was accepted by the holy Church, and it has been read and studied by everyone without the faintest hint of ally objection ever being conceived against its doctrines. Yet now that manifest experiences and necessary proofs have shown them to be well grounded, persons exist who would strip the author of his reward without so much as looking at his book, and add the shame of having him pronounced a heretic. All this they would do merely to satisfy their personal displeasure conceived without any cause against another man, who has no interest in Copernicus beyond approving his teachings.

Now as to the false aspersions which they so unjustly seek to cast upon me, I have thought it necessary to justify myself in the eyes of all men, whose judgment in matters of religion and of reputation I must hold in great esteem. I shall therefore discourse of the particulars which these men produce to make this opinion detested and to have it condemned not merely as false but as heretical. To this end they make a shield of their hypocritical zeal for religion. They go about invoking the Bible, which they would have minister to their deceitful purposes. Contrary to the sense of the Bible and the intention of the holy Fathers, if I am not mistaken, they would extend such authorities until even in purely physical matters—where faith is not involved—they would have us altogether abandon reason and the evidence of our senses in favor of some biblical passage, though under the surface meaning of its words this passage may contain a different sense.

I hope to show that I proceed with much greater piety than they do, when I argue not against condemning this book, but against condemning it in the way they suggest—that is, without understanding it, weighing it, or so much as reading it. For Copernicus never discusses matters of religion or faith, nor does he use arguments that depend in any way upon the authority of sacred writings which he might have interpreted erroneously. He stands always upon physical conclusions pertaining to the celestial motions, and deals with them by astronomical and geometrical demonstrations, founded primarily upon sense experiences and very exact observations. He did not ignore the Bible, but he knew very well that if his doctrine were proved, then it could not contradict the Scriptures when they were rightly understood. And thus at the end of his letter of dedication, addressing the pope, he said:

"If there should chance to be any exegetes ignorant of mathematics who pretend to skill in that discipline, and dare to condemn and censure this hypothesis of mine upon the authority of some scriptural passage twisted to their purpose, I value them not, but disdain their unconsidered judgment. For it is known that Lactantius—a poor mathematician though in other respects a worthy author—writes very childishly about the shape of the earth when he scoffs at those who affirm it to be a globe. Hence it should not seem strange to the ingenious if people of that sort should in turn deride me. But mathematics is written for mathematicians, by whom, if I am not deceived, these labors of mine will be recognized as contributing something to their domain, as also to that of the Church over which Your Holiness now reigns."

Such are the people who labor to persuade us that an author like Copernicus may be condemned without being read, and who produce various authorities from the Bible, from theologians, and from Church Councils to make us believe that this is not only lawful but commendable. Since I hold these to be of supreme authority, I consider it rank temerity for anyone to contradict them—when employed according to the usage of the holy Church. Yet I do not believe it is wrong to speak out when there is reason to suspect that other men wish, for some personal motive, to produce and employ such authorities for purposes quite different from the sacred intention of the holy Church.

Therefore I declare (and my sincerity will make itself manifest) not only that I mean to submit myself freely and renounce any errors into which I may fall in this discourse through ignorance of matters pertaining to religion, but that I do not desire in these matters to engage in disputes with anyone, even on points that are disputable. My goal is this alone; that if, among errors that may abound in these considerations of a subject remote from my profession, there is anything that may be serviceable to the holy Church in making a decision concerning the Copernican system, it may be taken and utilized as seems best to the superiors. And if not, let my book be torn and burnt, as I neither intend nor pretend to gain from it any fruit that is not pious and Catholic. And though many of the things I shall reprove have been heard by my own ears, I shall freely grant to those who have spoken them that they never said them, if that is what they wish, and I shall confess myself to have been mistaken. Hence let whatever I reply be addressed not to them, but to whoever may have held such opinions.

The reason produced for condemning the opinion that the earth moves and the sun stands still is that in many places in the Bible one may read that the sun moves and the earth stands still. Since the Bible cannot err, it follows as a necessary consequence that anyone takes an erroneous and heretical position who maintains that the sun is inherently motionless and the earth movable.

With regard to this argument, I think in the first place that it is very pious to say and prudent to affirm that the holy Bible can never speak untruth—whenever its true meaning is understood. But I believe nobody will deny that it is often very abstruse, and may say things which are quite different from what its bare words signify. Hence in expounding the Bible if one were always to confine oneself to the unadorned grammatical meaning, one might fall into error. Not only contradictions and propositions far from true might thus be made to appear in the Bible, but even grave heresies and follies. Thus it would be necessary to assign to God feet, hands, and eyes, as well as corporeal and human affections, such as anger, repentance, hatred, and sometimes even the forgetting of things past and ignorance of those to come. These propositions uttered by the Holy Ghost were set down in that manner by the sacred scribes in order to accommodate them to the capacities of the common people, who are rude and unlearned. For the sake of those who deserve to be separated from the herd, it is necessary that wise expositors should produce the true senses of such passages, together with the special reasons for which they were set down in these words. This doctrine is so widespread and so definite with all theologians that it would be superfluous to adduce evidence for it.

Hence I think that I may reasonably conclude that whenever the Bible has occasion to speak of any physical conclusion (especially those which are very abstruse and hard to understand), the rule has been observed of avoiding confusion in the minds of the common people which would render them contumacious toward the higher mysteries. Now the Bible, merely to condescend to popular capacity, has not hesitated to obscure some very important pronouncements, attributing to God himself some qualities extremely remote from (and even contrary to) His essence. Who, then, would positively declare that this principle has been set aside, and the Bible has confined itself rigorously to the bare and restricted sense of its words, when speaking but casually of the earth, of water, of the sun, or of any other created thing? Especially in view of the fact that these things in no way concern the primary purpose of the sacred writings, which is the service of God and the salvation of souls—matters infinitely beyond the comprehension of the common people.

This being granted, I think that in discussions of physical problems we ought to begin not from the authority of scriptural passages, but from sense-experiences and necessary demonstrations; for the holy Bible and the phenomena of nature proceed alike from the divine Word, the former as the dictate of the Holy Ghost and the latter as the observant executrix of God's commands. It is necessary for the Bible, in order to be accommodated to the understanding of every man, to speak many things which appear to differ from the absolute truth so far as the bare meaning of the words is concerned. But Nature, on the other hand, is inexorable and immutable; she never transgresses the laws imposed upon her, or cares a whit whether her abstruse reasons and methods of operation are understandable to men. For that reason it appears that nothing physical which sense-experience sets before our eyes, or which necessary demonstrations prove to us, ought to be called in question (much less condemned) upon the testimony of biblical passages which may have some different meaning beneath their words. For the Bible is not chained in every expression to conditions as strict as those which govern all physical effects; nor is God any less excellently revealed in Nature's actions than in the sacred statements of the Bible. Perhaps this is what Tertullian meant by these words:

"We conclude that God is known first through Nature, and then again, more particularly, by doctrine; by Nature in His works, and by doctrine in His revealed word."

From this I do not mean to infer that we need not have an extraordinary esteem for the passages of holy Scripture. On the contrary, having arrived at any certainties in physics, we ought to utilize these as the most appropriate aids in the true exposition of the Bible and in the investigation of those meanings which are necessarily contained therein, for these must be concordant with demonstrated truths. I should judge that the authority of the Bible was designed to persuade men of those articles and propositions which, surpassing all human reasoning, could not be made credible by science, or by any other means than through the very mouth of the Holy Spirit.

Yet even in those propositions which are not matters of faith, this authority ought to be preferred over that of all human writings which are supported only by bare assertions or probable arguments, and not set forth in a demonstrative way. This I hold to be necessary and proper to the same extent that divine wisdom surpasses all human judgment and conjecture.

But I do not feel obliged to believe that that same God who has endowed us with senses, reason, and intellect has intended to forgo their use and by some other means to give us knowledge which we can attain by them. He would not require us to deny sense and reason in physical matters which are set before our eyes and minds by direct experience or necessary demonstrations. This must be especially true in those sciences of which but the faintest trace (and that consisting of conclusions) is to be found in the Bible. Of astronomy, for instance, so little is found that none of the planets except Venus are so much as mentioned, and this only once or twice under the name of "Lucifer." If the sacred scribes had had any intention of teaching people certain arrangements and motions of the heavenly bodies, or had they wished us to derive such knowledge from the Bible, then in my opinion they would not have spoken of these matters so sparingly in comparison with the infinite number of admirable conclusions which are demonstrated in that science. Far from pretending to teach us the constitution and motions of the heavens and the stars, with their shapes, magnitudes, and distances, the authors of the Bible intentionally forbore to speak of these things, though all were quite well known to them. Such is the opinion of the holiest and most learned Fathers, and in St. Augustine we find the following words:

"It is likewise commonly asked what we may believe about the form and shape of the heavens according to the Scriptures, for many contend much about these matters. But with superior prudence our authors have forborne to speak of this, as in no way furthering the student with respect to a blessed life—and, more important still, as taking up much of that time which should be spent in holy exercises. What is it to me whether heaven, like a sphere, surrounds the earth on all sides as a mass balanced in the center of the universe, or whether like a dish it merely covers and overcasts the earth? Belief in Scripture is urged rather for the reason we have often mentioned; that is, in order that no one, through ignorance of divine passages, finding anything in our Bibles or hearing anything cited from them of such a nature as may seem to oppose manifest conclusions, should be induced to suspect their truth when they teach, relate, and deliver more profitable matters. Hence let it be said briefly, touching the form of heaven, that our authors knew the truth but the Holy Spirit did not desire that men should learn things that are useful to no one for salvation."

The same disregard of these sacred authors toward beliefs about the phenomena of the celestial bodies is repeated to us by St. Augustine in his next chapter. On the question whether we are to believe that the heaven moves or stands still, he writes thus:

"Some of the brethren raise a question concerning the motion of heaven, whether it is fixed or moved. If it is moved, they say, how is it a firmament? If it stands still, how do these stars which are held fixed in it go round from east to west, the more northerly performing shorter circuits near the pole, so that heaven (if there is another pole unknown to us) may seem to revolve upon some axis, or (if there is no other pole) may be thought to move as a discus? To these men I reply that it would require many subtle and profound reasonings to find out which of these things is actually so; but to undertake this and discuss it is consistent neither with my leisure nor with the duty of those whom I desire to instruct in essential matters more directly conducing to their salvation and to the benefit of the holy Church."

From these things it follows as a necessary consequence that, since the Holy Ghost did not intend to teach us whether heaven moves or stands still, whether its shape is spherical or like a discus or extended in a plane, nor whether the earth is located at its center or off to one side, then so much the less was it intended to settle for us any other conclusion of the same kind. And the motion or rest of the earth and the sun is so closely linked with the things just named, that without a determination of the one, neither side can be taken in the other matters. Now if the Holy Spirit has purposely neglected to teach us propositions of this sort as irrelevant to the highest goal (that is, to our salvation), how can anyone affirm that it is obligatory to take sides on them, and that one belief is required by faith, while the other side is erroneous? Can an opinion be heretical and yet have no concern with the salvation of souls? Can the Holy Ghost be asserted not to have intended teaching us something that does concern our salvation? I would say here something that was heard from an ecclesiastic of the most eminent degree: "That the intention of the Holy Ghost is to teach us how one goes to heaven, not how heaven goes."

But let us again consider the degree to which necessary demonstrations and sense experiences ought to be respected in physical conclusions, and the authority they have enjoyed at the hands of holy and learned theologians. From among a hundred attestations I have selected the following:

"We must also take heed, in handling the doctrine of Moses, that we altogether avoid saying positively and confidently anything which contradicts manifest experiences and the reasoning of philosophy or the other sciences. For since every truth is in agreement with all other truth, the truth of Holy Writ cannot be contrary to the solid reasons and experiences of human knowledge."

And in St. Augustine we read: "If anyone shall set the authority of Holy Writ against clear and manifest reason, he who does this knows not what he has undertaken; for he opposes to the truth not the meaning of the Bible, which is beyond his comprehension, but rather his own interpretation; not what is in the Bible, but what he has found in himself and imagines to be there."

This granted, and it being true that two truths cannot contradict one another, it is the function of wise expositors to seek out the true senses of scriptural texts. These will unquestionably accord with the physical conclusions which manifest sense and necessary demonstrations have previously made certain to us. Now the Bible, as has been remarked, admits in many places expositions that are remote from the signification of the words for reasons we have already given. Moreover, we are unable to affirm that all interpreters of the Bible speak by divine inspiration, for if that were so there would exist no differences between them about the sense of a given passage. Hence I should think it would be the part of prudence not to permit anyone to usurp scriptural texts and force them in some way to maintain any physical conclusion to be true, when at some future time the senses and demonstrative or necessary reasons may show the contrary. Who indeed will set bounds to human ingenuity? Who will assert that everything in the universe capable of being perceived is already discovered and known? Let us rather confess quite truly that "Those truths which we know are very few in comparison with those which we do not know."

We have it from the very mouth of the Holy Ghost that God delivered up the world to disputations, *so that man cannot find out the work that God hath done from the beginning even to the end.* In my opinion no one, in contradiction to that dictum, should close the road to free philosophizing about mundane and physical things, as if everything had already been discovered and revealed with certainty. Nor should it be considered rash not to be satisfied with those opinions which have become common. No one should be scorned in physical disputes for not holding to the opinions which happen to please other people best, especially concerning problems which have been debated among the greatest philosophers for thousands of years. One of these is the stability of the sun and mobility of the earth, a doctrine believed by Pythagoras and all his followers, by Heracleides of Pontus (who was one of them), by Philolaus the teacher of Plato, and by Plato himself according to Aristotle. Plutarch writes in his *Life of Numa* that Plato, when he had grown old, said it was most absurd to believe otherwise. The same doctrine was held by Aristarchus of Samos, as Archimedes tells us; by Seleucus the mathematician, by Nicetas the philosopher (on the testimony of Cicero), and by many others. Finally this opinion has been amplified and confirmed with many observations and demonstrations by Nicholas Copernicus. And Seneca, a most eminent philosopher, advises us in his book on comets that we should more diligently seek to ascertain whether it is in the sky or in the earth that the diurnal rotation resides.

Hence it would probably be wise and useful counsel if, beyond articles which concern salvation and the establishment of our Faith, against the stability of which there is no danger whatever that any valid and effective doctrine can ever arise, men would not aggregate further articles unnecessarily. And it would certainly be preposterous to introduce them at the request of persons who, besides not being known to speak by inspiration of divine grace, are clearly seen to lack that understanding which is necessary in order to comprehend, let alone discuss, the demonstrations by which such conclusions are supported in the subtler sciences. If I may speak my opinion freely, I should say further that it would perhaps fit in better with the decorum and majesty of the sacred writings to take measures for preventing every shallow and vulgar writer from giving to his compositions (often grounded upon foolish fancies) an air of authority by inserting in them passages from the Bible, interpreted (or rather distorted) into senses as far from the right meaning of Scripture as those authors are near to absurdity who thus ostentatiously adorn their writings. Of such abuses many examples might be produced, but for the present I shall confine myself to two which are germane to these astronomical matters. The first concerns those writings which were published against the existence of the Medicean planets recently discovered by me, in which many passages of holy Scripture were cited. Now that everyone has seen these planets, I should like to know what new interpretations those same antagonists employ in expounding the Scripture and excusing their own simplicity. My other example is that of a man who has lately published, in defiance of astronomers and philosophers, the opinion that the moon does not receive its light from the sun but is brilliant by its own nature. He supports this fancy (or rather thinks he does) by sundry texts of Scripture which he believes cannot be explained unless his theory is true; yet that the moon is inherently dark is surely as plain as daylight.

It is obvious that such authors, not having penetrated the true senses of Scripture, would impose upon others an obligation to subscribe to conclusions that are repugnant to manifest reason and sense, if they had any authority to do so. God forbid that this sort of abuse should gain countenance and authority, for then in a short time it would be necessary to proscribe all the contemplative sciences. People who are unable to understand perfectly both the Bible and the sciences far outnumber those who do understand. The former, glancing superficially through the Bible, would arrogate to themselves the authority to decree upon every question of physics on the strength of some word which they have misunderstood, and which was employed by the sacred authors for some different purpose. And the smaller number of understanding men could not dam up the furious torrent of such people, who would gain the majority of followers simply because it is much more pleasant to gain a reputation for wisdom without effort or study than to consume oneself tirelessly in the most laborious disciplines. Let us therefore render thanks to Almighty God, who in His beneficence protects us from this danger by depriving such persons of all authority, reposing the power of consultation, decision, and decree on such important matters in the high wisdom and benevolence of most prudent Fathers, and in the supreme authority of those who cannot fail to order matters properly under the guidance of the Holy Ghost. Hence we need not concern ourselves with the shallowness of those men whom grave and holy authors rightly reproach, and of whom in particular St. Jerome said, in reference to the Bible:

"This is ventured upon, lacerated, and taught by the garrulous old woman, the doting old man, and the prattling sophist before they have learned it. Others, led on by pride, weigh heavy words and philosophize amongst women concerning holy Scripture. Others—oh, shame!—learn from women what they teach to men, and (as if that were not enough) glibly expound to others that which they themselves do not understand. I forbear to speak of those of my own profession who, attaining a knowledge of the holy Scriptures after mundane learning, tickle the ears of the people with affected and studied expressions, and declare that everything they say is to be taken as the law of God. Not bothering to learn what the prophets and the apostles have maintained, they wrest incongruous testimonies into their own senses—as if distorting passages and twisting the Bible to their individual and contradictory whims were the genuine way of teaching, and not a corrupt one."

I do not wish to place in the number of such lay writers some theologians whom I consider men of profound learning and devout behavior, and who are therefore held by me in great esteem and veneration. Yet I cannot deny that I feel some discomfort which I should like to have removed, when I hear them pretend to the power of constraining others by scriptural authority to follow in a physical dispute that opinion which they think best agrees with the Bible, and then believe themselves not bound to answer the opposing reasons and experiences. In explanation and support of this opinion they say that since theology is queen of all the sciences, she need not bend in any way to accommodate herself to the teachings of less worthy sciences which are subordinate to her; these others must rather be referred to her as to their supreme empress, changing and altering their conclusions according to her statutes and decrees. They add further that if in the inferior sciences any conclusion should be taken as certain in virtue of demonstrations or experiences, while in the Bible another conclusion is found repugnant to this, then the professors of that science should themselves undertake to undo their proofs and discover the fallacies in their own experiences, without bothering the theologians and exegetes. For, they say, it does not become the dignity of theology to stoop to the investigation of fallacies in the subordinate sciences; it is sufficient for her merely to determine the truth of a given conclusion with absolute authority, secure in her inability to err.

Now the physical conclusions in which they say we ought to be satisfied by Scripture, without glossing or expounding it in senses different from the literal, are those concerning which the Bible always speaks in the same manner and which the holy Fathers all receive and expound in the same way. But with regard to these judgments I have had occasion to consider several things, and I shall set them forth in order that I may be corrected by those who understand more than I do in these matters—for to their decisions I submit at all times.

First, I question whether there is not some equivocation in failing to specify the virtues which entitle sacred theology to the title of "queen." It might deserve that name by reason of including everything that is learned from all the other sciences and establishing everything by better methods and with profounder learning. It is thus, for example, that the rules for measuring fields and keeping accounts are much more excellently contained in arithmetic and in the geometry of Euclid than in the practices of surveyors and accountants. Or theology might be queen because of being occupied with a subject which excels in dignity all the subjects which compose the other sciences, and because her teachings are divulged in more sublime ways.

That the title and authority of queen belongs to theology in the first sense, I think will not be affirmed by theologians who have any skill in the other sciences. None of these, I think, will say that geometry, astronomy, music, and medicine are much more excellently contained in the Bible than they are in the books of Archimedes, Ptolemy, Boethius, and Galen. Hence it seems likely that regal pre-eminence is given to theology in the second sense; that is, by reason of its subject and the miraculous communication of divine revelation of conclusions which could not be conceived by men in any other way, concerning chiefly the attainment of eternal blessedness.

Let us grant then that theology is conversant with the loftiest divine contemplation, and occupies the regal throne among sciences by dignity. But acquiring the highest authority in this way, if she does not descend to the lower and humbler speculations of the subordinate sciences and has no regard for them because they are not concerned with blessedness, then her professors should not arrogate to themselves the authority to decide on controversies in professions which they have neither studied nor practiced. Why, this would be as if an absolute despot, being neither a physician nor an architect but knowing himself free to command, should undertake to administer medicines and erect buildings according to his whim—at grave peril of his poor patients' lives, and the speedy collapse of his edifices.

Again, to command that the very professors of astronomy themselves see to the refutation of their own observations and proofs as mere fallacies and sophisms is to enjoin something that lies beyond any possibility of accomplishment. For this would amount to commanding that they must not see what they see and must not understand what they know, and that in searching they must find the opposite of what they actually encounter. Before this could be done they would have to be taught how to make one mental faculty command another, and the inferior powers the superior, so that the imagination and the will might be forced to believe the opposite of what the intellect understands. I am referring at all times to merely physical propositions, and not to supernatural things which are matters of faith.

I entreat those wise and prudent Fathers to consider with great care the difference that exists between doctrines *sub*ject to proof and those subject to opinion. Considering the force exerted by logical deductions, they may ascertain that it is not in the power of the professors of demonstrative sciences to change their opinions at will and apply themselves first to one side and then to the other. There is a great difference between commanding a mathematician or a philosopher and influencing a lawyer or a merchant, for demonstrated conclusions about things in nature or in the heavens cannot be changed with the same facility as opinions about what is or is not lawful in a contract, bargain, or bill of exchange. This difference was well understood by the learned and holy Fathers, as proven by their having taken great pains in refuting philosophical fallacies. This may be found expressly in some of them; in particular, we find the following words of St. Augustine: "It is to be held as an unquestionable truth that whatever the sages of this world have demonstrated concerning physical matters is in no way contrary to our Bibles; hence whatever the sages teach in their books that is contrary to the holy Scriptures may be concluded without any hesitation to be quite false. And according to our ability let us make this evident, and let us keep the faith of our Lord, in whom are hidden all the treasures of wisdom, so that we neither become seduced by the verbiage of false philosophy nor frightened by the superstition of counterfeit religion."

From the above words I conceive that I may deduce this doctrine: That in the books of the sages of this world there are contained some physical truths which are soundly demonstrated, and others that are merely stated; as to the former, it is the office of wise divines to show that they do not contradict the holy Scriptures. And as to the propositions which are stated but not rigorously demonstrated, anything contrary to the Bible involved by them must be held undoubtedly false and should be proved so by every possible means.

Now if truly demonstrated physical conclusions need not be subordinated to biblical passages, but the latter must rather be shown not to interfere with the former, then before a physical proposition is condemned it must be shown to be not rigorously demonstrated—and this is to be done not by those who hold the proposition to be true, but by those who judge it to be false. This seems very reasonable and natural, for those who believe an argument to be false may much more easily find the fallacies in it than men who consider it to be true and conclusive. Indeed, in the latter case it will happen that the more the adherents of an opinion turn over their pages, examine the arguments, repeat the observations, and compare the experiences, the more they will be confirmed in that belief. And Your Highness knows what happened to the late mathematician of the University of Pisa who undertook in his old age to look into the Copernican doctrine in the hope of shaking its foundations and refuting it, since he considered it false only because he had never studied it. As it fell out, no sooner had he understood its grounds, procedures, and demonstrations than he found himself persuaded, and from an opponent he became a very staunch defender of it. I might also name other mathematicians who, moved by my latest discoveries, have confessed it necessary to alter the previously accepted system of the world, as this is simply unable to subsist any longer.

If in order to banish the opinion in question from the world it were sufficient to stop the mouth of a single man—as perhaps those men persuade themselves who, measuring the minds of others by their own, think it impossible that this doctrine should be able to continue to find adherents—then that would be very easily done. But things stand otherwise. To carry out such a decision it would be necessary not only to prohibit the book of Copernicus and the writings of other authors who follow the same opinion, but to ban the whole science of astronomy. Furthermore, it would be necessary to forbid men to look at the heavens, in order that they might not see Mars and Venus sometimes quite near the earth and sometimes very distant, the variation being so great that Venus is forty times and Mars sixty times as large at one time as another. And it would be necessary to prevent Venus being seen round at one time and forked at another, with very thin horns; as well as many other sensory observations which can never be reconciled with the Ptolemaic system in any way, but are very strong arguments for the Copernican. And to ban Copernicus now that his doctrine is daily reinforced by many new observations and by the learned applying themselves to the reading of his book, after this opinion has been allowed and tolerated for those many years during which it was less followed and less confirmed, would seem in my judgment to be a contravention of truth, and an attempt to hide and suppress her the more as she revealed herself the more clearly and plainly. Not to abolish and censure his whole book, but only to condemn as erroneous this particular proposition, would (if I am not mistaken) be a still greater detriment to the minds of men, since it would afford them occasion to see a proposition proved that it was heresy to believe. And to prohibit the whole science would be but to censure a hundred passages of holy Scripture which teach us that the glory and greatness of Almighty God are marvelously discerned in all his works and divinely read in the open book of heaven. For let no one believe that reading the lofty concepts written in that book leads to nothing further than the mere seeing of the splendor of the sun and the stars and their rising and setting, which is as far as the eyes of brutes and of the vulgar can penetrate. Within its pages are couched mysteries so profound and concepts so sublime that the vigils, labors, and studies of hundreds upon hundreds of the most acute minds have still not pierced them, even after continual investigations for thousands of years. The eyes of an idiot perceive little by beholding the external appearance of a human body, as compared with the wonderful contrivances which a careful and practiced anatomist or philosopher discovers in that same body when he seeks out the use of all those muscles, tendons, nerves, and bones; or when examining the functions of the heart and the other principal organs, he seeks the seat of the vital faculties, notes and observes the admirable structure of the sense organs, and (without ever ceasing in his amazement and delight) contemplates the receptacles of the imagination, the memory, and the understanding. Likewise, that which presents itself to mere sight is as nothing in comparison with the high marvels that the ingenuity of learned men discovers in the heavens by long and accurate observation. And that concludes what I have to say on this matter.

Next let us answer those who assert that those physical propositions of which the Bible speaks always in one way, and which the Fathers all harmoniously accept in the same sense, must be taken according to the literal sense of the words without glosses or interpretations, and held as most certain and true. The motion of the sun and stability of the earth, they say, is of this sort; hence it is a matter of faith to believe in them, and the contrary view is erroneous.

To this I wish first to remark that among physical propositions there are some with regard to which all human science and reason cannot supply more than a plausible opinion and a probable conjecture in place of a sure and demonstrated knowledge; for example, whether the stars are animate. Then there are other propositions of which we have (or may confidently expect) positive assurances through experiments, long observation, and rigorous demonstration; for example, whether or not the earth and the heavens move, and whether or not the heavens are spherical. As to the first sort of propositions, I have no doubt that where human reasoning cannot reach—and where consequently we can have no science but only opinion and faith—it is necessary in piety to comply absolutely with the strict sense of Scripture. But as to the other kind, I should think, as said before, that first we are to make certain of the fact, which will reveal to us the true senses of the Bible, and these will most certainly be found to agree with the proved fact (even though at first the words sounded otherwise), for two truths can never contradict each other. I take this to be an orthodox and indisputable doctrine, and I find it specifically in St. Augustine when he speaks of the shape of heaven and what we may believe concerning that. Astronomers seem to declare what is contrary to Scripture, for they hold the heavens to be spherical, while the Scripture calls it "stretched out like a curtain." St. Augustine opines that we are not to be concerned lest the Bible contradict astronomers; we are to believe its authority if what they say is false and is founded only on the conjectures of frail humanity. But if what they say is proved by unquestionable arguments, this holy Father does not say that the astronomers are to be ordered to dissolve their proofs and declare their own conclusions to be false. Rather, he says it must be demonstrated that what is meant in the Bible by "curtain" is not contrary to their proofs. Here are his words:

"But some raise the following objection. 'How is it that the passage in our Bibles, *Who stretcheth out the heavens as a curtain,* does not contradict those who maintain the heavens to have a spherical shape?' It does contradict them if what they affirm is false, for that is true which is spoken by divine authority rather than that which proceeds from human frailty. But if, peradventure, they should be able to prove their position by experiences which place it beyond question, then it is to be demonstrated that our speaking of a curtain in no way contradicts their manifest reasons."

He then proceeds to admonish us that we must be no less careful and observant in reconciling a passage of the Bible with any demonstrated physical proposition than with some other biblical passage which might appear contrary to the first. The circumspection of this saint indeed deserves admiration and imitation, when even in obscure conclusions (of which we surely can have no knowledge through human proofs) he shows great reserve in determining what is to be believed. We see this from what he writes at the end of the second book of his commentary on Genesis, concerning the question whether the stars are to be believed animate:

"Although at present this matter cannot be settled, yet I suppose that in our further dealing with the Bible we may meet with other relevant passages, and then we may be permitted, if not to determine anything finally, at least to gain some hint concerning this matter according to the dictates of sacred authority. Now keeping always our respect for moderation in grave piety, we ought not to believe anything inadvisedly on a dubious point, lest in favor of our error we conceive a prejudice against something that truth hereafter may reveal to be not contrary in any way to the sacred books of either the Old or the New Testament."

From this and other passages the intention of the holy Fathers appears to be (if I am not mistaken) that in questions of nature which are not matters of faith it is first to be considered whether anything is demonstrated beyond doubt or known by sense-experience, or whether such knowledge or proof is possible; if it is, then, being the gift of God, it ought to be applied to find out the true senses of holy Scripture in those passages which superficially might seem to declare differently. These senses would unquestionably be discovered by wise theologians, together with the reasons for which the Holy Ghost sometimes wished to veil itself under words of different meaning, whether for our exercise, or for some purpose unknown to me.

As to the other point, if we consider the primary aim of the Bible, I do not think that its having always spoken in the same sense need disturb this rule. If the Bible, accommodating itself to the capacity of the common people, has on one occasion expressed a proposition in words of different sense from the essence of that proposition, then why might it not have done the same, and for the same reason, whenever the same thing happened to be spoken of? Nay, to me it seems that not to have done this would but have increased confusion and diminished belief among the people.

Regarding the state of rest or motion of the sun and earth, experience plainly proves that in order to accommodate the common people it was necessary to assert of these things precisely what the words of the Bible convey. Even in our own age, people far less primitive continue to maintain the same opinion for reasons which will be found extremely trivial if well weighed and examined, and upon the basis of experiences that are wholly false or altogether beside the point. Nor is it worth while to try to change their opinion, they being unable to understand the arguments on the opposite side, for these depend upon observations too precise and demonstrations too subtle, grounded on abstractions which require too strong an imagination to be comprehended by them. Hence even if the stability of heaven and the motion of the earth should be more than certain in the minds of the wise, it would still be necessary to assert the contrary for the preservation of belief among the all-too-numerous vulgar. Among a thousand ordinary men who might be questioned concerning these things, probably not a single one will be found to answer anything except that it looks to him as if the sun moves and the earth stands still, and therefore he believes this to be certain. But one need not on that account take the common popular assent as an argument for the truth of what is stated; for if we should examine these very men concerning their reasons for what they believe, and on the other hand listen to the experiences and proofs which induce a few others to believe the contrary, we should find the latter to be persuaded by very sound arguments, and the former by simple appearances and vain or ridiculous impressions.

It is sufficiently obvious that to attribute motion to the sun and rest to the earth was therefore necessary lest the shallow minds of the common people should become confused, obstinate, and contumacious in yielding assent to the principal articles that are absolutely matters of faith. And if this was necessary, there is no wonder at all that it was carried out with great prudence in the holy Bible. I shall say further that not only respect for the incapacity of the vulgar, but also current opinion in those times, made the sacred authors accommodate themselves (in matters unnecessary to salvation) more to accepted usage than to the true essence of things. Speaking of this, St. Jerome writes:

"As if many things were not spoken in the Holy Bible according to the judgment of those times in which they were acted, rather than according to the truth contained." And elsewhere the same saint says: "It is the custom for the biblical scribes to deliver their judgments in many things according to the commonly received opinion of their times." And on the words in the twenty-sixth chapter of Job, *He stretcheth out the north over the void, and hangeth the earth above* nothing, St. Thomas Aquinas notes that the Bible calls "void" or "nothing" that space which we know to be not empty, but filled with air. Nevertheless the Bible, he says, in order to accommodate itself to the beliefs of the common people (who think there is nothing in that space), calls it "void" and "nothing." Here are the words of St. Thomas: "What appears to us in the upper hemisphere of the heavens to be empty, and not a space filled with air, the common people regard as void; and it is usually spoken of in the holy Bible according to the ideas of the common people."

Now from this passage I think one may very logically argue that for the same reason the Bible had still more cause to call the sun movable and the earth immovable. For if we were to test the capacity of the common people, we should find them even less apt to be persuaded of the stability of the sun and the motion of the earth than to believe that the space which environs the earth is filled with air. And if on this point it would not have been difficult to convince the common people, and yet the holy scribes forbore to attempt it, then it certainly must appear reasonable that in other and more abstruse propositions they have followed the same policy.

Copernicus himself knew the power over our ideas that is exerted by custom and by our inveterate way of conceiving things since infancy. Hence, in order not to increase for us the confusion and difficulty of abstraction, after he had first demonstrated that the motions which appear to us to belong to the sun or to the firmament are really not there but in the earth, he went on calling them motions of the sun and of the heavens when he later constructed his tables to apply them to use. He thus speaks of "sunrise" and "sunset," of the "rising and setting" of the stars, of changes in the obliquity of the ecliptic and of variations in the equinoctial points, of the mean motion and variations in motion of the sun, and so on. All these things really relate to the earth, but since we are fixed to the earth and consequently share in its every motion, we cannot discover them in the earth directly, and are obliged to refer them to the heavenly bodies in which they make their appearance to us. Hence we name them as if they took place where they appear to us to take place; and from this one may see how natural it is to accommodate things to our customary way of seeing them.

* * * * *

The more we learn of science, the more we see that its wonderful mysteries are all explained by a few simple laws so connected together and so dependent upon each other, that we see the same mind animating them all.

Olympia Brown, *An Autobiography*, 1960

And now the announcement of Watson and Crick about DNA. This is for me the real proof of the existence of God.

Salvadore Dali

The cosmic religious experience is the strongest and the noblest driving force behind scientific research.

Albert Einstein

Raffiniert ist der Herr Gott, aber Boshaft ist der nicht. The Lord God is subtle, but he is not malicious.

Albert Einstein

SCIENCE AND RELIGION

ALBERT EINSTEIN

For a biographical sketch of Albert Einstein (1879-1955) see page 235. Einstein maintained no formal religious affiliation, but he was known to be a deeply religious person. He thought of God as the Being at work in the universe rather than in strictly personal terms. He saw the universe as subject to law and order and not simply as the result of chance or chaos.

1. How does Einstein resolve the apparent conflict between science and religion?
2. What does Einstein see as the respective functions of science and religion?
3. According to Einstein, what are the sources of present day conflicts between the spheres of religion and of science?
4. Does his argument shed any light on the current evolution-creation argument?
5. How might science and religion each be of assistance to the other?

SCIENCE AND RELIGION

ALBERT EINSTEIN

DURING THE LAST CENTURY and part of the one before, it was widely held that there was an unreconcilable conflict between knowledge and belief. The opinion prevailed among advanced minds that it was time that belief should be replaced increasingly by knowledge; belief that did not itself rest on knowledge was superstition, and as such had to be opposed. According to this conception, the sole function of education was to open the way to thinking and knowing, and the school, as the outstanding organ for the people's education, must serve that end exclusively.

One will probably find but rarely, if at all, the rationalistic standpoint expressed in such crass form; for any sensible man would see at once how one-sided is such a statement of the position. But it is just as well to state a thesis starkly and nakedly, if one wants to clear up one's mind as to its nature.

It is true that convictions can best be supported with experience and clear thinking. On this point one must agree unreservedly with the extreme rationalist. The weak point of his conception is, however, this, that those convictions which are necessary and determinant for our conduct and judgments, cannot be found solely along this solid scientific way.

For the scientific method can teach us nothing else beyond how facts are related to, and conditioned by, each other. The aspiration toward such objective knowledge belongs to the highest of which man is capable, and you will certainly not suspect me of wishing to belittle the achievements and the heroic efforts of man in this sphere. Yet it is equally clear that knowledge of what is does not open the door directly to what *should be.* One can have the clearest and most complete knowledge of what is, and yet not be able to deduct from that what should be the *goal* of our human aspirations. Objective knowledge provides us with powerful instruments for the achievements of certain ends, but the ultimate goal itself and the longing to reach it must come from another source. And it is hardly necessary to argue for the view that our existence and our activity acquire meaning only by the setting up of such a goal and of corresponding values. The knowledge of truth as such is wonderful, but it is so little capable of acting as a guide that it cannot prove even the justification and the value of the aspiration towards that very knowledge of truth. Here we face, therefore, the limits of the purely rational conception of our existence.

But it must not be assumed that intelligent thinking can play no part in the formation of the goal and of ethical judgments. When someone realizes that for the achievement of an end certain means would be useful, the means itself becomes thereby an end. Intelligence makes clear to us the interrelation of means and ends. But mere thinking cannot give us a sense of the ultimate and fundamental ends. To make clear these fundamental ends and valuations, and to set them fast in the emotional life of the individual, seems to me precisely the most important function which religion has to perform in the social life of man. And if one asks whence derives the authority of such fundamental ends, since they cannot be stated and justified merely by reason, one can only answer: they exist in a healthy society as powerful traditions, which act upon the conduct and aspirations and judgments of the individuals; they are there, that is, as something living, without its being necessary to find justification for their existence. They come into being not through demonstration but through revelation, through the medium of powerful personalities. One must not attempt to justify them, but rather to sense their nature simply and clearly.

The highest principles for our aspirations and judgments are given to us in the Jewish-Christian religious tradition. It is a very high goal which, with our weak powers, we can reach only very inadequately, but which gives a sure foundation to our aspirations and valuations. If one were to take that goal out of its religious form and look merely at its purely human side, one might state it perhaps thus: free and responsible development of the individual, so that he may place his powers freely and gladly in the service of all mankind.

There is no room in this for the divinization of a nation, of a class, let alone of an individual. Are we not all children of one father, as it is said in religious language? Indeed, even the divinization of humanity, as an abstract totality, would not be in the spirit of that ideal. It is only to the individual that a soul is given. And the high destiny of the individual is to serve rather than to rule, or to impose himself in any other way.

If one looks at the substance rather than at the form, then one can take these words as expressing also the fundamental democratic position. The true democrat can worship his nation as little as can the man who is religious, in our sense of the term.

What, then, in all this, is the function of education and of the school? They should help the young person to grow up in such a spirit that these fundamental principles should be to him as the air which he breathes. Teaching alone cannot do that.

If one holds these high principles clearly before one's eyes, and compares them with the life and spirit of our times, then it appears glaringly that civilized mankind finds itself at present in grave danger. In the totalitarian states it is the rulers themselves who strive actually to destroy that spirit of humanity. In less threatened parts it is nationalism and intolerance, as well as the oppression of the individuals by economic means, which threaten to choke these most precious traditions.

A realization of how great is the danger is spreading, however, among thinking people, and there is much search for means with which to meet the danger—means in the field of national and international politics, of legislation, of organization in general. Such efforts are, no doubt, greatly needed. Yet the ancients knew something which we seem to have forgotten. All means prove but a blunt instrument, if they have not behind them a living spirit. But if the longing for the achievement of the goal is powerfully alive within us, then shall we not lack the strength to find the means for reaching the goal and for translating it into deeds.

II

It would not be difficult to come to an agreement as to what we understand by science. Science is the century-old endeavor to bring together by means of systematic thought the perceptible phenomena of this world into as thoroughgoing an association as possible. To put it boldly, it is the attempt at the posterior reconstruction of existence by the process of conceptualization. But when asking myself what religion is I cannot think of the answer so easily. And even after finding an answer which may satisfy me at this particular moment I still remain convinced that I can never under any circumstances bring together, even to a slight extent, all those who have given this question serious consideration.

At first, then, instead of asking what religion is I should prefer to ask what characterizes the aspirations of a person who gives me the impression of being religious: A person who is religiously enlightened appears to me to be one who has, to the best of his ability, liberated himself from the fetters of his selfish desires and is preoccupied with thoughts, feelings, and aspirations to which he clings because of their super-personal value. It seems to me that what is important is the force of this super-personal content and the depth of the conviction concerning its overpowering meaningfulness, regardless of whether any attempt is made to unite this content with a divine Being, for otherwise it would not be possible to count Buddha and Spinoza as religious personalities. Accordingly, a religious person is devout in the sense that he has no doubt of the significance and loftiness of those super-personal objects and goals which neither require nor are capable of rational foundation. They exist with the same necessity and matter-of-factness as he himself. In this sense religion is the age-old endeavor of mankind to become clearly and completely conscious of these values and goals and constantly to strengthen and extend their effect. If one conceives of religion and science according to these definitions then a conflict between them appears impossible. For science can only ascertain what is, but not what *should be,* and outside of its domain value judgments of all kinds remain necessary. Religion, on the other hand, deals only with evaluations of human thought and action: it cannot justifiably speak of facts and relationships between facts. According to this interpretation the well-known conflicts between religion and science in the past must all be ascribed to a misapprehension of the situation which has been described.

For example, a conflict arises when a religious community insists on the absolute truthfulness of all statements recorded in the Bible. This means an intervention on the part of religion into the sphere of science; this is where the struggle of the Church against the doctrines of Galileo and Darwin belongs. On the other hand, representatives of science have often made an attempt to arrive at fundamental judgments with respect to values and ends on the basis of scientific method, and in this way have set themselves in opposition to religion. These conflicts have all sprung from fatal errors.

Now, even though the realms of religion and science in themselves are clearly marked off from each other, nevertheless there exist between the two strong reciprocal relationships and dependencies. Though religion may be that which determines the goal, it has, nevertheless, learned from science, in the broadest sense, what means will contribute to the attainment of the goals it has set up. But science can only be created by those who are thoroughly imbued with the aspiration towards truth and understanding. This source of feeling, however, springs from the sphere of religion. To this there also belongs the faith in the possibility that the regulations valid for the world of existence are rational, that is, comprehensible to reason. I cannot conceive of a genuine scientist without that profound faith. The situation may be expressed by an image: Science without religion is lame, religion without science is blind.

Though I have asserted above that in truth a legitimate conflict between religion and science cannot exist I must nevertheless qualify this assertion once again on an essential point, with reference to the actual content of historical religions. This qualification has to do with the concept of God. During the youthful period of mankind's spiritual evolution human fantasy created gods in man's own image, who, by the operations of their will were supposed to determine, or at any rate to influence the phenomenal world. Man sought to alter the disposition of these gods in his own favor by means of magic and prayer. The idea of God in the religions taught at present is a sublimation of that old conception of the gods. Its anthropomorphic character is shown, for instance, by the fact that men appeal to the Divine Being in prayers and plead for the fulfilment of their wishes.

Nobody, certainly, will deny that the idea of the existence of an omnipotent, just and omnibeneficent personal God is able to accord man solace, help, and guidance; also, by virtue of its simplicity it is accessible to the most undeveloped mind. But, on the other hand, there are decisive weaknesses attached to this idea in itself, which have been painfully felt since the beginning of history. That is, if this being is omnipotent then every occurrence, including every human action, every human thought, and every human feeling and aspiration is also His work; how is it possible to think of holding men responsible for their deeds and thoughts before such an almighty Being? In giving out punishment and rewards He would to a certain extent be passing judgment on Himself. How can this be combined with the goodness and righteousness ascribed to Him?

The main source of the present-day conflicts between the spheres of religion and of science lies in this concept of a personal God. It is the aim of science to establish general rules which determine the reciprocal connection of objects and events in time and space. For these rules, or laws of nature, absolutely general validity is required—not proven. It is mainly a program, and faith in the possibility of its accomplishment in principle is only founded on partial successes. But hardly anyone could be found who would deny these partial successes and ascribe them to human self-deception. The fact that on the basis of such laws we are able to predict the temporal behavior of phenomena in certain domains with great precision and certainty is deeply embedded in the consciousness of the modern man, even though he may have grasped very little of the contents of those laws. He need only consider that planetary courses within the solar system may be calculated in advance with great exactitude on the basis of a limited number of simple laws. In a similar way, though not with the same precision, it is possible to calculate in advance the mode of operation of an electric motor, a transmission system, or of a wireless apparatus, even when dealing with a novel development.

To be sure, when the number of factors coming into play in a phenomenological complex is too large scientific method in most cases fails us. One need only think of the weather, in which case prediction even for a few days ahead is impossible. Nevertheless no one doubts that we are confronted with a causal connection whose causal components are in the main known to us. Occurrences in this domain are beyond the reach of exact prediction because of the variety of factors in operation, not because of any lack of order in nature.

We have penetrated far less deeply into the regularities obtaining within the realm of living things, but deeply enough nevertheless to sense at least the rule of fixed necessity. One need only think of the systematic order in heredity, and in the effect of poisons, as for instance alcohol, on the behavior of organic beings. What is Still lacking here is a grasp of connections of profound generality, but not a knowledge of order in itself.

The more a man is imbued with the ordered regularity of all events the firmer becomes his conviction that there is no room left by the side of this ordered regularity for causes of a different nature. For him neither the rule of human nor the rule of divine will exists as an independent cause of natural events. To be sure, the doctrine of a personal God interfering with natural events could never be *refuted,* in the real sense, by science, for this doctrine can always take refuge in those domains in which scientific knowledge has not yet been able to set foot.

But I am persuaded that such behavior on the part of the representatives of religion would not only be unworthy but also fatal. For a doctrine which is able to maintain itself not in clear light but only in the dark, will of necessity lose its effect on mankind, with uncalculable harm to human progress. In their struggle for the ethical good, teachers of religion must have the stature to give up the doctrine of a personal God, that is, give up that source of fear and hope which in the past placed such vast power in the hands of priests. In their labors they will have to avail themselves of those forces which are capable of cultivating the Good, the True, and the Beautiful in humanity itself. This is, to be sure, a more difficult but an incomparably more worthy task. After religious teachers accomplish the refining process indicated they will surely recognize with joy that true religion has been ennobled and made more profound by scientific knowledge.

If it is one of the goals of religion to liberate mankind as far as possible from the bondage of egocentric cravings, desires, and fears, scientific reasoning can aid religion in yet another sense. Although it is true that it is the goal of science to discover rules which permit the association and foretelling of facts, this is not its only aim. It also seeks to reduce the connections discovered to the smallest possible number of mutually independent conceptual elements. It is in this striving after the rational unification of the manifold that it encounters its greatest successes, even though it is precisely this attempt which causes it to run the greatest risk of falling a prey to illusions. But whoever has undergone the intense experience of successful advances made in this domain, is moved by profound reverence for the rationality made manifest in existence. By way of the understanding he achieves a far-reaching emancipation from the shackles of personal hopes and desires, and thereby attains that humble attitude of mind towards the grandeur of reason incarnate in existence, and which, in its profoundest depths, is inaccessible to man. This attitude, however, appears to me to be religious, in the highest sense of the word. And so it seems to me that science not only purifies the religious impulse of the dross of its anthropomorphism but also contributes to a religious spiritualization of our understanding of life.

The further the spiritual evolution of mankind advances, the more certain it seems to me that the path to genuine religiosity does not lie through the fear of life, and the fear of death, and blind faith, but through striving after rational knowledge. In this sense I believe that the priest must become a teacher if he wishes to do justice to his lofty educational mission.

The whole of science is nothing more than a refinement of everyday thinking.
Albert Einstein, *Out of My Later Years*, 1950

Extinguished theologians lie about the cradle of every science as the strangled snakes beside that of Hercules.
Thomas Huxley, *Darwiniana* (1860)

www.ingramcontent.com/pod-product-compliance
Ingram Content Group UK Ltd.
Pitfield, Milton Keynes, MK11 3LW, UK
UKHW041142300926
15471UKWH00011B/107

9 780761 808299